CONTENTS

PREFACE

The Process Industries Challenge

In the early 1990s, the process industries recognized that they would face a major manpower shortage due to the large number of employees retiring. Industry partnered with community colleges, technical colleges and universities to provide training for their process technicians, recognizing that substantial savings on training and traditional hiring costs could be realized. In addition, the consistency of curriculum content and exit competencies of process technology graduates could be ensured if industry collaborated with education.

To achieve this consistency of graduates' exit competencies, the Gulf Coast Process Technology Alliance and the Center for the Advancement of Process Technology identified a core technical curriculum for the Associate Degree in Process Technology. This core, consisting of eight technical courses, is taught in alliance member institutions throughout the United States. This textbook is intended to provide a common standard reference for the *Introduction to Process Technology* course that serves as part of the core technical courses in the degree program.

Purpose of the Textbook

Instructors who teach the process technology core curriculum, and who are recognized in the industry for their years of experience and their depth of subject matter expertise, requested that a textbook be developed to match the standardized curriculum. Reviewers from a broad array of process industries and education institutions participated in the production of these materials so that the widest audience possible would be represented in the presentation of the content.

The textbook is intended for use in community colleges, technical colleges, universities and corporate settings in which process technology is taught. However, educators in many disciplines will find these materials a complete reference for both theory and practical application. Students will find this textbook to be a valuable resource throughout their process technology career.

Organization of the Textbook

This textbook has been divided into 29 chapters. Chapters 1-9 provide an overview of the various process industries. Chapters 10-15 describe the various concepts and skills that process technicians must understand. Chapters 16-29 explain some common equipment that process technicians will work with as part of their job.

Each chapter is organized in the following way:

- Learning objectives
- Key terms

- Introduction
- Key topics
- Summary
- Checking your knowledge questions
- Activities

The **Learning Objectives** for a chapter may cover one or more sessions in a course. For example, Chapter 2 may take two weeks (or two sessions) to complete in the classroom setting.

The **Key Terms** are a listing of important terms and their respective definitions that students should know and understand before proceeding to the next chapter.

The **Introduction** may be a simple introductory paragraph, or may introduce concepts necessary to the development of the content of the chapter itself.

Any of the **Key Topics** can have several subtopics. Although these topics and subtopics do not always follow the flow of the learning objectives as stated at the beginning, all learning objectives are addressed in the chapter.

The **Summary** is a restatement of the learning outcomes of the chapter.

The **Checking Your Knowledge** questions are designed to help students to self-test on potential learning points from the chapter.

The **Activities** section contains activities that can be performed by a student on their own or with other students in small groups, and activities that should be performed with instructor involvement.

Chapter Summaries

CHAPTER 1: PROCESS TECHNOLOGY: AN OVERVIEW

This chapter defines process industries and what they do. In addition, process technician duties, responsibilities, expectations and working conditions are discussed.

CHAPTER 2: OIL AND GAS INDUSTRY OVERVIEW

This chapter explains the growth and development of the oil and gas industries. The different segments of the oil and gas industries, process technician duties, responsibilities and expectations, and future trends in the oil and gas industry are discussed.

CHAPTER 3:
CHEMICAL INDUSTRY OVERVIEW

This chapter explains the growth and development of the chemical processing industry. Process technician duties, responsibilities and expectations, and future trends in the chemical processing industry are also included in this chapter.

CHAPTER 4:
MINING INDUSTRY OVERVIEW

This chapter explains the growth and development of the mining industry. In addition, process technician duties, responsibilities and expectations are discussed, as well as future trends.

CHAPTER 5:
POWER GENERATION INDUSTRY OVERVIEW

This chapter explains the growth and development of power generation in the process industry. In addition, the different segments of the power generation industry, process technician duties, responsibilities and expectations, and future trends in the power generation industry are discussed.

CHAPTER 6:
PULP AND PAPER INDUSTRY OVERVIEW

This chapter explains the growth and development of pulp and paper industries. Furthermore, the different segments of the pulp and paper industries are discussed, along with process technician duties, responsibilities and expectations, and future trends in the pulp and paper industries.

CHAPTER 7:
WATER AND WASTE WATER TREATMENT INDUSTRY OVERVIEW

This chapter explains the history of wastewater treatment and its importance to the process industry and the community. In addition, the water treatment process, process technician duties, responsibilities and expectations, and environmental statutes associated with water treatment are discussed.

CHAPTER 8:
FOOD AND BEVERAGE INDUSTRY OVERVIEW

This chapter explains the history of the food manufacturing industry and the food manufacturing process. The various types of equipment used in the food manufacturing industry are also identified, as well as the most common products produced by the industry and who the food manufacturing industry serves. Process technician duties, responsibilities, and expectations are discussed, as well as the responsibilities of various regulatory agencies and their impact on the food manufacturing industry.

CHAPTER 9: PHARMACEUTICAL INDUSTRY OVERVIEW

This chapter explains the history of the pharmaceutical industry and the drug manufacturing process. In addition, process technician duties, responsibilities, and expectations are discussed, as well as future trends in the pharmaceutical industry.

CHAPTER 10: BASIC PHYSICS

This chapter explains how physics is applied in the process industry. The states of matter, methods of heat transfer, characteristics of matter under different conditions, and the scientific laws that explain the relationships between various physical phenomena are also explained.

CHAPTER 11: BASIC CHEMISTRY

This chapter explains how chemistry is applied in the process industry. The relationship between molecules, atoms, and the components of atoms are also explained, along with the differences between organic and inorganic chemistry, chemical versus physical properties, and acids versus bases.

CHAPTER 12: SAFETY, HEALTH, ENVIRONMENT AND SECURITY

This chapter identifies safety, health and environmental hazards found in the process industry. Various regulatory agencies and the regulations they impose are discussed, as well the consequences of non-compliance. The role of the process technician with regard to safety is explained, along with key safety concepts and equipment.

CHAPTER 13: QUALITY

This chapter explains the concept of quality as it pertains to the process industry. The quality movement and its pioneers are also discussed, along with the components of Total Quality Management, the International Organization for Standardization, Statistical Process Control, and the roles and responsibilities of process technicians with regard to quality improvement in the workplace.

CHAPTER 14: TEAMS

This chapter explains the difference between working groups and teams, and identifies the different types of teams encountered in the process industry. In addition, characteristics of high performing teams, stages through which teams evolve, factors that contribute to the failure of teams, and the impact of diversity on workplace relations are discussed.

CHAPTER 15: PROCESS DRAWINGS

This chapter explains the purpose of process system drawings in the process industry. Common drawing types and their components are identified.

CHAPTER 16: PIPING AND VALVES

This chapter explains various types of piping and valves, and how they are used in the process industry. Materials of construction, hazards associated with improper operation, monitoring and maintenance activities, and symbology are also discussed.

CHAPTER 17: VESSELS

This chapter explains various types of vessels and how they are used in the process industry. Construction characteristics, the purpose of containment walls, hazards associated with improper operation, monitoring and maintenance activities, and symbology are also discussed.

CHAPTER 18: PUMPS

This chapter explains various types of pumps, and how they are used in the process industry. Types of pumps, pump components, hazards associated with improper operation, monitoring and maintenance activities, and symbology are also discussed.

CHAPTER 19: COMPRESSORS

This chapter explains various types of compressors, and how they are used in the process industry. Types of compressors, compressor components, hazards associated with improper operation, monitoring and maintenance activities, and symbology are also discussed.

CHAPTER 20: TURBINES

This chapter explains various types of turbines and how they are used in the process industry. Turbine operating principles, hazards associated with improper operation, monitoring and maintenance activities, and symbology are also discussed.

CHAPTER 21: ELECTRICITY AND MOTORS

This chapter explains the purpose of motors in the process industry. The components of a typical motor are also discussed, along with the differences between alternating current (AC) and direct current (DC), hazards associated with improper operation, monitoring and maintenance activities, and symbology.

CHAPTER 22: HEAT EXCHANGERS

This chapter explains how heat exchangers are used in the process industry. Methods of heat transfer, heat exchanger components, hazards associated with improper operation, monitoring and maintenance activities, and symbology are also discussed.

CHAPTER 23: COOLING TOWERS

This chapter explains the principles of operation for cooling towers and how they are used in the process industry. Cooling tower components, hazards associated with improper operation, monitoring and maintenance activities, and symbology are also discussed.

CHAPTER 24: FURNACES

This chapter explains the principles of operation for furnaces and how they are used in the process industry. Different types of furnaces and fuel types, furnace components, hazards associated with improper operation, monitoring and maintenance activities, and symbology are also discussed.

CHAPTER 25: BOILERS

This chapter explains the principles of operation for boilers and how they are used in the process industry. Different types of boilers and fuel types, furnace components, hazards associated with improper operation, monitoring and maintenance activities, and symbology are also discussed.

CHAPTER 26: DISTILLATION

This chapter explains the principles of distillation and how it is used in the process industry. The components of a distillation column, the purpose of packing, hazards associated with improper operation, monitoring and maintenance activities, and symbology are also discussed.

CHAPTER 27: PROCESS UTILITIES

This chapter explains the different types of process utilities and their applications in the process industry. The different types of equipment used in these utility systems are also described.

CHAPTER 28: PROCESS AUXILIARIES

This chapter explains the different auxiliary systems used in the process industry, as well as their purpose and functions. The equipment associated with each of these systems is also discussed.

CHAPTER 29: INSTRUMENTATION

This chapter explains the purpose of process control instrumentation in the process industry. The different process variables and instruments used to measure them are discussed. Process control instrumentation, generic control loops, and distributive control systems are also discussed, along with hazards associated with improper operation, monitoring and maintenance activities, and symbology.

ACKNOWLEDGEMENTS

The following organizations and their dedicated personnel voluntarily participated in the production of this textbook. Their contributions to making this a successful project are greatly appreciated. Perhaps our gratitude for their involvement can best be expressed by this sentiment:

> "The credit belongs to those people who are actually in the arena...who know the great enthusiasms, the great devotions to a worthy cause; who at best, know the triumph of high achievement; and who, at worst, fail while daring greatly...so that their place shall never be with those cold and timid souls who know neither victory nor defeat." – Theodore Roosevelt

Process technicians both current and future will utilize the information within this textbook as a resource to more fully understand the process industries. This knowledge will strengthen these paraprofessionals by helping to make them better prepared to meet the ever challenging roles and responsibilities within their specific process industry.

INDUSTRY CONTENT DEVELOPERS AND REVIEWERS

Christine Archer, TAP Safety Services, Texas
William Anderson, Eastman, Texas
Johnny Arevalos, Valero, Texas
Ronnie Baker, ExxonMobil, Texas
Thomas J. Blacklock, Novartis, New Jersey
Ted Borel, Equistar Chemical, Texas
Linda Brown, Pasadena Refining System, Inc., Texas
Douglas Daniels, Heads Up Systems, Virginia
F.D. (Bubba) Diaz, Mississippi Power, Mississippi
Lisa Arnold Diederich, Marathon Ashland, Texas
Danny Dunagan, Formosa Plastics Corp., HDPE II, Texas
Errol Dupre, BP, Louisiana
Diane Dykes, BP, Texas
Bob Everard, Shell Chemical LP, Louisiana
Kathleen Garey, Shell Chemical LP, Louisiana
Eileen Gurney, ExxonMobil, New Jersey
James Guynes, Basell USA, Inc., Louisiana
Wes Haworth, BP, Washington
Linda Hebert, ExxonMobil, Louisiana
George (Tommy) Henry, Chevron Texaco, Mississippi
Ky Holland, Alaska Process Industry Careers Consortium, Alaska
Richard Honea, The Dow Chemical Company, Texas
Glenn Johnson, Huish Detergents, Inc., Texas
Gina Jones, Westlake Chemical Corporation, Louisiana

Bharat Kamdar, INGENIOUS, INC., Texas
Martha McKinley, Eastman Chemical, Texas
Roy Murdock, Shell Oil Products, Washington
Eric Newby, BASF, Texas
Don Parsley, Valero Refining, Texas
Ray Player, Eastman Chemical Company, Texas Operations, Texas
Ed Podhirny, The Dow Chemical Company, Texas
Erv Rohde, The Dow Chemical Company, Texas
Pete Rygaard, Union Carbide, Texas
Vonya Sanders, BP, California
Tim Seeger, Eastman Chemical Company, Texas Operations, Texas
Oran Sonnier, BP, Texas
Steve Shreiber, Anchorage Water and Wastewater Utility, Alaska
Ken Stuchly, Kraft Foods, Texas
Paul Summers, The Dow Chemical Company, Texas
Mike Tucker, Eastman Chemical Company, Texas Operations, Texas
Robert Walls, Sherwin Alumina, Texas
Ray Wuertz, ConocoPhillips, New Jersey

EDUCATION CONTENT DEVELOPERS AND REVIEWERS

Louis Babin, ITI Technical College, Louisiana
Jack Baggett, ITI Technical College, Louisiana
Lon Bedney, Louisiana Technical College, Louisiana
Tommie Broome, Mississippi Gulf Coast Community College, Mississippi
Anita Brunsting, Victoria College, Texas
Lou Caserta, Alvin Community College, Texas
Michael Cobb, College of the Mainland, Texas
Michael Connella III, McNeese State University, Louisiana
Richard Cox, Baton Rouge Community College, Louisiana
Mark Demark, Alvin Community College, Texas
Ernest Duhon, Sowela, Louisiana
Jerry Duncan, College of the Mainland, Texas
George Foret, Louisiana Technical College, Louisiana
Charles Gaffen, Elizabeth High School, New Jersey
John K. Galiotos, Houston Community College, Texas
George Golombowski, San Juan College, New Mexico
Ronald Good, Delta College, Michigan
Demont (Monty) Henderson, Kennebec College, Maine
Ken Hernandez, Houston Community College, Texas
Gary Hicks, Brazosport College, Texas
William Hodge, City University of New York, New York
Clarence Hughes, Sowela, Louisiana
Mike Kukuk, College of the Mainland, Texas
Jerry Layne, Baton Rouge Community College, Louisiana

Linton LeCompte, Sowela Technical Community College, Louisiana
Roger Murphy, Linden High School, New Jersey
Richard Ortloff, Bellingham Technical College, Washington
Dorothy Ortego, McNeese University, Louisiana
Tony Otero, San Juan College, New Mexico
Lyndon Pousson, Louisiana Technical College, Louisiana
Jonathan Prater, University of Alaska, Alaska
Bill Raley, College of the Mainland, Texas
Denise Rector, Del Mar College, Texas
Robert Robertus, Montana State University - Billings, Montana
Paul Rodriguez, Lamar Institute of Technology, Texas
Vicki Rowlett, Lamar Institute of Technology, Texas
Dan Schmidt, Bismarck State College, North Dakota
Robert (Bobby) Smith, Texas State Technical College - Marshall, Texas
Dale Smith, Alabama Southern Community College, Alabama
Michael Speegle, San Jacinto College, Texas
Mark Stoltenberg, Brazosport College, Texas
Keith Tolleson, Louisiana Technical College, Louisiana
Diane Trainor, Middlesex County College, New Jersey
Walter Tucker, Lamar Institute of Technology, Texas
Lane Warner, Alameda High School, Colorado
Steve Wethington, College of the Mainland, Texas
Bennett Willis, Brazosport College, Texas

CENTER FOR THE ADVANCEMENT OF PROCESS TECHNOLOGY STAFF:

Tina Burkhalter, Secretary
Melissa Collins, Senior Instructional Designer
John Dees, Senior Instructional Designer
Joanna Kile, Principal Investigator
Joanna Perkins, Outreach Coordinator
Bill Raley, Co-Principal Investigator
Debi Shoots, Public Relations Coordinator
Merv Treigle, Assistant Director
Scott Turnbough, Graphic Artist
Cindy Washington, Program Assistant

This material is based upon work supported, in part, by the National Science Foundation under Grant No. DUE 0202400. Any opinions, findings, and conclusions or recommendations expressed in this material are those of the author(s) and do not necessarily reflect the views of the National Science Foundation.

Chapter 1

Process Technology: An Overview

OBJECTIVES

Upon completion of this chapter you will be able to:

1. Define what process industries do.
2. Define what process technology is.
3. Describe what a process technician does.
4. Identify the duties, responsibilities, and expectations of the process technician.
5. Describe working conditions in the process industries.

KEY TERMS

- **Process**—a system of people, methods, equipment, and structures that create products from other materials.
- **Process Industries**—a broad term for industries that convert raw materials, using a series of actions or operations, into products for consumers.
- **Process Technology**—a controlled and monitored series of operations, steps, or tasks that converts raw materials into a product.
- **Process Technician**—a worker in a process facility that monitors and controls mechanical, physical and/or chemical changes, throughout many processes, to produce either a final product or intermediate product, made from raw materials.
- **Raw Materials**—also called feedstock or input. The material sent to a processing unit to be converted into a different material or materials.
- **Product**—also called output. The desired end components from a particular process.
- **Production**—output, such as material made in a plant, oil from a well, or chemicals from a processing plant.
- **Facility**—also called a plant. Something that is built or installed to serve a specific purpose.
- **Unit**—an integrated group of process equipment used to produce a specific product or products. All equipment contained in a department.

INTRODUCTION

This chapter provides you with a description of what process industries do and their impact on people, the economy, and the environment. This chapter also describes what a process technician does, and the tasks and working conditions of process technicians.

Throughout this textbook, the term "process industries" is used. "Process industries" is a broad term for industries that convert raw materials, using a series of actions or operations, into products for consumers.

Generally speaking, the process industries are concerned with processes that take quantities of raw materials and transform them into other products. The result might be an end product for a consumer, or an intermediate product which is used to make an end product. Each company in the process industries uses a system of people, methods, equipment, and structures to create products.

Process industries share some basic processes and equipment. For example, the distillation process (which is discussed later on in this textbook) can be used to make anything from alcoholic beverages to petroleum products. It just depends on the materials used. However, most of these industries have their own unique processes. For example, how water is used in paper making paper differs greatly from how water is treated and made suitable for drinking.

The process industries are some of the largest industries in the world, employing hundreds of thousands of people in almost every country. These industries, either directly or indirectly, create and distribute thousands of products that affect the daily lives of almost everyone on the planet.

WHAT PROCESS INDUSTRIES DO

There are a variety of industries classified as process industries. These include:

- **Oil and gas**

  The exploration and production segment locates oil and gas then extracts them from the ground using drilling equipment and production facilities.

  The transportation segment moves petroleum from where it is found, to the refineries and petrochemical facilities, then takes finished products to markets.

  The refining segment of the oil and gas process industry takes quantities of hydrocarbons and transforms them into finished products, such as gasoline and jet fuel, or into feedstocks (a component used to make something else, like plastics).

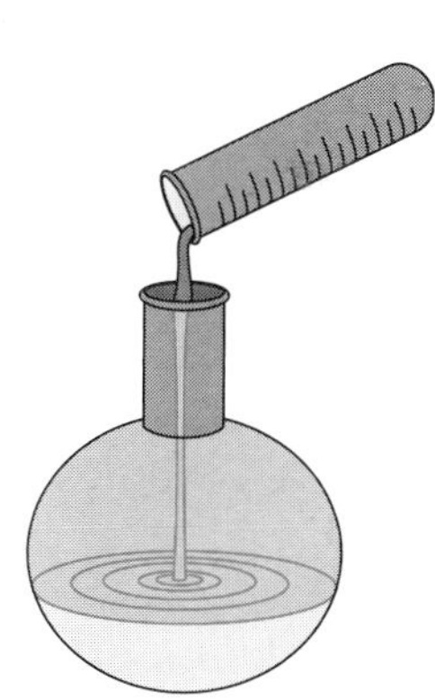

- **Chemicals**

  Chemicals play a vital role in a wide range of manufacturing processes, resulting in products such as plastic, fertilizers, dyes, detergent, explosives, film, paints, food preservatives and flavors, synthetic lubricants, and so on.

- **Mining**

  Mining is a complex process, which involves the extraction and processing of rocks and minerals from the ground. Mining products are integral to a wide range of industries, serving as base materials for utilities and power generation, construction, transportation, agriculture, electronics, food production, pharmaceuticals, personal hygiene, consumer products, and so on.

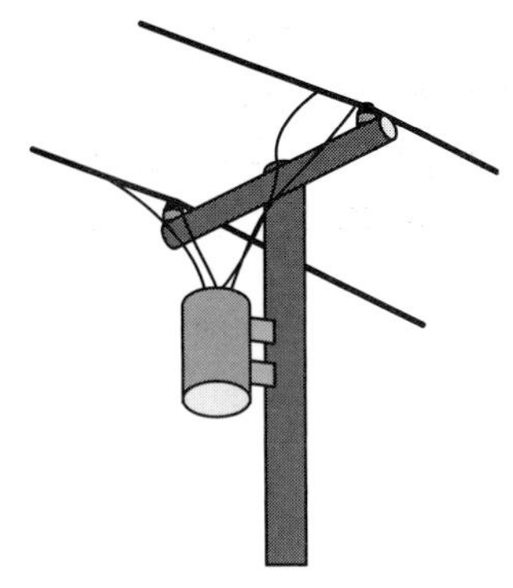

- **Power generation**

  Power generation involves the production and distribution of electrical energy in large quantities to industries, businesses, residences and schools. The role electricity plays in everyday life is enormous, supplying lighting, heating and cooling, and power to everything from coffee pots to refineries. There are three main segments to the power generation industry: generation, transmission, and distribution. Power can be generated in a variety of ways. These include burning fuels, splitting atoms, using water, and so on.

- **Water and Waste Water Treatment**

  Clean water is essential for life and many industrial processes. It is through water treatment facilities that process technicians are able to process and treat water so it is safe to drink, and safe to return to the environment.

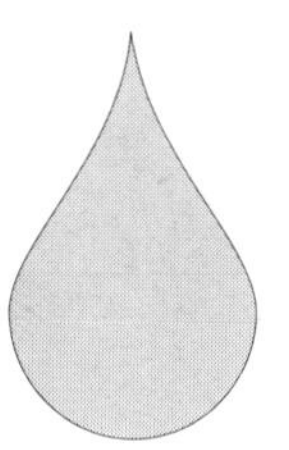

- **Food and Beverage**

  The food manufacturing industry links farmers to consumers through the production of finished food products. The products created by this industry can vary dramatically, and can range anywhere from fresh meats and vegetables, to processed foods that must simply be heated in the microwave.

- **Pharmaceuticals**

  Modern drug manufacturing establishments produce a variety of products, including finished drugs, biological products, bulk chemicals and botanicals used in making finished drugs, and diagnostic substances such as pregnancy and blood glucose kits.

  Modern drugs save lives and improve the well-being of countless patients, while improving health and quality of life, and reducing healthcare costs.

- **Pulp and Paper**

  Paper plays a huge role in everyday life. If, along with paper, you include items made from natural wood chemicals, then the pulp and paper industry creates and distributes thousands of products used daily around the world. The products include items such as packaging, documents, bandages, insulation, textbooks, playing cards, money and so on.

While this list of industries is not all-inclusive, it covers a broad range of industries which are responsible for producing or contributing to a wide range of items. These items include:

- DVD and CD media
- Electronic parts
- Telephones and cell phones
- Antihistamines, aspirin and antibiotics
- Carpets and wallpaper
- Parts for cars, boats and planes

- Gasolines and fuels
- Foods and drinks
- Boxes and packaging
- Jewelry
- Electricity
- Clean water
- Tires
- Checks, notepads, newspapers and books
- Credit cards
- Detergents and soaps
- Paints and glues
- Sporting goods
- Cosmetics and cologne
- Cameras and film
- Toothpaste
- Cement, steel and other building materials

It is easy to see the importance of the process industries and how they impact everyday life.

HOW PROCESS INDUSTRIES OPERATE

Although there are a wide ranges of processes and products associated with each type of process industry, they have some common operations:

1. Raw materials (sometimes called input or feedstock) are made available to a process facility (or plant), a place that is built to perform a specific function.
2. The raw materials are sorted by the proper process. Most facilities have different units that perform a specific process. A unit is an integrated group of process equipment used to produce a specific product.
3. Units can be referred to by the processes they perform, or named after their end products.
4. The raw materials are processed using people, equipment and methods. Process technicians monitor and control the mechanical, physical and/or chemical changes that occur during the process. Safety, quality, and efficiency are key elements to the process.
5. A product (output), or desired component, is the result of a particular process. The product can either be an end product for consumers, or an intermediate product used as part of another process to make an end product.
6. The product is then distributed to consumers.

WHAT A PROCESS TECHNICIAN DOES

A process technician is a key member of a team responsible for planning, analyzing, and controlling the production of products from the acquisition of raw materials through the production and distribution of products to customers in a variety of process industries.

A process technician monitors and controls mechanical, physical and/or chemical changes throughout many processes to produce a final product or intermediate product, made from raw materials.

Process technician duties can also include the acquisition and test of raw materials, monitoring the production and distribution stages and ultimately the shipment of products to customers in a variety of industries.

It is essential for a process technician to have the ability to work effectively in a team-based environment. Productivity within industries is accomplished by teams of people from many different backgrounds and with many different levels of education. Strong computer, oral, and written communication skills are essential for a process technician to operate within the organizational structure of the company as well as describe activities for relief personnel, maintain data logs, prepare reports, and perform other needed tasks.

The duties of a process technician include maintaining a safe work environment, controlling, monitoring and troubleshooting equipment, analyzing, evaluating and communicating about data and training others while also continuing his or her own learning process.

A process technician applies quality principles to all activities performed. Particular emphasis is placed on process control applied to production operations and the continuous improvement of those operations. While applying these principals, the process technician wears safety equipment, uses industrial safety devices and/or promotes safety among co-workers.

The life of a process technician must be flexible since he/she will work shiftwork in all types of weather. This career provides a variety of experiences for an individual looking for a challenging occupation.

Process technician duties include but are not limited to

- Maintaining a safe work environment
- Controlling, monitoring and troubleshooting equipment in processes
- Analyzing, evaluating and communicating data obtained through the use of technology.

Process technician ensures customer satisfaction by applying quality principles to all activities performed with an emphasis on production, operations, and continuous improvement of those operations.

Employees in the process industries are generally rewarded for job excellence through salary increases, promotions, and bonuses. Job benefits usually include health and dental insurance, profit sharing and retirement plans.

DUTIES, RESPONSIBILITIES, AND EXPECTATIONS OF PROCESS TECHNICIANS

Today's process technicians must deal with more automation and computerization than workers in the past. Fewer people are now required to work as process technicians due to advances in technology. Today's workers must have more education and be more highly skilled. A two-year Associates degree is recommended.

Process technician jobs are more complex and cross functional (meaning workers are required to perform a wider variety of tasks) than in the past. Technicians are responsible for entire complex processes, not just simple tasks.

Companies have an increased emphasis on keeping costs low, by improving safety, performing preventative maintenance, optimizing processes and making efforts to increase efficiency.

Along with technology changes, process technicians must keep up with constantly changing governmental regulations pertaining to safety and the environment.

This section covers general responsibilities, since there are a wide range of processes involved with the process industries and companies have different ways of organizing job duties.

Skills

The skills and traits companies look for when hiring process technicians include:

- An education (preferably a two year Associates degree), including a foundation in science and math.
- Technical knowledge and skills.
- Interpersonal skills (relating to and working with others).
- Communication skills (reading, writing, speaking).
- Computer skills.
- Physical capabilities.
- Ability to deal with change.
- Aptitude for technology.
- Ability to stay current in work skills.
- Flexible, self-directed team player.

- Ability to work on a team.
- Ability to appreciate a diverse workplace.
- Ability to follow safety, health, environment, and security procedures and policies.
- Sense of responsibility.
- Strong work ethic.
- Positive attitude.
- Respectful of others.
- Ability to accept criticism and feedback.

Job Duties

Process technicians may be responsible for a wide variety of tasks, including:

- Monitoring and controlling processes
    - Sample processes.
    - Inspect equipment.
    - Start, stop and regulate equipment.
    - View instrumentation readouts.
    - Analyze data.
    - Evaluate processes for optimization.
    - Make process adjustments.
    - Respond to changes, emergencies and abnormal operations.
    - Document activities, issues and changes.
- Assisting with equipment maintenance
    - Change or clean filters or strainers.
    - Lubricate equipment.
    - Monitor and analyze equipment performance.
    - Prepare equipment for repair.
    - Return equipment to service.
- Troubleshooting and problem solving
    - Apply troubleshooting processes such as root cause analysis.
    - Participate in corrective action teams.
    - Use statistical tools.
- Communicating and working with others
    - Write reports.
    - Perform analysis.
    - Write and review procedures.
    - Document incidents.
    - Help train others.
    - Learn new skills and information.
    - Work as part of a team.

- Performing administrative duties
  - Do housekeeping (make sure the work area is clean and organized).
  - Perform safety checks.
  - Perform environmental checks.
- Performing with a focus on safety, health, environment, and security
  - Keep safety, health, environmental and security regulations in mind at all time.
  - Look for unsafe acts.
  - Watch for signs of potentially hazardous situations.

Equipment

Process technicians work with many different types of equipment, including:

- Pipes and valves
- Vessels (tanks, drums, reactors)
- Pumps
- Compressors
- Furnaces
- Boilers
- Heat exchangers
- Cooling towers
- Distillation columns
- Turbines
- Motors
- Instrumentation
- Auxiliary systems
- Utilities

Later chapters in this textbook explain what this equipment does, how to monitor and maintain it, potential hazards, and other information.

Workplace Conditions and Expectations

The following are some workplace conditions that process technicians can expect:

- All types of weather, including extreme conditions.
- Being outside some or most of the time.
- A drug and alcohol free environment.
- Being a member of a team.

- Shift work in a facility that operates 24 hours a day, 7 days per week.
- Using tools and lifting some heavy objects.

Companies expect process technicians to:

- Avoid incidents and errors (i.e., work safely, in compliance with government and industry regulation).
- Help with environmental compliance.
- Work smarter and focus on business goals.
- Look for ways to reduce waste and improve efficiency.
- Keep up with industry trends and constantly improve skills.

Safety

A major expectation of employers is that employees have a proactive attitude regarding safety. Two crucial safety points that process technicians should remember are that accidents are preventable and they must be proactive (thinking ahead) before a problem occurs.

Government regulations are in place to protect worker health and safety, the community and the environment. A range of federal government agencies and their regulations impact the process industries: OSHA, EPA, Department of Transportation, and so on. State agencies also regulate certain elements within their jurisdiction.

Companies use physical and cyber security measures to protect assets and workers from internal and external threats. Process technicians must be trained and be able to recognize safety and security threats, and understand the impact on themselves and the facility where they work.

Quality

Quality is an important part of the process industries. Quality has two major definitions: (1) A product or service free of deficiencies; and (2) the characteristics of a product or service that bear on its ability to satisfy stated or implied needs.

Without quality measures, products and services could be deficient or unsatisfactory. Unsatisfactory products lead to unhappy customers, increased waste, inefficiencies, increased costs, reduced profits, and an inability to maintain a competitive edge.

Teams

A team usually consists of a small number of people. People are picked for a team because they have skills that complement the other team members' skills. Everyone on the team is committed to a common purpose. All team members hold the other members mutually accountable for the success of the team.

When working as part of a team, it is important to recognize and appreciate others for their contributions to the workplace, while not discounting them because of their differences. It is vital to understand diversity and practice its principles in the workplace.

Shift Work

Since most process facilities operate 24 hours a day, seven days a week, 365 days a year, process technicians are typically shift workers. Each plant or process facility is unique in the way shifts and work days are arranged.

Most shift work involves 8 hour or 12 hour rotations, with numerous variations on the way non-working days are arranged. Some examples of working day arrangements are:

- Four days on, four days off—the process technician works four consecutive days and then is off four consecutive days.
- EOWO (every other weekend off)—the process technician works one weekend and is off the next weekend.

August

	1	2	3	4	5	6
7	8	9	10	11	12	13
14	15	16	17	18	19	20
21	22	23	24	25	26	27
28	29	30	31			

Process technicians must to be able to adjust to the schedules that a plant or process facility operates under. Because of this, it is vital that process technicians understand the impacts of shift work.

A person operates on a natural time clock that is different from work hours. Two mental low points can occur every 24 hours, typically between 2-6 a.m. and 2-6 pm. The "Sun up effect" may also occur, where a person wakes up when the sun rises, no matter how sleep deprived the person is.

Shift work has been compared to having permanent jetlag. People who work long, irregular hours tend to:

- Experience fatigue.
- Have a reduced attention span.
- Have a slowed reaction time.
- Experience conflicting body clock and work schedule.
- Be less attentive.

- Think and remember less clearly.
- Have more accidents.

Shift work can affect:

- **Physical health**—resulting in high rates of alcohol, drug, and tobacco use, overeating, lack of exercise, and long-term sleep disturbances.
- **Emotional health**—resulting in increased irritability and a tendency toward depression and a lack of social life or healthy leisure activities.
- **Family matters**—resulting in higher divorce rates; little time with children and spouse, few shared family activities, and missing out on social outings.

Process technicians can reduce the impact of shift work by taking care of themselves. To maintain physical and mental health, a process technician should:

- Establish as regular a schedule as possible.
- Create a day-sleeping environment.
- Take naps when possible.
- Avoid stimulants, alcohol, and caffeine.
- Eat only light snacks in the 2-6 a.m./p.m. period.
- Compensate for lower awareness.

Union versus Non-Union Work Environment

Many companies in the process industries hire employees that belong to labor organizations called unions. A goal of a union is to help employees establish better working environments and benefits for themselves. Process technicians should understand how working conditions can vary based on union versus non-union work environments.

However, not every employee in the process industries is a union member. Companies with employees that belong to unions often have different labor practices and operating conditions than those companies with non-union employees.

Unions use collective bargaining to negotiate various work conditions and policies for its members. Unions usually select representatives to negotiate with employers. These negotiations seek to create collective bargaining agreements (work rules) that govern company practices ranging from wages and benefits, to unit staffing levels, to job assignments based on seniority.

Non-union employees, on the other hand, do not negotiate labor practices with their employers as union members do. Instead, non-union employees follow the labor policies that the employer has established.

The labor policies that employers provide differ from company to company. Companies provide standard benefits and compensation for their employees. Adopting such policies helps to ensure employee retention and job satisfaction.

SUMMARY

Process industries are industries that convert raw materials into products for consumers using a series of actions or operations. Each company in the process industries uses a system of people, methods, equipment, and structures to create products.

Process industries share some basic processes and equipment. However, most of these industries have their own unique processes.

The process industries are some of the largest industries in the world, employing hundreds of thousands of people in almost every country. These industries, either directly or indirectly, create and distribute thousands of products that affect the daily lives of almost everyone on the planet.

Process technician jobs are more complex and cross functional than in the past. Technicians are responsible for entire complex processes, not just simple tasks.

There is an increased emphasis on keeping costs low, by improving safety, performing preventative maintenance, optimizing processes and making efforts to increase efficiency.

Along with technology changes, process technicians must keep up with constantly changing governmental regulations pertaining to safety and the environment.

CHECKING YOUR KNOWLEDGE

1. Define the following key terms:
   a. Process
   b. Process Industries
   c. Process Technology
   d. Process Technician
   e. Raw Materials
   f. Product

g. Production

h. Facility

i. Unit

2. A process is a system of people, methods, equipment, and structures that create products from other materials.

3. Name at least five process industries.
Oil & Gas, Mining, Waste Water, Power Plant, Food & Beverage

4. Which of the following is the best definition of a unit in the process industries?

   a. 144 items grouped together

   b. A distribution center

   c. A plant

   d. An integrated group of process equipment used to produce a specific product or products

5. Process technicians Monitor and control the mechanical, physical and/or chemical changes that occur during a process.

6. *(True or False)* Process technicians do not need an understanding of math and science for their jobs.

   a. True

   b. False

7. Name at least three skills that companies look for when hiring process technicians. 2 year degree, Computer Skills, Communication

8. Name at least five traits that companies look for when hiring process technicians. Responsibility, Positive attitude, Safety, Ability to Work on a team, Respectfulness

9. *(True or False)* Many facilities operate 24 hours a day, seven days a week.

10. When are two mental low periods during the day?

   a. 12:00 a.m.–2:00. p.m.

   b. 12:00 a.m.–4:00 p.m.

   c. 2:00 a.m.–6:00 p.m.

   d. 5:00 a.m.–9:00 p.m.

11. What can you do to combat the fatigue associated with shift work? (Select all that apply)
    a. Drink plenty of coffee or tea
    b. Take naps when possible
    c. Avoid stimulants and alcohol
    d. Eat only light snacks in the 2-6 a.m. or p.m. period
    e. Work an irregular schedule.

ACTIVITIES

1. Which process industry sounds the most interesting to you? Why? Write a one page paper in response.

2. What do you think are the benefits of working as a process technician? What are the drawbacks? Make a chart with one column for benefits and one for drawbacks. Write down you responses, then compare the benefits and drawbacks.

3. Think about the impact shift work would have on your life. Make a list of how it would affect your daily life.

Chapter 2

Oil and Gas Industry Overview

OBJECTIVES

Upon completion of this chapter you will be able to:

1. Explain the growth and development of the oil and gas process industry.
2. Explain the different segments of the oil and gas industry.
3. Identify the duties, responsibilities, and expectations of the process technician in the oil and gas industry.
4. Describe changes and future trends in the role of the process technician in the oil and gas industry.

KEY TERMS

- **Exploration**—the process of locating oil and gas reservoirs by conducting surveys and studies, and drilling wells.
- **Hydrocarbon**—organic compounds that contain only carbon and hydrogen that are most often found occurring in petroleum, natural gas and coal.
- **Petroleum**—a substance found in the earth, such as oil or gas, composed of chemical compounds consisting primarily of hydrogen and carbon.
- **Production**—the process of bringing oil and gas to the surface then preparing it for transport.
- **Refining**—the process of purifying a crude substance into other products, such as petroleum being separated into gasoline, kerosene, gas oil.
- **Transportation**—the oil and gas industry segment responsible for moving petroleum from wells to processing facilities and finished products to consumers. Transportation methods include pipelines, watercraft, railways and trucks.

INTRODUCTION

This chapter provides you with a history of the oil and gas process industry and its impact on people, the economy, and the environment. This chapter also describes the tasks and working conditions of process technicians, as well as futures trends and how they affect these technicians.

In this chapter, we refer to the oil and gas industry. Many segments make up this industry, including exploration and oilfield production, transportation (e.g. pipelines, trucks), and refining. In this text, then primary focus will be on the refining segment, although transportation and exploration and production will be discussed some.

The exploration and production segment locates oil and gas then extracts it from the ground using drilling equipment and production facilities.

The transportation segment moves petroleum from where it is found, to the refineries and petrochemical facilities, then take finished products to markets.

The refining segment of the oil and gas process industry takes quantities of hydrocarbons and transforms them into finished products, such as gasoline and jet fuel, or in to feedstocks (a component used to make something else, like plastics).

A closely related industry is the chemical industry. Sometimes you will hear the term petrochemical, which refers to chemicals derived from petroleum products. See ***Chemical Industry Overview*** for details.

GROWTH AND DEVELOPMENT OF THE OIL AND GAS INDUSTRY

Petroleum is the name given to the liquid oils found in the earth. Petroleum is a substance made almost exclusively of **hydrocarbons** (organic compounds that contain only carbon and hydrogen).

Petroleum can be either a solid, liquid, or gas. The main benefits of petroleum are that it contains a considerable amount of energy, and that it can be used to create many products (e.g., fuels, plastics, adhesives, paints, explosives, and pharmaceuticals).

Although no one is exactly sure how petroleum was formed, the most accepted theory is that it came from organic materials (the remains of carbon-based plants and animals that fell to the sea floor from millions of years ago and were buried by sediments like silt and sand.

Because oxygen could not penetrate the layers of sediment covering these remains, the material did not decay properly. Over time, many layers of organic material and sediment were created. The heat and pressure of all the layers stacked on top of one another caused the organic material to change into petroleum, and the surrounding sediments to change into sedimentary rock. The end product of these changes was an oil-saturated rock, similar to a kitchen sponge.

As geological forces changed the face of the Earth over time, seas became land and land became covered with water. During these shifts, petroleum was forced into different formations both onshore and offshore. These formations lay dormant for many years until they were discovered by ancient civilizations.

DID YOU KNOW?

The word petroleum comes from the Latin word *petra,* which means rock, and the Greek word *oleum* which means oil.

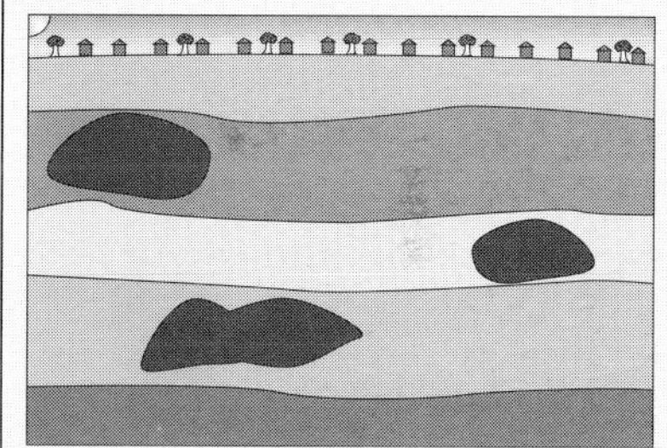

Literally, the term *petroleum* means "rock oil," or oil that comes from rocks.

The Early Days of Refining

> **DID YOU KNOW?**
> - The ancient Egyptians used petroleum to help preserve mummies.
> - Pitch, a tarry petroleum substance, was used extensively in ancient construction to seal off the pyramids and to waterproof Egyptian reed boats.
> - The ancient Chinese sometimes found oil when they were digging for brine (salt water). They captured this oil and burned it as fuel to heat their homes.
> - In ancient Persia, the streets of Babylon were said to be paved with asphalt, a solid form of petroleum.
> - The ancient Persians, Sumatrans and American Indians believed that oil had medicinal properties.
> - In the 1800's, before people knew what could be done with petroleum, it was bottled throughout the U.S. and sold as a miracle tonic that people would rub on their skin or drink.

The chemical processing industry can be traced back thousands of years to many different civilizations, including the ancient Egyptians, Chinese, and Native Americans. In the early days, however, the ability to harness and use raw petroleum products was very limited.

In 1821 a gunsmith named William Aaron Hart drilled the first natural gas well in Fredonia, New York. The gas from this twenty-seven feet deep well was piped to nearby buildings and was used for fuel and lighting. In the years following Hart's experiment, many scientists and researchers continued investigating additional applications for coal, shale, crude oil and natural gas.

In 1850 Samuel Kier built the first petroleum refinery in Pittsburgh, Pennsylvania. This refinery consisted of a one-barrel still that was used to distill petroleum. Later Kier added a larger 5-barrel still.

In 1853 kerosene was extracted from petroleum for the first time. This was an important discovery because whale oil (the first source of lamp oil) was becoming scarce.

In 1855 Benjamin Sillman Jr., a Yale University professor, began researching crude oil. He theorized that the components of the crude mixture could be separated out using heat and the distillation process. As it turned out, Sillman was correct. During his investigation he discovered that each component in the mixture had a unique boiling point and could, therefore, be separated out into a number of products including naphtha, solvents, kerosene, heavy oils and tars. However, the heavy oils and tars produced during this process were considered useless and were dumped into nearby rivers and streams.

Figure 2-1: Colonel Edwin Drake
Courtesy of The Pennsylvania Historical and Museum Commission, Drake Well Museum, Titusville, PA.

From ancient times until the mid 1800s, petroleum was collected when it seeped to the surface or from very shallow wells. While this method produced a quantity of oil, it was not satisfactory for large-scale applications. So, in 1859, four years after Sillman's discovery, Colonel Edwin Drake began exploring for oil with an old steam engine.

In order to facilitate his exploration, Drake adapted the engine to fit a drill and then selected a spot near Titusville, Pennsylvania as a drill site. On that spot Drake drilled a well seventy feet deep and was met with almost immediate success. Drake's well was considered the first commercially successful oil well.

In response to Drake's success, other oil drillers began setting down wells. Before long the beautiful Pennsylvania landscape was changed into an industrial community teeming with roughnecks, wooden oil derricks and wagons waiting to carry the raw crude to riverboats so it could be further transported to a handful of refineries located along the east coast.

With the introduction of commercial oils wells came the need for more refineries. By 1860, there were 15 refineries in the United States, all of which primarily produced kerosene.

These early refineries consisted mainly of large iron drums with a long tube that acted as a condenser. A coal fire heated the drum and three parts of the crude (called fractions) were boiled off. The first fraction to boil off was naphtha. The second was kerosene, and the third were the heavy oils and tar, which were considered useless and were dumped into the river.

Oil was transported from wells to refineries in wooden barrels (usually old whiskey barrels). Wagons or barges hauled these barrels. Barrels used to transport oil ranged greatly in size, but were finally standardized to 42 gallons around 1870. This 42-gallon standard is still the unit of measure used today.

Early attempts at transporting oil were often unsuccessful. It was not uncommon for oil to seep out of the barrels. As the oil seeped out, the smell of raw petroleum became nauseating to anyone working near the barrels. Furthermore, the volatility of the oil also made it dangerous to transport because of the likelihood of fire or explosion.

Figure 2-2: Early Oil Tank Wagon

Courtesy of PetroleumHistory.com.

In the early days of refining it was difficult to move the crude materials from the drilling site to the refinery for processing. Horse-drawn wagons were employed to carry countless loads from the drilling site to nearby refineries, railroads, or barges. The wagon drivers were called teamsters, because they managed the team of horses need to pull the wagon. Sometimes the wagons would take the barrels directly to the refineries. However, in most cases, water and barges were used to complete the transport.

Early oil barges were pulled by horse or depended on the river current to move. Eventually, however, these barges and riverboats were transformed into tanker boats specially designed to haul oil.

Hauling oil in barrels through horse-drawn wagons and riverboats was a slow and tedious process, so in 1865 a pipeline was built from Drake's Titusville well to a nearby railroad. This new transportation system, however, was not without problems. Teamsters, angry at losing transport business to the pipeline, attacked it with pickaxes, causing significant damage.

Early pipelines were made of wood or iron. Most designs tried to use gravity flow to push oil through the pipes. In 1865, Samuel Van Syckel built a wrought iron pipeline with steam pumps to push the oil. It also had a telegraph line running next to it, so people on each end of the pipeline could communicate about oil shipments, leaks, and more. This pipeline is considered one of the first successful oil pipelines in the U.S.

DID YOU KNOW?

Karl Benz preceded Henry Ford with his invention of the gasoline engine in 1885.

Courtesy of Brandon Classics, Atlanta, GA.

The Ford Motor Company was not founded until 1903. The first mass-produced automobile, the Model T, was not manufactured until 1908.

In 1869 the Transcontinental Railroad was completed. This railroad stretched from the east coast of the United States to the west coast. Special tanker cars were built to hold the petroleum. By 1869, a wooden, horizontal style tank car with a dome to allow the expansion of dangerous gases was in use. This basic style remains the industry standard today, although wood has been phased out and replaced with metal.

The completion of the Transcontinental Railroad helped improve the transportation of refinery products to all parts of the United States. That same year, new products made from refinery "wastes" (e.g., petroleum jelly, lubricants, candle way and chewing gum) were introduced into the U.S. market. However, there was still no market for the major waste product, gasoline.

In the early days, gasoline was tried as a local anesthetic and was also used in place of kerosene but with extremely limited success since it had a tendency to set homes ablaze or, in more severe instance, caused them to explode.

Finally, in 1885, Karl Benz invented a use for gasoline – the gasoline automobile. This was followed in 1892 by the diesel engine.

Between 1885 and 1886, two Germans, Gottlieb Daimler and Karl Benz, separately invented gasoline-powered automobiles. With this technology, a number of companies in France, Germany, and the U.S. began to produce automobiles. However, they tended to be costly since only one car could be manufactured at a time.

In the U.S., Henry Ford created a way to mass-produce automobiles that were more reasonably priced. In 1908, the Model T rolled off the assembly line and changed the transportation industry. Automobile transportation became affordable for many Americans. Tank trucks were developed to haul petroleum products where rail, watercraft and pipelines could not go.

With the introduction of the automobile, gasoline, previously considered a waste product, was now considered valuable. With this new market, better processes for making gasoline were needed.

In 1897, H.L. Williams drilled the first offshore oil well in California. Williams built a wharf extending 300 feet into the Pacific Ocean, on which he erected a drilling rig. When the well began to successfully produce, other wharves were built, with some extending over 1,200 feet over the ocean.

World War I and World War II

In 1914, war broke out in Europe. "The Great War" (now called "World War I") was the first mechanized, chemical war. Tanks, airplanes, and chemical weapons were used to change the nature of warfare. This brought renewed focus to the importance of the oil, gas, and chemical industries to military power.

Figure 2-3: World War I Tank

Courtesy of Library of Congress.

Twenty-five years later, in 1939, war broke out again in Europe and around the world. World War II spurred a huge growth in the oil, gas, and chemical industries, stimulating new advances in processing.

During the war, tankers were the lifeblood of both the Allied (e.g. the U.S., Britain) and Axis (e.g. Germany, Italy, Japan) forces. Unfortunately, German U-boats sank many Allied oil tankers as they traveled to East Coast refineries. Looking for a safe alternative, the federal government built two pipelines, the 24-inch diameter "Big Inch" Pipeline and the 20-inch diameter "Little Big Inch" Pipeline, to deliver crude oil from Texas to the East Coast.

Transporting oil and gas by rail and truck also was integral during both the World Wars. The outcome of many battles was determined by the supply and performance of petroleum products.

In 1940, Standard Oil developed a new process to increase the octane in gasoline, which helped Allied transportation outperform the Axis powers. The Allies' logistical superiority in moving fuels around the world is also credited with winning some victories. Some German offensives stalled when they ran out of fuel.

At the end of the war, millions of veterans returned home. This return unleashed several "booms" (dramatic increases in activity): a consumer buying boom, a housing boom, an education boom, and a baby boom. During this time the oil and gas industry enjoyed solid growth and profitability.

Increases in automobile ownership and new interstate highway systems increased the demand for gasoline products. Consumer goods manufacturing also increased to meet the demand for household goods, such as plastics, synthetic fibers, and other petrochemical products.

The 1950s to Present Day

In the 1950s, American and European companies with manufacturing locations all over the world dominated the petroleum industry. Other countries profited little when the U.S. and Europe tapped their oil resources.

In the 1960s and 1970s, foreign oil producing countries realized they could go into business for themselves. There was a wave of nationalization and the emergence of state-owned petroleum companies. An increase in production of oil caused prices to fall.

In 1973, the Organization of Petroleum Exporting Countries (OPEC) decided to boost prices with an oil embargo (trade restriction). The result was skyrocketing prices, lines at the gas stations, national trends towards conservation (e.g., no Christmas lights on houses), and purchases of smaller, more fuel-efficient cars.

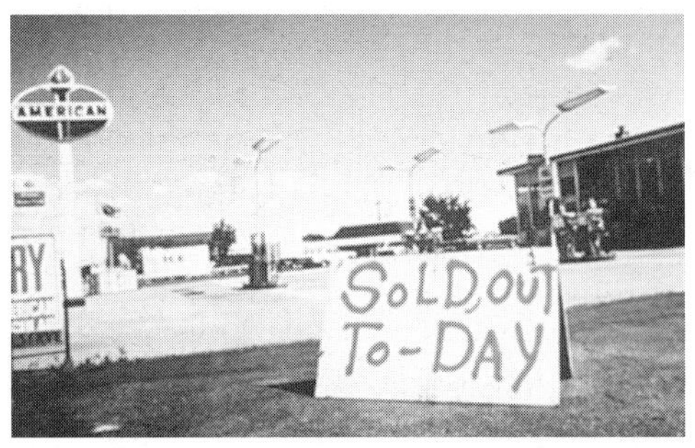

Figure 2-4: Cars lined up to buy gas during the 1970s oil embargo
Courtesy of U.S. Department of Energy.

As a result of the embargo, interest increased in alternative energy sources, such as: hydroelectricity, nuclear power plants and solar energy. In addition, exploration and production expanded into areas with harsh environments, such as Alaska, the North Sea, and the deep waters of the Gulf of Mexico.

Increased prices encouraged other nations (e.g., South America, Africa, China, and Russia) to begin producing their own oil. Eventually, the supply of oil increased and prices dropped.

Concern for the environment and worker safety boomed in the 1960s and 1970s. Governments enacted legislation to provide for cleaner air and water, along with safer working environments.

In the 1980s, the price of oil dropped, greatly affecting the oil and gas industry. Exploration and production dropped off, and refining operations were optimized for better performance. During the 1980s and 1990s, increased global competition, governmental regulations, a reduction in work force, and other factors contributed to tight markets. Many companies downsized and/or merged with others to remain in business.

OIL AND GAS INDUSTRY SEGMENTS

The oil and gas industry is broken into three branches or segments. These segments include: exploration and production, transportation and refining.

Exploration and Production

Exploration is the process of locating oil and gas reservoirs by conducting surveys and studies, and drilling wells. During the exploration process, scientists called geologists and geophysicists examine the Earth and locate potential petroleum reservoirs on land or under water using seismic studies and other tools. Once a location is identified, surveys, legal issues, and drilling rights are resolved to allow a company to drill. Once a potential reservoir is detected, a well is drilled to verify the presence of hydrocarbons. Only a small percentage of wells in a new prospect are commercially viable.

Once the company is ready to drill, specialized drilling equipment (called a rig) is set up at the location. Land rigs can be transported by truck, barge, or sometimes by helicopter, while offshore rigs are built into barges, moveable platforms or ships. Figure 2-5 shows an example of an offshore rig.

A rig consists of a derrick (also called a tower), a power system, a mechanical hoisting system, and rotating equipment.

Figure 2-5: Pompano offshore oil rig

Courtesy of BP. © 2005 BP Photo Resources.

Through the rig, workers drill a hole in the ground using a rotating bit. The rotating equipment on the rig provides the bit with the motion it needs to cut the hole (called a well bore). Sections of drilling pipe are attached to the bit, which is raised and lowered by the derrick as the well bore gets deeper.

As the well bore is drilled, fluid (either water or drilling fluid called mud) is circulated into the hole to cool the drill bit and remove cuttings. Casing, or pipe, is placed in the hole to keep it from collapsing. A Blow-Out Preventer is a device placed on the surface above the well bore to prevent an uncontrolled rush of oil or gas to the surface.

Drilling continues in stages. The well is drilled deeper and then more casing is added. Testing processes (samples and electronic sensors) are used to check for the presence of petroleum. When the well's depth reaches the petroleum reservoir, a device blows holes in the casing to let the petroleum flow. Tubing is inserted in the well and a valve structure is placed at the top of the well. This valve structure allows the flow of petroleum to be controlled.

Once the well is drilled, the production phase starts. Facilities and equipment are brought in to extract the petroleum to the surface and transport it to another location for processing.

Figure 2-6: Oil drilling pipe and downhole tools
Courtesy of CAPT.

A variety of other tasks are associated with exploration and production, but are outside the scope of this textbook.

There are many state and federal agencies involved with the regulation of oil and gas exploration and production. These agencies include: OSHA, EPA, Department of Transportation, U.S. Coast Guard, Minerals Management Service, and more.

Transportation

Transportation is the oil and gas industry segment responsible for moving petroleum from wells to processing facilities and finished products to consumers. Transportation methods include pipelines, watercraft, railways and trucks.

Pipelines are the prime method for transporting petroleum products. They move high volumes of product at low cost. Pipelines account for almost two-thirds of the petroleum quantities moved in the U.S. In general, pipelines tend to run east-west since a majority of the oil is in states such as Texas, Louisiana, Oklahoma, and Alaska, while demand and some of the refining capabilities are on the East Coast.

Figure 2-7: Oil tanker
Courtesy of Corbis Images.

Gas pipelines and oil pipelines are traditionally separate systems. A gas pipeline uses compressors to generate pressure as a driving force for the gas, while an oil pipeline uses pumps to generate pressure to drive liquids. When compared with other transportation methods, pipelines are one of the safest methods for transporting oil and gas.

The second most used transport method for oil and gas is watercrafts (e.g., oil tankers). The costs for this type of transportation are moderate.

Tankers transport products around the world, from large oil producing regions to consumer countries. Within North America, barges and other crafts transport petroleum products through inter-coastal waterways and to coastal ports. After the boom of the 1970's and the following downturn, the fleet of oil tankers began to reduce in size, and new vessel construction remains slow as the fleet continues to age.

Pipelines replaced railways as the major method of transport during the 20th century. The cost of rail delivery is moderate. However, rail delivery is limited to the existing system of rails. The current use of rail transportation for petroleum products is fairly static, remaining at less than 5% of the total volume of transportation. Since rails tend to run north-south rather than the east-west, they prove useful in transporting petroleum products to northern and southern locales.

Trucks are the most flexible method of petroleum products delivery. They can be used to transport products to smaller distribution centers and from fields that do not have a pipeline system yet. However, because the amount of product that can be transported at one time is limited, the cost of using trucks for petroleum product transportation is high.

Figure 2-8: The beginning of the Alaska pipeline
Courtesy of CAPT.

There are many state and federal agencies involved with the regulation of oil and gas transportation. These agencies include: OSHA, EPA, Department of Transportation, U.S. Coast Guard, Minerals Management Service, and more.

A variety of other topics are associated with transportation, but are outside the scope of this textbook.

Refining

Refining is the process of purifying a crude substance into other products, such as petroleum being separated into gasoline, kerosene, gas, and oil.

Figure 2-9: Refinery
Courtesy of EyeWire/Getty Images, Inc.

A refinery is a facility used to separate various petroleum components into fuel products or feedstocks for other processes, such as plastics or pharmaceuticals. Refineries involve a series of processes to separate and change petroleum components.

The refining process relies on differences between chemicals (e.g., boiling points) to separate petroleum components. The following is an overview at the refining process:

1. **Distillation** (also called fractionation)—the refining process begins with the distillation of petroleum into separate components. This process uses the different boiling points of liquids to separate components (called fractions). Atmospheric and vacuum towers are used during this process.
2. **Conversion**—a process that changes the size and/or structure of the petroleum components. One method called "cracking" breaks down large molecules into smaller ones. If heat is used, the process is called thermal cracking. If a catalyst (a substance that is used to affect the rate of a chemical reaction) is used, the process is called catalytic cracking. Other methods include combining components and rearranging components.
3. **Treatment**—a process that prepares the components for additional processing. This process also creates some final products. Various

components or contaminants may be removed or separated at this point. Chemical or physical separation can be used along with a variety of other methods and treatments.

4. **Formulating and blending**—a process that combines or mixes components and additives to produce finished products with specific performance requirements.
5. **Additional processes**—processes such as recovering components, treating wastes and water, cooling, handling, moving and storing.

There are many state and federal agencies involved with the regulation of oil and gas refining. These agencies include: OSHA, EPA, Department of Transportation, U.S. Coast Guard, Minerals Management Service, and more.

DUTIES, RESPONSIBILITIES, AND EXPECTATIONS OF PROCESS TECHNICIANS

This section deals primarily with process technicians in the refining industry. The chemical industry is covered in the Chemical Industry Overview chapter of this textbook. Exploration, production, and transportation are outside the scope of this textbook.

A Declining Workforce

The process industries are one of the largest industries in the world, employing over 500,000 refining and chemical processing workers alone. However, this industry is predicting a severe shortage of skilled workers in the relatively near future.

According to statistics, it is estimated that some industries (namely the chemical processing industry) will be forced to replace as much as 80-90% of the existing workforce by the year 2015, primarily because the "baby boom" generation is reaching retirement age.

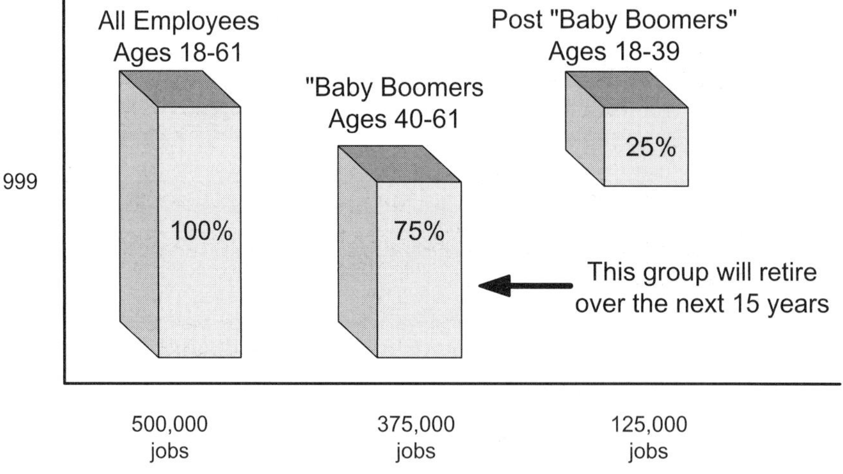

Figure 2-10: Projected worker shortage once the "baby boomers" retire

If these statistics are accurate, industry will need to hire 4,800 to 6,000 process technicians each year to support industry demands (Source: Petrochem Magazine, October 1997). Unfortunately, records indicate that as many as 70% of current high school students do not have the basic knowledge and skills required to fill these industry positions. Because of this, education beyond the high school level will be required for all process technicians.

In the past, most processing jobs involved operating machinery. Today, much of the machinery is automated and computerized, so fewer people are required to make it run. Those who run this machinery often find that their jobs are more complex and cross–functional than traditional employees, since workers today are responsible for entire processes, not just simple tasks.

Employees of today are required to be more technologically savvy and have better reading, writing and verbal communication skills than their predecessors. In addition, they must be able to keep up with constantly changing technology and regulations.

With the threat of global competition always looming in the background, the employees of today must be highly skilled and able to work quickly, safely, and efficiently using modern technology and processes. Because of this, a two-year Associates degree is preferred for process technicians.

Skills

The skills companies look for most when searching for potential applicants include:

- Education (preferably an Associates degree)
- Technical knowledge and skills
- Interpersonal skills
- Computer skills
- Physical abilities

Good common sense and the ability to troubleshoot are also desirable skills.

Job Duties

Process technicians may be asked to perform a wide variety of job duties. The duties may include:

- Monitoring and controlling processes and equipment.
- Sampling process streams and analyzing data.
- Inspecting, operating and maintaining equipment.
- Troubleshooting processes and equipment.

- Making process adjustments.
- Responding to changes, emergencies and abnormal operations.
- Documenting activities, issues and changes.
- Communicating and working with others as part of a team.
- Learning new skills and information.
- Helping train others.
- Performing administrative and housekeeping duties (e.g., making sure the work area is clean and organized).
- Performing safety and environmental checks and looking for unsafe acts or potentially hazardous situations.
- Keeping safety, health, environment and security regulations in mind at all time.

Equipment

Process technicians in the oil and gas industry will work with many different types of equipment, including:

- Pumps
- Compressors
- Pipes
- Valves
- Vessels
- Furnaces
- Boilers
- Heat exchangers
- Cooling towers
- Distillation columns
- Turbines
- Motors
- Instrumentation

These types of equipment are explained in more detail later on in this textbook.

Workplace Conditions and Expectations

The following are some of the workplace conditions that process technicians can expect:

- All types of weather, including extreme conditions.
- Potential of working away from home for long periods of time (e.g. offshore rigs, Alaska pipeline)
- Working outside at least part of the time.
- A drug and alcohol free environment.

- Being a member of a team.
- Working shift work in a facility that operates 24 hours a day, 7 days per week.
- Using tools and doing some heavy lifting.

In addition to these conditions, companies also expect process technicians to:

- Avoid accidents and errors (work safely, in compliance with government and industry regulation).
- Help with environmental compliance.
- Work smarter and focus on business goals.
- Look for ways to reduce waste and improve efficiency.
- Keep up with industry trends and constantly improve skills.

CHANGES AND FUTURE TRENDS IN THE ROLE OF PROCESS TECHNICIANS

The role of process technicians in the oil and gas industry is changing. Process technicians should expect to see the following trends in the future:

- Increased use of computers, automation, and advanced technology.
- New and updated government and industry regulations.
- More involvement in process control and business decision.
- Increased workplace diversity.
- Less traditional methods of supervision.
- Continuous upgrading of skills and knowledge.
- New training methods and technologies, including virtual facilities, advanced simulators, and partnerships between education and industry.

SUMMARY

Many segments make up the oil and gas industry, including oilfield exploration, production, transportation, and refining. The exploration and production segment locates oil and gas then extracts it from the ground using drilling equipment and production facilities. The transportation segment moves petroleum from where it is found, to the refineries and petrochemical facilities, then take finished products to markets. The refining segment of the oil and gas process industry takes quantities of hydrocarbons and transforms them into finished products, such as gasoline and jet fuel, or in to feedstocks (components used to make something else, like plastics).

The history of the oil and gas process industry is long and varied. Ancient civilizations discovered many uses for petroleum. They used it for everything from heating to medicine to waterproofing. In the 19th century, early refineries and oil wells starting appearing in the U.S. The main use of petroleum then was kerosene, for heating and lighting.

The invention of the automobile and two world wars caused gasoline and similar petroleum products to gain importance. A population boom after World War II lead to expansion of consumer markets and increased use of petroleum products.

Environmental and safety regulations grew in importance during the 1960s and 1970s. During the oil crisis of the 1970s, petroleum products increased. This lead to exploration by countries that previously depended on others for petroleum, and new technologies such as wind and solar power were explored.

In the 1980s, oil prices dropped due to excess production and many companies merged or downsized to stay in business. In the 1990s, companies sought to operate more efficiently and effectively for lower costs.

Process technicians must deal with more automation and computerization than past workers in the industry. Fewer people are now required to run processes due to advances in technology. However, these workers must have more education and must be more highly skilled than workers in the past. A two-year Associates degree is preferred for process technicians.

Process technician jobs are more complex and cross functional (meaning workers are required to perform a wider variety of tasks) than in the past. Technicians are responsible for complex processes, not just simple tasks.

There is an increased emphasis on keeping costs low by improving safety, performing preventative maintenance, optimizing processes and making efforts to increase efficiency.

Along with technology changes, process technicians must keep up with constantly changing governmental regulations, which address safety issues and the environment.

Process technicians must have strong communication skills, as well as solid computer skills and an understanding of math, physics and chemistry.

Futures trend such as increased automation and the use of new technologies will affect the duties of a typical process technician.

CHECKING YOUR KNOWLEDGE

1. Define the following key terms:
    a. Exploration
    b. Hydrocarbon
    c. Petroleum
    d. Production
    e. Refining
    f. Transportation

2. _______________ is a substance found in the earth, such as oil or gas, composed of chemical compounds consisting primarily of hydrogen and carbon.

3. Which of the following was not an ancient use of petroleum?
    a. Heating homes in China
    b. Making chariot wheels and shoes for Romans
    c. Using it as a medicine among Native Americans
    d. Preserving mummies in Egypt

4. Put the following events in order, from earliest to latest:
    a. Automobile invented
    b. First refinery built in the U.S.
    c. Improvements in refining during World War II
    d. First successful commercial oil well drilled in Pennsylvania

5. Name three segments of the oil and gas industry.

6. What did the earliest refineries produce?
    a. Kerosene
    b. Propane
    c. Gasoline
    d. Diesel

7. What is the most common method of transporting petroleum?
    a. Watercraft
    b. Pipelines
    c. Rail cars
    d. Trucks

8. What is the process of purifying a crude substance into other products called?
   a. Exploration
   b. Transportation
   c. Fractional concatenating
   d. Refining

9. Select the item below that is not a major criterion that you must meet to work as a process technician.
   a. Perfect hearing and vision
   b. Education
   c. Physical capabilities
   d. Technical knowledge and skills

10. Companies that hire process technicians now prefer them to have:
    a. Bachelor of Science
    b. GED
    c. Associate degree
    d. Industry certification

11. What tasks are associated with monitoring of a process? Select all that apply.
    a. Sample process streams
    b. Use statistical tools
    c. Analyze data
    d. Make process adjustments
    e. Do housekeeping (make sure the work area is clean and organized)

ACTIVITIES

1. Which event or period in time had the biggest impact on the oil and gas industry? Why? Write at least five paragraphs in response.

2. Tell which of the following oil and gas segments sound the most interesting to you and explain why. Write several paragraphs in response.
   a. Exploration and Production.
   b. Transportation.
   c. Refining.

3. Pick three of the following tasks and explain what would be tough about performing those tasks. Then pick the three remaining tasks and explain what would be easy about performing those.
   a. Monitoring and controlling process
   b. Maintaining equipment
   c. Troubleshooting and problem solving
   d. Performing administrative activities
   e. Performing assigned duties with a focus on safety, health, environment and production
   f. Returning equipment to service

4. Think about the skills required of process technicians. Make a list of the skills you think you already have and the skills you will need to learn.

Chapter 3

Chemical Industry Overview

OBJECTIVES

Upon completion of this chapter you will be able to:

1. Explain the growth and development of the chemical process industry.
2. Identify the duties, responsibilities, and expectations of the process technician in the chemical industry.
3. Describe changes and future trends in the role of the process technician in the chemical industry.

KEY TERMS

- **Alchemy**—a medieval practice that combined occult mysticism and chemistry.
- **Chemical**—a substance with a distinct composition that is used in, or produced by, a chemical process.
- **Chemistry**—a science that studies the composition, structure and properties of a substance, along with the changes they undergo.
- **Commodity Chemicals—**basic chemicals that are typically produced in large quantities, and in large facilities. Most of these chemicals are inexpensive and are used as intermediates.
- **Intermediates—**substances that are not made to be used directly, but are used to produce other useful compounds.
- **Petrochemical**—refers to chemicals derived from fossil fuels or petroleum products.
- **Specialty Chemicals—**chemicals that are produced in smaller quantities, are more expensive, and are used less frequently than commodity chemicals.
- **Synthetic**—a substance resulting from combining components, instead of being naturally produced.

INTRODUCTION

This chapter provides you with a history of the chemical process industry and its impact on people, the economy, and the environment. This chapter also describes the tasks and working conditions of process technicians. It also covers futures trends and how they affect process technicians.

WHAT ARE CHEMICALS?

A **chemical** is a substance with a distinct composition that is used in, or produced by, a chemical process. Chemicals play a vital role in a wide range of manufacturing processes. Consider plastic, one of the most pervasive chemical products. Plastics are product found in almost every aspect of modern society. They are used in computers, cars, hospitals, houses, appliances, electronics, food packaging, games, toys, manufacturing, aircraft, schools, businesses, sporting goods, and more.

But plastics are not the only items produced through chemistry. The following are just a few of the items produced by the chemical industry:

TABLE 3-1: Examples of products produced through chemical manufacturing

Industry	Examples of chemical products
Food and Beverage	sweeteners, preservatives, flavorings, packaging
Medical	pharmaceuticals, plastics (e.g., syringes, tubes)
Construction	paints, coatings, adhesives, colored glass
Agriculture	fertilizers, pesticides
Clothing	dyes, fibers
Transportation	artificial rubber, improved fuels, oils, lightweight materials
Cleaning	detergents, bleaches
Military	explosives, armor plating
Personal Hygiene	soaps, perfumes, beauty aids
Amusement and Hobbies	photographic materials, plastic toys and games, sporting goods

Chemical manufacturing requires an understanding of chemistry, manufacturing processes, and the special equipment and structures used to process the chemicals.

Some chemical companies produce products that sell directly to consumers, although most create feedstocks that are used to create other products. Some chemicals are considered commodity chemicals and others are considered specialty chemicals.

Commodity chemicals are basic chemicals, typically produced in large quantities, and in large facilities. Most of these chemicals are inexpensive and are used as **intermediates** (substances that are not made to be used directly, but are used to produce other useful compounds).

Specialty chemicals are chemicals that are produced in smaller quantities, are more expensive, and are used less frequently than commodity chemicals.

To learn more about chemistry, refer to the *Chemistry* chapter in this textbook.

GROWTH AND DEVELOPMENT OF THE CHEMICAL INDUSTRY

The chemical industry is generally viewed as having seven different segments:

1. Basic chemicals
2. Synthetic materials (e.g., resins, artificial rubber, fibers)
3. Agricultural chemicals
4. Paints, coating and adhesives
5. Cleaning materials
6. Pharmaceuticals and medicines
7. Various chemical products

All of the industry segments, except for pharmaceuticals, are addressed generally in this chapter. Pharmaceuticals are covered in the ***Pharmaceutical Industry Overview*** chapter of this textbook.

DID YOU KNOW?

The word chemistry is supposedly derived from the word "khemeia," a variation of the Greek word "khumos", or juice.

Khemeia is said to relate to the art of extracting juices, which was how early chemistry was viewed.

An industry closely related to the chemical industry is the oil and gas industry. Within the industries the term petrochemical is often used. The term **petrochemical** refers to chemicals derived from fossil fuels or petroleum products. Petroleum is discussed in more detail in the ***Oil and Gas Industry Overview*** chapter of this textbook.

In the Beginning

The use of chemicals can be traced back to early civilizations that existed thousands of years ago. For example, in 7,000 B.C., Persian artisans produced glass using refined alkali and limestone. Roughly 1,000 years later, Phoenicians produced soap. Early Greek, Chinese and Egyptian cultures also contributed greatly to advances in chemistry and chemicals. These cultures sought ways to cure diseases, prolong life and change common metals into precious metals like gold and silver. Through their studies and experiments, a pseudo-science called alchemy was born. **Alchemy** was a medieval practice that combined occult mysticism and chemistry (e.g., some alchemists tried to blend non-precious metals to create gold, while others tried to create elixirs of immortality).

Arabs added to the study and practice of alchemy. But during this learning process, they also created small amounts of chemical, medicinal elixirs and a compound similar to plaster of Paris that could be used to heal broken bones.

1000s–1600s

Europeans picked up the practice of alchemy from the Arabs during the Crusades, around the 11th century. Around the same time, the Chinese invented black powder, an early explosive.

Between the 11th and 6th centuries, many alchemy practices, which previously were closely guarded secrets passed on from a mentor to a student verbally, were written down by scholars and scientists. In 1597, the German alchemist Andreas Libau published what is considered the first chemistry textbook, *Alchemia.*

1600s–1800s

In the 1630s, Pilgrim in Massachusetts were using chemicals to tan leather and making saltpeter, a crucial component in gunpowder.

In the 18th century, alchemy began to make a transition from mysticism to a serious science. A Flemish physician, Jan Baptista Van Helmont, described how air and vapors were similar in appearance but had different properties. He called air and vapors "Chaos," a Greek word for the material used to create the universe. In Flemish, Chaos is pronounced similarly to "gas", which is what we still collectively call air, vapors and similar substances.

Large-scale chemical manufacturing began around the time of the Industrial Revolution, between the late 1700s through the late 1800s, when England and the U.S. transformed from farming and trading economies into large scale manufacturing economies.

In the 1800s, consumer demand rose for items such as glass, paper, explosives, matches, fertilizers, soaps, and textiles. Manufacturing of these items required large quantities of chemicals, including soda ash, sulfuric acid, potash, and others. In the 1820s, James Muspratt began mass-producing soda ash in Britain. Soda ash is used for making soap, glass and other products. Some years later, synthetic dyes for textiles were produced from coal tars.

As early as 1839, English citizens were petitioning to have production of certain chemicals such as soda ash banned because the manufacturing emissions killed the trees, vegetation, and animal life around the factories and caused serious ailments.

Around 1890, German companies starting producing sulfuric acid in large quantities. Other companies began to mass-produce caustic soda and chlorine.

DID YOU KNOW?

Women's' nylon stockings were in short supply during World War II due to nylon shortages (the nylon was needed for parachutes and other equipment).

Because of this, nylon stockings became a "black market" item and the price skyrocketed.

World War I

At the start of the 1900s, synthetic fertilizers revolutionized agriculture by improving crop yields. By the late 19th and early 20th centuries, the U.S. and Britain had become dominant players in chemical manufacturing, but Germany was more advanced in basic chemistry research.

In 1914 "The Great War" (now called World War I) broke out in Europe. WWI was the first major conflict that involved mechanized and chemical warfare. The countries involved in this war all used chemical weapons to kill the enemy, primarily in the form of mustard gas. The war brought renewed focus on the importance of chemical and refining industries to military power. Extensive chemical research was conducted on chemical weapons during WWI.

Rayon (dubbed "artificial silk") was the first synthetic fabric. Nylon was the first synthetic fiber made solely from petrochemicals. Nylon was first produced by Du Pont in 1928 for women's stockings.

World War II

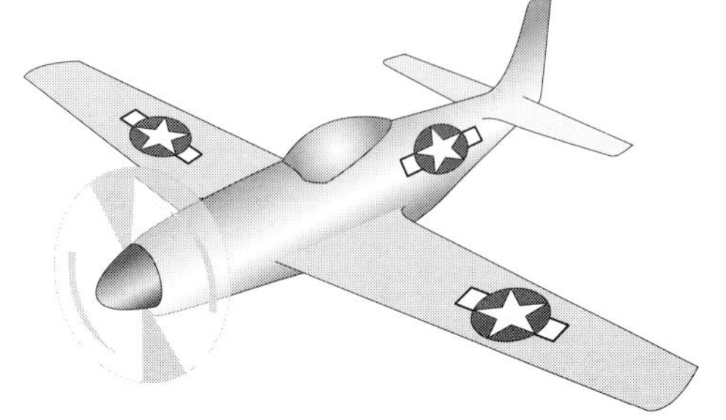

In the late 1930s and early 40s, another World War (WWII), broke out and once again chemicals played an important role in the conflict. The war spurred a huge growth in the refining and chemical processing industries, and stimulated many new process advances.

The chemical and refining industries made significant contributions to the U.S. victories during World War II. By the time the U.S. entered the war in 1941, Japan had captured 90% of the rubber-producing nations in Asia. This created a need for rubber substitutes for a variety of products (e.g., for tires, shoes, raingear, and tents). In 1940, the first synthetic rubber tire was produced in the U.S. from butadiene and styrene. Also during the war, the nylon industry was entirely taken over for the war effort, for parachutes, tents, rope, and other uses.

1960–Present

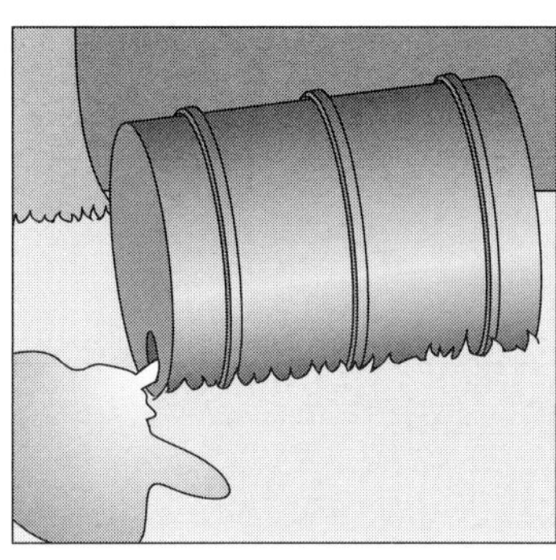

Concern for the environment and worker safety boomed in the 1960s and 1970s. Governments enacted legislation to provide for cleaner air and water, along with safer working environments.

A range of federal government agencies and their regulations impact the chemical industry. These agencies include OSHA, EPA, Department of Transportation, US Coast Guard, and more. State agencies also regulate certain elements within their jurisdiction.

During the 1980s, the U.S. chemical industry was greatly affected by increased global competition and governmental regulations. The price of oil dropped, impacting the petrochemical business. Chemical operations were optimized for better performance and many companies downsized and/or merged with others to remain in business.

The need to minimize risks from chemical manufacturing and processing has driven many of the advancements in chemical manufacturing processes, equipment, and construction of today's chemical plants.

DUTIES, RESPONSIBILITIES, AND EXPECTATIONS OF PROCESS TECHNICIANS

Process industries are one of the largest industries in the world. Over 500,000 workers are employed in just the refining and chemical processing industries. With a large number of workers expected to retire over the coming years (as people from the Baby Boom era reach retirement age), the oil and gas and chemical industries estimate that between 4,000 to 6,000 process technicians will be hired every year.

There are a limited number of fundamental processes used in chemical manufacturing, such as storing materials, transporting materials, heating and cooling fluids, filtering, and so on. This means that no matter what part of the chemical industry you may work in, there are many common skills and technical knowledge required. Each production process utilizes similar systems comprised of machines, equipment, and structures including tanks, pipes, valves and pumps.

Process technicians must deal with more automation and computerization than past workers in the industry. Although fewer people are now required to work as process technicians due to advances in technology, these workers must have more education and be more highly skilled. A two-year Associates degree is preferred for process technicians.

Process technician jobs are more complex and cross functional (meaning workers are required to perform a wider variety of tasks) than in the past. Technicians are responsible for entire complex processes, not just simple tasks.

There is an increased emphasis on keeping costs low, by improving safety, performing preventative maintenance, optimizing processes and making efforts to increase efficiency.

Along with technology changes, process technicians must keep up with constantly changing governmental regulations addressing safety and the environment.

Process technicians are required to have strong communication skills (i.e. reading, writing, and verbal) along with solid computer skills and an understanding of math, physics and chemistry.

Skills

The skills companies look for most when searching for potential applicants include:

- Education (preferably an Associates degree)
- Technical knowledge and skills
- Interpersonal skills
- Computer skills
- Physical abilities

Job Duties

Process technicians may be asked to perform a wide variety of job duties. The duties can include:

- Monitoring and controlling processes and equipment.
- Sampling process streams and analyzing data.
- Inspecting, operating and maintaining equipment.
- Troubleshooting processes and equipment.
- Making process adjustments.
- Responding to changes, emergencies and abnormal operations.
- Documenting activities, issues and changes.
- Communicating and working with others as part of a team.
- Learning new skills and information
- Helping train others.
- Performing administrative and housekeeping duties (e.g., making sure the work area is clean and organized).
- Performing safety and environmental checks and looking for unsafe acts or potentially hazardous situations.
- Keeping safety, health, environment and security regulations in mind at all times.

Equipment

Process technicians in the chemical industry will work with many different types of equipment, including:

- Pumps
- Compressors
- Pipes
- Valves
- Vessels
- Furnaces
- Boilers
- Heat exchangers
- Cooling towers
- Distillation columns
- Turbines
- Motors
- Instrumentation

These types of equipment are explained in more detail later on in this textbook.

Workplace Conditions and Expectations

The following are some of the workplace conditions that process technicians can expect:

- All types of weather, including extreme conditions.
- Working outside at least part of the time.
- A drug and alcohol free environment.
- Being a member of a team.
- Working shift work in a facility that operates 24 hours a day, 7 days per week.
- Using tools and doing some heavy lifting.

In addition to these conditions, companies also expect process technicians to:

- Avoid accidents and errors (work safely, in compliance with government and industry regulation).
- Help with environmental compliance.
- Work smarter and focus on business goals.
- Look for ways to reduce waste and improve efficiency.
- Keep up with industry trends and constantly improve skills.

CHANGES AND FUTURE TRENDS IN THE ROLE OF PROCESS TECHNICIANS

The role of process technicians in the chemical industry is changing. Process technicians should expect to see the following trends in the future:

- Increased use of computers, automation, and advanced technology.
- New and updated government and industry regulations.
- More involvement in process control and business decision.
- Increased workplace diversity.
- Less traditional methods of supervision.
- Continuous upgrading of skills and knowledge.
- New training methods and technologies, including virtual facilities, advanced simulators, and partnerships between education and industry.

SUMMARY

Chemicals play a vital role in a wide range of manufacturing processes. Following are just a few of the many ways that the chemical industry has impacted everyday life: food and beverage, medical, construction, agriculture, clothing, transportation, cleaning, military, personal hygiene, amusements and hobbies, and so on.

Chemical manufacturing requires an understanding of chemistry, manufacturing processes, and the special equipment and structures used to process the chemicals.

The chemical industry is generally viewed as having seven different segments: basic chemicals; synthetic materials; agricultural chemicals; paints, coating and adhesives; cleaning materials; pharmaceutical and medicines; and various chemical products

The history of the chemical industry is long and varied. Ancient civilizations discovered many uses for chemicals, including glass-making, soap and black powder. The pseudo-science of alchemy developed, with a focus on ways to change less valuable metals into valuable ones. Many cultures (e.g., Greek, Egyptian, Chinese and Arabic cultures) contributed to the study of alchemy and the use of chemicals.

In the late 17th and 18th centuries, alchemy became a true science, chemistry, with many scholars and scientists contributing to its advancement. In the 1800s, consumer demand rose for items such as glass, paper, explosives, matches, fertilizers, soaps, and textiles.

World War I was the first major conflict that involved mechanized and chemical warfare. Extensive chemical research on chemical weapons was conducted during the war. In the late 1930s and early 40s, World War, II broke out and once again chemicals played an important role in the conflict. The war spurred a huge growth in refining and chemical process industries and stimulated many new advances in processes.

Concern for the environment and worker safety boomed in the 1960s and 1970s. Governments enacted legislation to provide for cleaner air and water, along with safer working environments. A range of federal government agencies and their regulations impact the chemical industry: OSHA, EPA, Department of Transportation, US Coast Guard, and so on. State agencies also regulate certain elements within their jurisdiction.

During the 1980s, the U.S. chemical industry was greatly affected by increased global competition and governmental regulations. The price of oil dropped, impacting the petrochemical business. Chemical operations were optimized for better performance and many companies downsized and/or merged with others to remain in business.

Process technicians must deal with more automation and computerization than past workers in the industry. Although fewer people are now required to work as process technicians due to advances in technology, these workers must have more education and be more highly skilled. A two-year Associates degree is preferred for process technicians.

Process technician jobs are more complex and cross-functional (meaning workers are required to perform a wider variety of tasks) than in the past. Technicians are responsible for entire complex processes, not just simple tasks.

There is an increased emphasis on keeping costs low, by improving safety, performing preventative maintenance, optimizing processes and making efforts to increase efficiency.

Along with technology changes, process technicians must keep up with constantly changing governmental regulations addressing safety and the environment.

Process technicians are required to have strong communication skills (i.e. reading, writing, and verbal) along with solid computer skills and an understanding of math, physics and chemistry.

Futures trend such as increased automation and use of new technology affect process technicians' duties.

CHECKING YOUR KNOWLEDGE

1. Define the following key terms:
   a. Alchemy
   b. Chemical
   c. Chemistry
   d. Intermediate
   e. Petrochemical
   f. Commodity chemicals
   g. Specialty chemicals

2. _______________ refers to chemicals derived from petroleum products.

3. Which of the following civilizations is generally credited with creating black powder?
   a. Russians
   b. Phoenicians
   c. Greeks
   d. Chinese

4. Put the following events in order, from earliest to latest:
   a. Nylon is invented
   b. Phoenicians create soap
   c. English citizens petition to have production of certain chemicals banned
   d. Arabs create medicinal elixirs

5. Name four segments of the chemical industry.

6. What chemical weapon was used extensively in World War I?
   a. Anthrax
   b. Sarin
   c. German gas
   d. Mustard gas

7. Which of the following was the first synthetic fiber?
   a. Silk
   b. Rayon
   c. Nylon
   d. Dacron

8. Which of the following is NOT a troubleshooting and problem solving skill?
   a. Applying troubleshooting processes such as root cause analysis
   b. Participating in corrective action teams
   c. Using statistical tools
   d. Lubricating equipment

9. Name four skills that companies like process technicians to have.

10. Companies expect process technicians to perform with a focus on ________, health, security, and environment (choose the term below that fits best).
    a. Effectiveness
    b. Statistics
    c. Safety
    d. Technology

ACTIVITIES

1. Use the Internet to find out more about the history of the chemical industry, including discoveries, scientists, and impact. Write a one-page summary of your findings.

2. Make a list of common items around your home, school or workplace that are chemical products. See how many you can identify.

3. Discuss how you as a process technician can prepare to handle future trends in the chemical industry. Write at least five paragraphs.

Chapter 4
Mining Industry Overview

OBJECTIVES

Upon completion of this chapter you will be able to:

1. Explain the growth and development of the mining process industry.
2. Identify the duties, responsibilities, and expectations of the process technician in the mining industry.
3. Describe changes and future trends in the role of the process technician in the mining industry.

KEY TERMS

- **Deposit**—a natural accumulation of ore.
- **Geology**—the study of the earth and its history as recorded in rocks.
- **Metal**—chemical elements that have luster (ability to reflect light) and can conduct heat and electricity (e.g., copper, bauxite, iron, lead, gold, silver, zinc, nickel, and uranium).
- **Mine**—a pit or excavation from which minerals are extracted.
- **Minerals**—naturally occurring, inorganic substances, which have a definite chemical composition and a characteristic crystalline structure.
- **Mining**—the extraction of valuable minerals or other geological materials from the earth.
- **Non-metal**—substances that conduct heat and electricity poorly, are brittle, waxy or gaseous, and cannot be hammered into sheets or drawn into wire (e.g., gems and precious stones, coal, gravel, sand, lime, stone, soda ash, phosphate rock, and clay).
- **Ore**—a metal bearing mineral that is valuable enough to warrant mining (e.g., iron or gold).
- **Quarry**—an open excavation from which stones are extracted.

INTRODUCTION

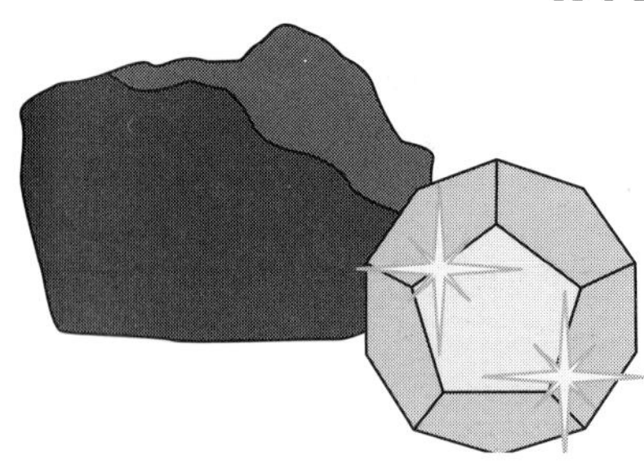

This chapter provides you with a history of the mining industry and its impact on people, the economy, and the environment. This chapter also describes the tasks and working conditions of process technicians, as well as future trends and how they affect process technicians.

Mining products are integral to a wide range of industries, serving as base materials for utilities and power generation, construction, transportation, agriculture, electronics, food production, pharmaceuticals, personal hygiene, consumer products, and so on. The following are some examples of the many ways products from mining processes impact everyday life:

- Coal for power generation
- Zinc for pharmaceuticals

- Salt for food
- Cement for construction
- Iron for steel (e.g., for cars, ships, buildings)
- Copper for electronics
- Gold for communications
- Graphite for pencils
- Fluoride for toothpaste
- Silver for jewelry
- Soda ash for fertilizers
- Feldspars for ceramics

HOW MINERALS ARE FORMED

Many minerals are formed when magma (molten rock beneath the Earth's surface) in the earth, interacts with other liquids and gases. As the magma cools, deposits of the mineral (called veins or lodes) collect according to the mineral's chemical properties. Rocks are usually consolidations of minerals combined with stone. Coal, an important energy-producing mineral, is an organic material consisting primarily of hydrogen and carbon, formed millions of years ago by biological and geological forces.

Geology is the study of the earth and its history as recorded in rocks. Process technicians in the mining industry should study and understand geology. Key concepts in geology include forces such as:

- **The Rock Cycle**—the process by which the different types of rocks are formed, exposed to weather and other forces that erode or break them down, and then are re-formed by geological forces (such as heat and pressure).
- **Plate tectonics**—a theory that explain how the earth moves, as semi-rigid plates under the earth (in its crust) drift or flow, causing geological changes like volcanoes, mountains and earthquakes.
- **Rock formations**—the study of the earth's landscape and its features.
- **Rocks**—minerals and gems are intertwined with the history of ancient civilizations, specifically with regard to buildings, food and cooking, weapons, religion, tools, and medicine.

OVERVIEW OF THE MINING INDUSTRY

Mining is the extraction of valuable minerals or other geological materials from the earth. **Minerals** are naturally occurring, inorganic substances, which have a definite chemical composition and a characteristic crystalline structure. To date, more than 3,000 different types of minerals have been identified. Minerals are the building blocks of rocks. Rocks can be composed of a single mineral, or a combination of minerals. Rocks and minerals are located in the earth, either just below the surface or buried deeply.

Minerals generally fall into two categories: metallic (metals) and non-metallic (non-metals):

- **Metals** are chemical elements that have luster (ability to reflect light) and can conduct heat and electricity (e.g., copper, bauxite, iron, lead, gold, silver, zinc, nickel, and uranium).
- **Non-metals** are substances that conduct heat and electricity poorly, are brittle, waxy or gaseous, and cannot be hammered into sheets or drawn into wire (e.g., gems and precious stones, coal, gravel, sand, lime, stone, soda ash, phosphate rock, and clay).

DID YOU KNOW?

Over 2/3 of the world's magnesium supply is mined from seawater.

12
Magnesium
Mg
24.31

Like coal, oil and gas (which are discussed in more detail in the ***Oil and Gas Industry Overview*** chapter of this textbook) are also considered energy-producing minerals.

Minerals can be extracted from either land or water. A **mine** is a pit or excavation from which minerals are extracted. A **quarry** is an open excavation from which stones are extracted. Mining operations also take place offshore. For example, electrolysis and other techniques may be use to remove minerals from seawater.

Ore is a metal bearing mineral that is valuable enough to warrant mining (e.g., iron or gold). A **deposit** is a natural accumulation of ore. Ore is extracted through mining.

Mining involves three fundamental elements: earth, air, and water. For example, during the mining process:

- Large amounts of earth are moved.
- Air (ventilation) in provided to enclosed mines.
- Water is pumped in or out, as the process requires.

DID YOU KNOW?

The Fort Knox gold mine in Alaska only retrieves about .028 ounces of gold per ton of rocks moved.

Courtesy of CAPT.

Mining uses powerful equipment and forces to handle these elements. Explosives and massive machines are used to move earth. Large compressors and fans are used to provide ventilation. Huge pumps are used to move water.

In general, the mining process involves the following steps:

1. Searching for rock and mineral deposits.
2. Extracting the rocks and minerals.
3. Processing the rocks and minerals.
4. Providing the rocks and minerals to consumers.

Mining Economics and Types of Mining

The economics of mining are challenging. First, companies must search for rock and mineral sites, and then obtain mining rights. Then, a considerable amount of extracting work must be done, often to obtain only small amounts of minerals. Finally, physical and/or chemical processing must be performed to prepare and refine the rocks and minerals. To add to these challenges, this process, which is often expensive and time consuming, occurs in an economic environment where prices can fluctuate rapidly.

Mining operations can vary greatly depending on the type of rock or mineral being mined, the location of the deposit (above ground, near the surface, or deep below the surface), the environment, and many other factors. For this reason, this text describes mining operations in very general terms.

There are two basic types of mining: surface and underground. Surface mining, also called open-pit or strip mining, is used when the rocks or minerals are located near the earth's surface. Underground mining is used when mineral deposits are located deep below the earth's surface.

SURFACE MINING

Surface mining, also called open-pit or strip mining, is used when the rocks or minerals are located near the earth's surface. Surfacing mining is typically more cost-effective than underground mining, since it requires fewer workers to produce the same amount of ore.

Figure 4-1: Example of a surface mine
Courtesy of CAPT.

The following are the general tasks associated with surface mining:

1. Explosives are used to move quantities of earth away from the deposit.
2. Workers use large earth-moving equipment to move away layers of dirt and rock covering the deposit.
3. When the deposit is exposed, workers use smaller earth-moving equipment to lift and load the ore from the deposit into trucks. If necessary, explosives may be used to break down the deposit further.
4. The ore retrieved is often crushed on-site, and is then transported to a nearby facility for processing.
5. When all of the ore that can be practically extracted is removed, the mine and its surrounding area are restored to original condition.

DID YOU KNOW?

Arsenic is an element that is naturally found in geological formations, especially granites. As long as this arsenic remains bound to the granite, it causes no harm.

However, when the granite is disturbed, either through mining or natural erosion, the arsenic may be released into the soil, thereby contaminating groundwater. That is why mining companies must monitor water quality during and after mining operations.

Courtesy of CAPT.

Quarrying operations are similar to mining operations in that stones (e.g. marble, granite, sandstone) are split into blocks from a massive rock surface.

UNDERGROUND MINING

There are different types of underground mining. These include:

1. **Conventional**—the oldest method, which involves cutting beneath the ore to control the direction of its fall, and using explosives to blast the ore free.
2. **Longwall**—similar to conventional mining; uses a machine with a rotating drum to shear the ore and load it onto a conveyor; hydraulic jacks are used to reinforce the roof.
3. **Continuous**—involves a machine called a continuous miner that rips ore out of the ground and loads it directly onto a transportation device.

The following is an overview of the tasks generally associated with underground mining:

1. Miners dig openings (tunnels) in the earth. Depending on where the deposit is in relation to the surface, these tunnels are vertical, horizontal, or sloped. Tunnels serve one of two purposes: 1) to allow miners to reach the deposit, extract the mineral, and transport it to the surface or 2) as ventilation, allowing fresh air into the mine.
2. Supports and pillars of unmined ore are used to support the mine roof.
3. Additional interconnecting passages are dug.
4. The ore is removed from the deposit and moved to surface using shuttle carts, rail cars, conveyor belts or other methods.
5. The ore is crushed (usually at an on-site location), and is then is hauled away for processing.
6. When all of the ore that can be practically extracted is removed, the mine and its surrounding area are restored to original condition.

Once underground mining operations have ceased, the mining company must check groundwater to make sure it is not contaminated, and must take actions to ensure that abandoned mines do not collapse.

Processing Operations

Once the ore is mined, it must be processed. Processing plants are usually located next to or near a mine or quarry. The job of the processing plant is to remove the target metals from the ore, then process it to meet certain specifications.

The following are some common tasks associated with processing:

Figure 4-2: Rock sorting machine

Courtesy of CAPT.

- **Separating and classifying**—a method of sorting minerals for processing. Machines called cyclones typically use centrifugal force to sort minerals.
- **Crushing**—a mechanical method which uses physical impact to reduce the size of ore. Crushing is typically a dry operation performed in three stages.
- **Grinding**—a mechanical method which uses impact and abrasion to further reduce the size of crushed rock and ore. Grinding can be dry or wet, and can take place in three stages.
- **Sizing**—a method used to divide crushed and ground minerals into different size categories.
- **Chemical processing**—a method that uses different solutions to process the minerals. For example, during leaching ore is mixed with chemical solutions, solvents or other liquids in order to separate the different material. Thickeners and clarifiers are added to condition the minerals.
- **Filtering**—a method which uses a porous medium to separate solids from liquids. In this method, a filter retains solids while allowing liquids to pass through.
- **Washing**—a method that uses water or other fluids to rinse the minerals and carry away impurities.

Figure 4-3: Rock sorting screen

Courtesy of CAPT.

GROWTH AND DEVELOPMENT OF THE MINING INDUSTRY

Many early civilizations were familiar with metals and other minerals. Because of this, the early history of civilization is often described in different ages, based on the types of minerals or metals they used (e.g., stone, bronze, iron, steel).

- **Stone Age** (around 2 million years ago)—the earliest technological period in human culture when metal was unknown and tools were made of stone, wood, bone, or antlers.
- **Copper Age** (around 5000 BC)—a phase in the development of human culture in which the use of early metal tools (made of copper) appeared alongside the use of stone tools.

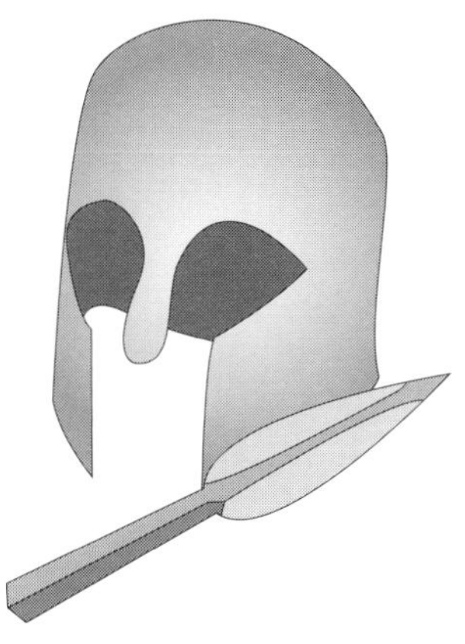

- **Bronze Age** (around 2500 BC)—the period in history characterized by the development of bronze and its use in the creation of weapons and tools.
- **Iron Age** (around 1500 BC)—the period in history characterized by the development of iron and its use in the creation of weapons and tools.
- **Steel Age** (around 200 AD)—the period in history characterized by the manufacture and use of steel in tools, weapons, and construction materials.

A variety of other metals and minerals were used during these ages as well. For example, the Egyptian pharaohs used gold, silver, and quarried stones to create jewelry and ornamentation for buildings, furnishing and monuments. The Chinese used jade and ceramics, and the Greeks had marble statues and cosmetics made from minerals.

Initially, these early civilizations found metals and minerals on the earth's surface, but soon discovered ways to obtain metals from the ground once surface supplies became scarce. During these ancient times the techniques for mining and processing minerals changed very little.

In the 16th century, cola was discovered as a good source of energy (e.g., heat). Because of this, the demand for coal increased. Within a few hundred years, it would provide the fuel source for the Industrial Revolution.

Industrial Revolution

The Industrial Revolution, which started in England the 1800s and spread to America shortly thereafter, changed the nature of manufacturing and sparked an increase in consumer products. Factories, powered by coal, began mass-producing products that previously were made by hand. Workers flocked to cities where factories provided jobs.

Coal also helped the growing transportation industry, which was required to open products to new markets. Railroads and ships benefited from the power of coal.

The discovery of gold and other precious minerals also spurred population shifts, such as gold rushes in California and Alaska.

World Wars I and II

In the early 1900s and later in the mid 1900s, two world wars increased the need for mining and minerals. Coal, along with oil and gas, provided fuel to the war efforts, while metals like steel (iron with low carbon content) and brass (essentially copper and zinc) were used to make vehicles and weapons.

Following World War II, a boom in the economy meant more minerals were needed for consumer products. For example, the construction business skyrocketed. This meant more cement, stones and metals were needed. Also, manufacturing increased, which meant a rise in the production of products such as electronics and appliances. The face of transportation changed, as more automobiles were produced, air travel became common, and ships transported goods around the world.

1960s to Present

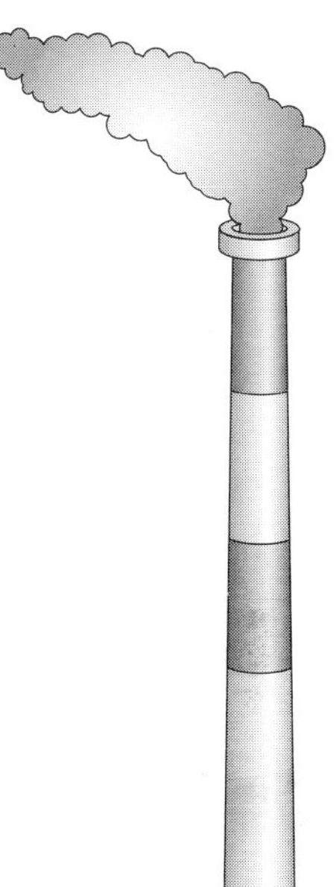

In the 1960s, coal provided about half the world's energy needs. By the 1970s, the need for coal had decreased and coal provided only one third of the world's energy needs. The rest was supplemented with oil and gas.

Concern for the environment and worker safety boomed in the 1960s and 1970s. Governments enacted legislation to provide for cleaner air, water, and land, along with safer working environments.

During the 1990s, production of both minerals and coal increased. Metal prices were volatile, leading to fluctuating production. Non-metallic mineral prices were more stable, leading to less production changes.

Today, changes in technology have made mining techniques safer and more automated (e.g., some mining operations even use lasers and robotics in their processes). Productivity has increased. Old sources of minerals, thought to be unprofitable to mine, are now being revisited due to more efficient recovery techniques.

DUTIES, RESPONSIBILITIES, AND EXPECTATIONS OF PROCESS TECHNICIANS

Process technicians in the mining industry can work in one of two primary segments: mining operations and processing operations.

Process technicians in these segments are exposed to more automation and computerization than past workers in the industry. Although fewer people are now required to work as process technicians due to advances in technology, these workers must have more education and be more highly skilled than workers in the past. A two-year Associates degree is preferred for process technicians.

Process technician jobs are more complex and cross-functional (meaning workers are required to perform a wider variety of tasks) than in the past. Technicians are responsible for entire complex processes, not just simple tasks.

There is an increased emphasis on keeping costs low, improving safety, optimizing processes and making efforts to improve efficiency.

Along with technology changes, process technicians must also keep up with constantly changing governmental regulations pertaining to safety and the environment.

Skills

Process technicians are required to have strong communication skills (i.e. reading, writing, and verbal); an understanding of geology, math, physics, chemistry and other physical sciences, along with solid computer skills.

So, companies are looking for skills such as:

- Education (preferably a two year Associates degree)
- Technical knowledge and skills
- Interpersonal skills
- Computer skills
- Physical abilities

Since there are a wide range of processes involved in mining and mineral processing, and since companies have different ways of organizing job duties, this section will only cover general responsibilities.

Job Duties

Process technicians in the mining industry may be responsible for a wide variety of tasks, including:

- Monitoring and controlling processes.
- Assisting with equipment maintenance.
- Troubleshooting and problem solving.
- Communicating and working with others.
- Performing administrative duties.
- Performing with a focus on safety, health, security, and environment.

Equipment

Process technicians may also be asked to work with a wide variety of equipment including:

- Pumps
- Compressors
- Pipes and chutes

- Valves
- Vessels
- Excavation equipment
- Supports and jacks
- Motors
- Instrumentation

Later chapters in this textbook explain what some of this equipment does, how to monitor and maintain it, potential hazards, and other information.

Process technicians may also work with explosives, jackhammers and other power tools, heavy lifting equipment, earth-moving equipment and various transportation equipment (e.g., carts, conveyors and rail cars), cutting tools and other related equipment.

Work Conditions and Expectations

Mining technicians will be expected to work in a variety of working conditions including:

- All types of weather, including extreme conditions.
- Spending a considerable amount of time outside.
- Working around noisy equipment and dust.
- A facility that operates 24 hours per day, as part of shift work.

Technicians will also be required to:

- Use tools.
- Perform duties that require physical exertion and stamina.
- Maintain a drug and alcohol free environment.
- Being a member of a team.

Companies expect process technicians to:

- Avoid accidents and errors (work safely, in compliance with government and industry regulation).
- Help with environmental compliance.
- Work smarter and focus on business goals.
- Look for ways to reduce waste and improve efficiency.
- Keep up with industry trends and constantly improve skills.

A range of federal government agencies and their regulations impact the mining industry: Minerals Management Service, OSHA, EPA, Department of Transportation, Mine Safety and Health Administration, Bureau of Land Management, and so on.

CHANGES AND FUTURE TRENDS IN THE ROLE OF PROCESS TECHNICIANS

Environmental regulations and issues will continue to affect the mining industry. Government regulations will continue to add and modify restrictions on types of mining and access to land. The government will also continue to impose stringent water, air and environmental quality standards, all of which impact the mining industry.

Many mining operations are expanding around the world, where reduced labor costs and lower environmental standards result in decreased production costs. International competition and consolidation have impacted the industry as well. Markets can fluctuate based on metal prices and needs from other industries (e.g. construction, transportation, and manufacturing), so the strength of the economy really determines the outlook for minerals.

Innovations in mining equipment have led to more efficient and safer operations. Production methods are becoming more automated and efficient. These technology advances mean fewer workers are required for operations, since many machines are computer-operated and can perform self-diagnostics if problems occur.

New, cleaner burning coal technologies are boosting the use of coal. These technologies not only burn coal more efficiently, they reduce emissions. Demand and production of coal is expected to increase in coming years. Coal provides about half of the electricity production in the U.S. Increased energy needs will also contribute to the demand for coal.

The role of process technicians in the mining industry is changing. Process technicians should expect to see these following trends:

- Increased use of computers, automation, and advanced technology.
- New and updated government and industry regulations.
- More involvement in process control and business decision.
- Increased workplace diversity.
- Less traditional methods of supervision.
- Continuous upgrading of skills and knowledge.
- New training methods and technologies, including virtual facilities, advanced simulators, and partnerships between education and industry.

SUMMARY

Mining products are integral to a wide range of industries, serving as base materials for utilities and power generation, construction, transportation, agriculture, electronics, food production, pharmaceuticals, personal hygiene, consumer products, and more.

Mining is a complex process, involving extracting and processing rocks and minerals from the ground.

Minerals are inorganic substances that occur naturally, with different chemical compositions and properties. Over 3,000 different types of minerals have been identified. Minerals are the building blocks of rocks. Minerals can combine with different types of minerals and other materials to form rocks, while some rocks of composed of just one type of mineral. Often, minerals are located under the earth, either just below the surface or buried deep.

Minerals generally fall into two categories: metallic (e.g. iron, copper) and non-metallic (e.g. coal, stone).

A mine is a pit or excavation from which minerals are extracted. A quarry is an open excavation from which stones are extracted. Mining operations also take place offshore.

The general mining process involves: searching for rock and mineral deposits, extracting the rocks and minerals, processing the rocks and minerals, and providing the rocks and minerals to consumers.

Mining operations can vary greatly depending on factors such as the type of rock or mineral being mined, the location of the deposit (above ground, near the surface, or deep below the surface), and the environment.

Two basic types of mining are surface and underground. Surface mining, also called open-pit or strip mining, is used when the rocks or minerals are located near the earth's surface. Underground mining is used when mineral deposits are located deep below the earth's surface.

Once the ore is mined, it must be processed. Processing plants are usually located near a mine or quarry. The job of a processing plant is to remove rock and other impurities from ore, then process it to meet certain specifications.

Rocks, minerals and gems are intertwined with the history of ancient civilizations. Initially, these early civilizations found metals and minerals on the earth's surface, but soon discovered ways to mine the metals when they became scarce on the surface. During these ancient times, not much changed about the way these minerals were excavated and processed.

The Industrial Revolution, which started in England the 1800s and spread to America shortly thereafter, changed the nature of manufacturing and sparked an increase in consumer products. Factories, powered by coal, began mass-producing products that previously were made by hand. In the early 1900s and later in the mid 1900s, two world wars increased the need for mining and minerals. Coal, along with oil and gas, provided fuel to the war efforts, while metals like steel (iron with low carbon content) and brass (essentially copper and zinc) were used to make vehicles and weapons.

Concern for the environment and worker safety boomed in the 1960s and 1970s. Governments enacted legislation to provide for cleaner air, water, and land, along with safer working environments. Changes in technology have made mining techniques safer and more automated. As a result, productivity has increased.

A range of federal government agencies and their regulations impact the mining industry. These agencies include: Minerals Management Service, OSHA, EPA, Department of Transportation, Mine Safety and Health Administration, and the Bureau of Land Management.

Process technicians in the mining industry today must interface with automation and computerization more than past workers in the industry. Although fewer people are now required to work as process technicians due to advances in technology, these workers must have more education and be more highly skilled. A two-year Associates degree is preferred for process technicians.

Process technician jobs are more complex and cross-functional (meaning workers are required to perform a wider variety of tasks) than in the past. Technicians are responsible for entire complex processes, not just simple tasks. There is an increased emphasis on keeping costs low, by improving safety, performing preventative maintenance, optimizing processes and making efforts to increase efficiency.

Along with technology changes, process technicians must keep up with constantly changing governmental regulations addressing safety and the environment. Process technicians are required to have strong communication skills (i.e. reading, writing, and verbal); an understanding of geology, math, physics, chemistry and other physical sciences, along with solid computer skills.

Environmental regulations and issues will continue to affect the mining industry. Government regulations continue to ad restrictions on types of mining and access to land. Stringent water, air and other environmental quality standards impact the industry. Also, futures trend such as increased automation and use of new technology affect process technicians' duties.

CHECKING YOUR KNOWLEDGE

1. Define the following key terms:
   a. Deposit
   b. Geology
   c. Metal
   d. Mine
   e. Mineral
   f. Mining
   g. Non-metal
   h. Ore
   i. Quarry

2. ______________ are inorganic substances that occur naturally, with different chemical compositions and properties.

3. What are the two primary types of minerals?
   a. Coal and metal
   b. Hard and soft
   c. Sedimentary and igneous
   d. Metallic and non-metallic

4. What is an ore? Select the best answer
   a. Made from hydrogen and carbon
   b. A concentration of minerals of sufficient quantity and quality that can be mined profitably.
   c. Another name for a mine
   d. A special type of coal.

5. What are the two main types of mining?
   a. Surface and strip
   b. Strip and open pit
   c. Surface and underground
   d. Crushing and stripping

6. What event sparked the use of coal?
   a. The Crusades
   b. The Industrial Revolution
   c. World War II
   d. The 1970s energy crisis

7. In the 1970s, coal provided about what percentage of the world's energy needs?
   a. One half
   b. Seventy five percent
   c. Ten percent
   d. One third

8. Which of the following are requirements for process technicians in the mining industry?
   a. Technical knowledge
   b. Computer skills
   c. Physical abilities
   d. Interpersonal skills
   e. All of the above

9. Which trend impacts coal demand? (select the best answer)
   a. Cleaner coal burning technologies
   b. Rise in automobile production
   c. A drop in the price of metals
   d. A reduction in workforce

ACTIVITIES

1. Explain which types of mineral impacts modern living the most, metals or non-metals? Write at least five paragraphs in response.

2. Which of the following mining segments would you be interested in and why? Which of the following would you be least interested in and why? Write a one-page paper in response.
   - Surface mining
   - Underground mining
   - Process operations.

3. Pick one of the following tasks and research some specific details about it for the mining industry. Use the Internet, library, or other resources for your research. Write at least five paragraphs in response.
   a. Monitoring and controlling process
   b. Maintaining equipment
   c. Troubleshooting and problem solving
   d. Performing administrative activities
   e. Performing assigned duties with a focus on safety, health, environment and production
   f. Returning equipment to service

4. Think about the skills required of process technicians in the mining industry. Make a list of the skills you think you already have and the skills you will need to learn.

Chapter 5

Power Generation Industry Overview

OBJECTIVES

Upon completion of this chapter you will be able to:

1. Explain the growth and development of the power generation process industry.
2. Explain the different segments of the power generation industry.
3. Identify the duties, responsibilities, and expectations of the process technician in the power generation industry.
4. Describe changes and future trends in the role of the process technician in the power generation industry.

KEY TERMS

- **Alternating Current (AC)**—an electric current that reverses direction periodically. This is the primary type of electrical current used in the process industries.
- **Atom**—the smallest particle of an element that still retains the properties and characteristics of that element.
- **Coal**—an organic, energy-producing mineral, consisting primarily of hydrogen and carbon.
- **Cogeneration Station**—A utility plant that produces both electricity and steam (for heating).
- **Direct Current (DC)**—an electrical current that always travels in the same direction.
- **Electricity**—a flow of electrons from one point to another along a pathway, called a conductor.
- **Fission**—the process of splitting the nucleus, the positively charged central part of an atom, that results in the release of large amounts of energy.
- **Generator**—a device that converts mechanical energy into electrical energy.
- **Geothermal**—a power generation source that uses steam produced by the earth to generate electricity.
- **Hydroelectric**—a power generation source that uses flowing water to generate electricity.
- **Nuclear**—a power generation source that uses the heat from splitting atoms to generate electricity.
- **Solar**—a power generation source that uses the power of the sun to heat water and generate electricity.
- **Turbine**—a machine for producing power. Activated by the expansion of a fluid (e.g., steam, gas, air) on a series of curved vanes on an impeller attached to a central shaft, which is used to create mechanical energy.

INTRODUCTION

This chapter provides you with a history of the power generation industry and its impact on people, the economy, and the environment. This chapter also describes the tasks and working conditions of process technicians, as well as future trends and how they affect these technicians.

Overview of Electricity

Electricity is a flow of electrons from one point to another along a pathway, called a conductor. Electricity flows in a continuous current from a high potential point (i.e. the power source) to a point of lower potential (i.e. homes or businesses), through a conductor (e.g., a wire). High voltage electricity is transmitted from the power plant to the power grid, a system of wires that channel the electricity to where it is needed.

When electricity leaves a power plant it is usually very high voltage. Substations are used to step the power down to a lower, safer voltage. The substation then distributes the electricity through feeder wires to a **transformer**, a device that steps down the voltage again. A service drop (or wire) runs from the transformer to the homes or businesses.

Once inside the home or business, electricity is used as energy to do work (e.g., light a bulb or operate a motor). Insulators, circuit breakers, fuses and other safety devices are used to make the whole power transmission process as safe as possible.

Overview of the Industry

Power generation involves producing and distributing electrical energy in large quantities to serve the requirements of industry, businesses, residences, schools and more. Electrically powered equipment harnesses the natural forces of electricity. We use electricity to power everything from coffee pots to refineries.

Every process industry requires power to operate. Because of this, process technicians working in segments outside of power generation should at least understand the fundamentals of this vital industry.

There are three main segments in the power generation industry:

- Generation
- Transmission
- Distribution

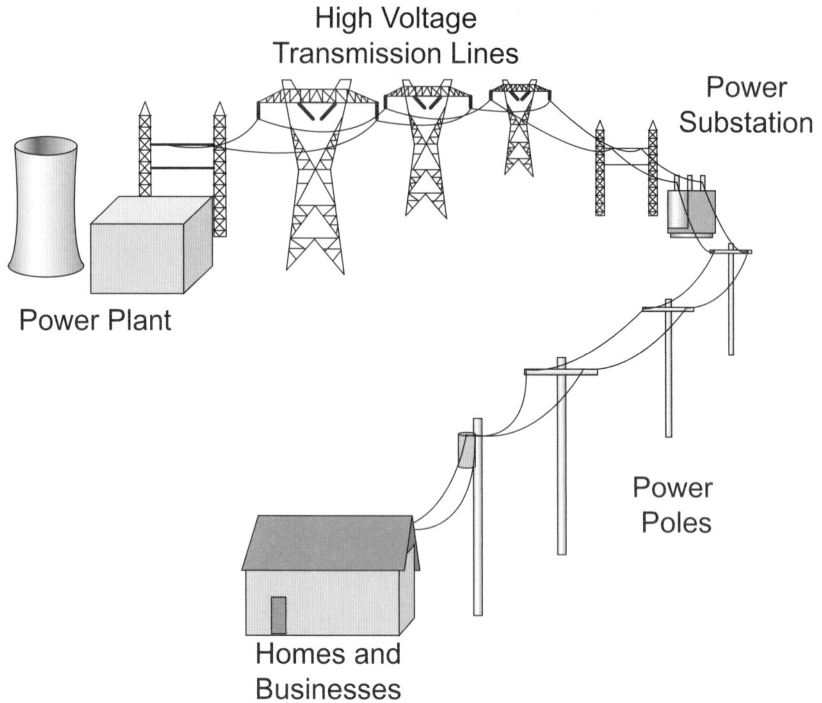

Generating plants produce electricity by converting mechanical energy or heat using a variety of methods. These methods include: burning a fuel (coal, oil, or gas), using water (hydroelectric power), splitting atoms (nuclear power), or harnessing resources such as the sun, wind or heat from the earth.

These sources of power are used to operate turbines and **generators,** which are machines that generate electricity. From the generating plant, the electricity travels to a transmission facility that moves the power over high voltage lines to certain regions. Distribution facilities then carry the power to end users, such as industrial or residential consumers.

For more information about electricity, refer to the ***Electricity and Motors*** chapter in this textbook.

GROWTH AND DEVELOPMENT OF THE POWER GENERATION INDUSTRY

Electricity occurs naturally in the world, a fact that is demonstrated during a lightning storm or when you shuffle your feet across the carpet and then touch a doorknob and receive a shock. Electrically powered equipment is just a way of harnessing these natural forces of electricity. Since the early days of civilization, people have been working to harness electricity and other sources of power.

Early Civilizations

The basis of power generation began with early civilizations sought to control and use various energy sources to do useful work. Civilizations over 2,000 years old used the power of water to turn wheels that ground wheat into flour. The ancient Greeks conducted experiments in static electricity by rubbing amber against a cloth. And a Greek inventor, Heron, created a simple amusement device that used the power of steam to rotate an object.

Eventually, around the 16th century, people started burning coal, an energy-releasing mineral, to provide heat. Windmills were also used to tap the power of wind and convert it to mechanical energy to grind wheat into flour and operate pumps.

Between the 17th and 20th centuries, many famous discoveries pertaining to the properties of electricity were made. For example, in the mid 1700s Benjamin Franklin performed experiments with kites to prove that lightning was electricity. Contrary to popular belief, Franklin did not "discover" electricity. Rather, he showed that it existed in nature.

DID YOU KNOW?

The terms Amp, Volt, Ohm, and Watt are all named after scientists who studied power and electricity.

Term	Named After
Amp	Marie Ampere
Volt	Alessandro Volta
Ohm	Georg Simon Ohm
Watt	James Watt

1700s

In 1763, James Watt created the first practical steam engine. His invention helped spur the Industrial Revolution in England and the U.S., since manufacturing could create products more quickly due to the use of machines like the steam engine. The electrical term watt is named after James Watt.

In the late 1700s, Alessandro Volta discovered that, when moisture comes between two different metals, electricity is generated. Using this discovery, he created a battery called the voltaic pile. This battery was constructed from thin sheets of copper and zinc separated by moist pasteboard. During his experiments, Volta demonstrated that electricity flows like water current. He also showed that electricity could be made to travel from one place to another, through conductors.

1800s

In the early 1800s Andre Marie Ampere, a French scientist, experimented with electromagnetism and discovered a way to measure electrical currents. The term Ampere is used as to describe units of electric current.

Also in the early 1800s, Georg Simon Ohm, a German physicist, discovered a law relating to the intensity of an electrical current, electromotive force, and resistance. His name, Ohm, is used to describe electrical resistance.

DID YOU KNOW?

In 1882, Edison provided lighting to a street in New York.

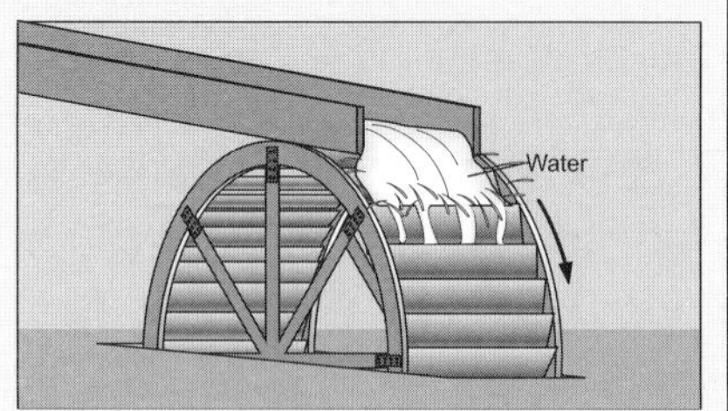

That same year in Wisconsin, a waterwheel on a river was used to supply hydroelectric power to two paper mills and a house.

In the mid 1800s Michael Faraday, an English scientist, devised a way to generate electrical current on a larger, more practical scale. Faraday expanded on earlier experiments that showed electricity could produce magnetism. During his experiments, Faraday demonstrated that magnetism could produce electricity through motion. He showed that moving a magnet inside a coil of copper wire produced an electrical current. His experiment resulted in the first electric generator.

In the latter part of the 1800s, incandescent lighting was developed by an English scientist name Joseph Swan. Thomas Edison, the American inventor, made a similar discovery in the U.S. several months after Swan. Swan and Edison created a company together to produce the first practical, incandescent lighting.

Also in the late 1800s, Thomas Edison created a Direct Current generator to power the lights in his lab. **Direct Current (DC)** is an electrical current that always travels in the same direction. Edison's generator, powered by Watt's steam engine, signaled the start of practical, large-scale electrical generation.

However, Edison's DC power had problems with long distance generation and voltage regulation, so Nikola Tesla (who had worked for Edison) devised another system for electrical generation: Alternating Current. **Alternating Current (AC)** is an electric current that reverses direction periodically.

George Westinghouse, an American inventor and industrialist, developed Tesla's Alternating Current power system as an option to DC power. The AC system had no problem sending electricity for hundreds of miles with little power loss. Eventually, AC won the battle and became the standard.

1900s

In 1916 the German-American physicist Albert Einstein published his ***Theory of Relativity***, which revolutionized physics. His theory also spurred research into developing atomic energy. **Atomic energy** is the energy released during a nuclear reaction. During a nuclear reaction, **atoms** (the smallest particle of an element that still retain the properties and characteristics of that element) are split and heat is produced.

During the 1920 and 1930s, a great financial depression swept around the world. Franklin Delano Roosevelt, the U.S. president, instituted sweeping programs, such as building the Hoover Dam and establishing the Tennessee Valley Authority. These projects created jobs and brought power to regions throughout the U.S.

During the 1930s, three German physicists, Hahn, Meitner and Strassman, performed the first human-initiated successful nuclear fission. **Fission** is the process of splitting the nucleus of the atom (the positively charged central part), which results in the release of large amounts of energy.

During the late 1930s and early 1940s, many nations raced to develop nuclear energy, first by attempting to build nuclear reactors, then by developing atomic weapons. Although German and Japanese programs came close, the U.S. was the first to produce atomic weapons.

In 1951, the U.S. began generating energy with a nuclear reactor. The Soviet Union opened its first commercial nuclear reactor in 1954.

Concern for the environment and worker safety boomed in the 1960s and 1970s. Governments enacted legislation to provide for cleaner air and water, along with safer working environments.

In 1973, OPEC (the Organization of Petroleum Exporting Countries) decided to adjust oil and gas prices with an embargo. The result was skyrocketing prices, lines at the gas stations, national trends towards conservation (e.g., no Christmas lights on houses), and purchases of smaller, more fuel-efficient cars. This spurred interest in, and development of, alternative energy sources, such as hydroelectricity, wind, nuclear power plants and solar energy.

In the late 1970s and mid 1980s, the nuclear energy industry suffered two highly visible accidents: Three Mile Island in the U.S., and Chernobyl in the Soviet Union.

In the 1990s, power generation companies faced regulations that impacted the structure and competition within the industry. Companies were forced to reorganize their operations and reduce costs, in order to compete more effectively.

DID YOU KNOW?

According to the U.S. Nuclear Regulatory Commission, the 1979 accident at the Three Mile Island nuclear power plant near Middletown, Pennsylvania, was one of the most serious accidents in U.S. commercial nuclear power plant operating history.

Courtesy of U.S. Department of Energy.

Fortunately, there were no injuries from this event. However, it did bring about major changes in emergency response planning, reactor operator training, and other areas associated with nuclear power generation.

SEGMENTS OF THE POWER GENERATION INDUSTRY

A variety of power sources are used to generate electricity. Generating plants produce electricity by converting mechanical or heat energy into electricity using a variety of methods which include:

- Burning fuels (coal, oil, gas, fuel oil)
- Using water (hydroelectric)

- Splitting atoms (nuclear)
- Harnessing other resources (the sun, wind, or heat from the earth)

Once electricity is generated, the transmission and distribution facilities are the same, regardless of the power source.

Coal, Oil and Gas

Figure 5-1: Fossil fuel burning power plant

Courtesy of www.OSHA.gov.

Coal is used most often as the fuel source for generating electricity in the U.S. Many power generation facilities burn coal, oil, or gas to heat water and turn it into steam. The steam causes a turbine to rotate. A **turbine** is a rotary machine that converts a fluid (e.g. water, steam, gas) into mechanical energy. The turbine has a wheel or rotor, which turns as a fluid flows past it. The wheel or rotor turns a shaft attached to a generator. The generator uses rotating magnets inside coils of copper wire to generate electricity.

One type of power generation facility is called a **cogeneration station**. This type of facility provides both heat and electricity to consumers. When energy is created, heat is formed as a by-product. Cogeneration facilities use this heat to provide industrial or residential heating, instead of cooling the heat and venting it to the atmosphere like other power plants do. Although more efficient than conventional power generation operations, cogeneration facilities must be located near an area that can the heat, since heat does not travel well over long distances.

Hydroelectric

Figure 5-2: Hydroelectric power plant

Courtesy of www.OSHA.gov.

Water is used to generate electricity for almost one-quarter of the world's electricity requirements. Hydroelectric power plants harness the energy of rushing water and use it to turn turbines that are connected to generators, creating electricity.

Typically, in a hydroelectric system, a dam is built across a river. However, some facilities use tidal forces from the ocean as the water source.

Hydroelectric turbines are very large, weighing upwards of 170 tons or more. In a dam-based hydroelectric facility, water is channeled into a narrow passageway (called penstocks) in order to increase the force of the water's movement. Water pressure builds up as it is channeled to the turbines. The power of the water is transferred to the turbine blades. The water is then released through an outtake (called a tailrace) downstream of the dam.

Hydroelectric power is considered a renewable power source, since the water is not consumed or used up during the energy generating process. However, the limitation is in finding a waterway suitable for damming.

Nuclear

Figure 5-3: Nuclear power plant

Courtesy of www.OSHA.gov.

Nuclear power plants produce heat by splitting atoms in a device called a **reactor**. Within a reactor, atoms are split using a process called fission.

Fission is the process of splitting the nucleus, the positively charged central part of an atom, which results in the release of large amounts of energy. Fission produces a tremendous amount of heat. This heat is used to boil water and create steam. Steam is then used to rotate a turbine which is connected to the shaft of a generator. When the generator turns, electricity is created.

Uranium and plutonium are radioactive metallic elements used as fuel for nuclear reactors. During nuclear reactions, a small amount of fuel produces large amounts of energy. For example, on some Navy ships (e.g., nuclear powered aircraft carriers and submarines) a pound of uranium is roughly equivalent to one million gallons of gasoline.

In a nuclear reactor, the fuel is arranged in long rods connected into bundles. These bundles are placed in water contained in a pressurized vessel. The water also acts as a coolant during this reaction.

Figure 5-4: Solar power plant

Courtesy of www.OSHA.gov.

As the bundles undergo a slow, controlled, chain reaction (i.e., a splitting of the atoms) heat is released. The heat from the chain reaction turns the water into radioactive steam. Some types of reactors use this steam to directly drive a turbine, while other types use this steam to heat a separate water loop, which is used to drive a turbine (this prevents the turbine from coming into direct contact with the radioactive steam).

Shielding around the reactor and an outer building made of concrete (built strong enough to withstand the crash of a jet airplane), prevent the escape of radiation and radioactive steam.

Nuclear power plant operations are much cleaner than coal, oil or gas powered plants. However, they produce radioactive wastes which must be stored and monitored, and which can remain hazardous for centuries.

Other Resources

Other resources, such as the sun, wind, and heat of the earth, can be used to generate electricity. Although these alternative energy sources come from renewable sources (in other words, the resource is not consumed during the power generation process), they account for only a small percentage of the world's electricity production.

Figure 5-5: Power generating windmills
Courtesy of www.OSHA.gov.

Solar power uses large reflective panels to collect and focus the sun's energy. This energy can be used to heat water or create electricity directly using special devices called solar cells.

Wind power uses flowing air currents to push against the blades of giant windmill-like **turbines** (rotary machines that convert fluid pressure into mechanical energy). These turbines are usually gathered together on a site called a farm. As the turbine rotates, it turns a shaft on a generator that converts mechanical energy into electrical energy.

Geothermal power is like a steam-powered facility, except it uses the naturally occurring steam from the earth to power turbines. Deep wells are drilled to tap underground sources of water that are superheated by hot geological forces (like lava). The water is then cooled and re-injected into the ground.

DUTIES, RESPONSIBILITIES, AND EXPECTATIONS OF PROCESS TECHNICIANS

Process technicians working in the power generation industry control the equipment that generates and distributes the electricity. These process technicians must deal with more automation and computerization than past workers in the industry. Although fewer people are now required to work as process technicians due to advances in technology, these workers must have more education and be more highly skilled. A two-year Associates degree is preferred for process technicians.

Figure 5-6: Geothermal power plant
Courtesy of www.OSHA.gov.

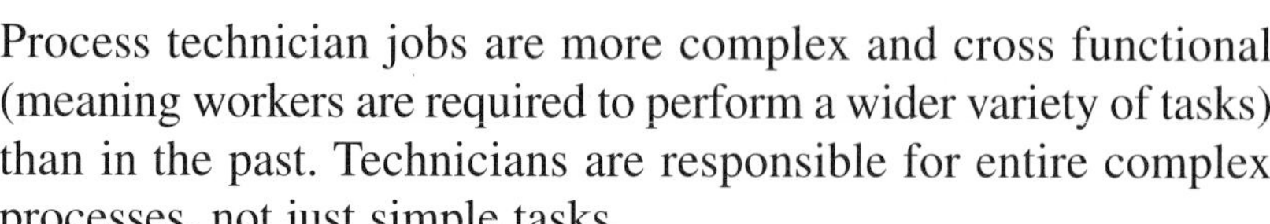

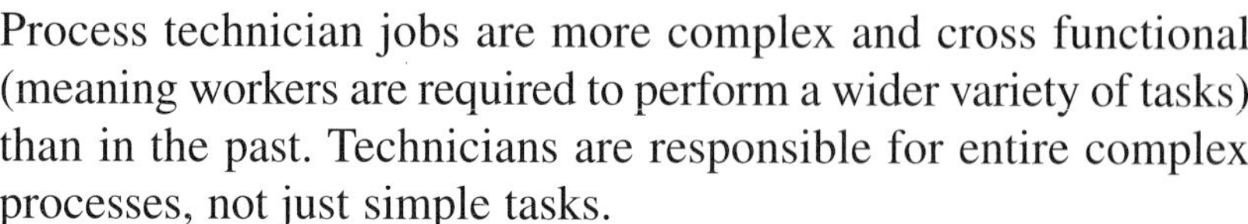

Process technician jobs are more complex and cross functional (meaning workers are required to perform a wider variety of tasks) than in the past. Technicians are responsible for entire complex processes, not just simple tasks.

There is an increased emphasis on keeping costs low, by improving safety, performing preventative maintenance, optimizing processes and making efforts to increase efficiency.

Along with technology changes, process technicians must keep up with constantly changing governmental regulations addressing safety and the environment.

Process technicians are required to have strong communication skills (i.e. reading, writing, and oral) along with solid computer skills and an understanding of math, physics and chemistry.

Some process technicians in this industry must obtain licensing from a government agency based on the job duties they perform (e.g. reactor operators at a nuclear power plant).

Skills

When companies search for potential applicants, are looking for skills such as:

- Education (preferably a two year Associates degree)
- Technical knowledge and skills
- Interpersonal skills
- Computer skills
- Physical abilities

Job Duties

This section will covers general responsibilities, since there are a wide range of processes involved with power generation and companies have different ways of organizing job duties.

The process technician can be responsible for a wide variety of tasks, including:

- Monitoring and controlling processes
  - Inspect equipment.
  - Start, stop and regulate equipment.
  - View instrumentation readouts.
  - Analyze data.
  - Make process adjustments.
  - Respond to changes, emergencies and abnormal operations.
  - Document activities, issues and changes.
- Assisting with equipment maintenance
  - Change or clean filters or strainers.
  - Lubricate equipment.
  - Monitor and analyze equipment performance.
  - Prepare equipment for repair.
  - Return equipment to service.
- Troubleshooting and problem solving
  - Apply troubleshooting processes such as root cause analysis.
  - Participate in corrective action teams.
  - Use statistical tools.
- Communicating and working with others
  - Write reports.
  - Perform analysis.
  - Write and review procedures.
  - Document incidents.
  - Help train others.

- Learn new skills and information.
- Work as part of a team.

- Performing administrative duties
  - Do housekeeping (make sure the work area is clean and organized).
  - Perform safety checks.
  - Perform environmental checks.
- Performing with a focus on safety, health, security, and environment
  - Keeps safety, health, environmental and security regulations in mind at all time.
  - Looks for unsafe acts.
  - Watches for signs of potentially hazardous situations.

Equipment

As a process technician in the power generation industry, you will work with all different types of equipment including:

- Pumps
- Compressors
- Pipes
- Valves
- Vessels
- Boilers
- Turbines
- Cooling towers
- Motors, engines and generators
- Instrumentation

Later chapters in this textbook explain what this equipment does, how to monitor and maintain it, potential hazards, and other information.

Workplace Conditions and Expectations

The following are some workplace conditions that process technicians can expect:

- All types of weather, including extreme conditions.
- Being outside part of the time.
- A drug and alcohol free environment.
- Being a member of a team.
- A facility that operates 24 hours a day, as part of shift work and occasionallyremote facility operation.
- Using tools and doing some heavy lifting.
- Working around hot and noisy conditions.

Companies expect process technicians to:

- Avoid hazardous conditions and errors (work safely, in compliance with government and industry regulation).
- Help with environmental compliance.
- Work smarter and focus on business goals.
- Look for ways to reduce waste and improve efficiency.
- Keep up with industry trends and constantly improve skills.

A range of federal government agencies and their regulations impact the power generation: OSHA, EPA, Department of Energy, Nuclear Regulatory Commission, US Coast Guard, and so on. State agencies also regulate certain elements within their jurisdiction.

CHANGES AND FUTURE TRENDS IN THE ROLE OF PROCESS TECHNICIANS

The role of process technicians in the power generation industry is changing. Process technicians should expect to see these following trends:

- Restructured organizations looking to reduce costs and compete more effectively.
- Increased use of computers, automation, and advanced technology.
- Increased reliance on communication skills.
- New and updated government and industry regulations.
- More involvement in process control and business decision.
- Increased workplace diversity.
- Less traditional methods of supervision.
- Continuous upgrading of skills and knowledge.
- New training methods and technologies, including virtual facilities, advanced simulators, and partnerships between education and industry.

SUMMARY

For thousands of years, civilizations have harnessed the forces of nature, such as water, steam and heat.

Between the 17th and 20th centuries, many famous discoveries about the properties of electricity were made. Benjamin Franklin, Alessandro Volta, Andre Marie Ampere and many others contributed to advancements in the development of electricity.

Michael Faraday, an English scientist, is credited for generating electrical current on a practical scale. He showed that relative motion between and a coil of copper wire produced an electrical current. His experiment resulted in the first electric generator. In 1763, James Watt created the first practical steam engine.

In the latter part of the 1800s, incandescent lighting was developed by an English scientist, Joseph Swan. Thomas Edison, the American inventor, made a similar discovery in the U.S. several months after Swan. Edison also created a DC generator to power the lights in his lab. DC is an electrical current that always travels in the same direction through a conductor. Edison's generator, coupled with a steam engine invented by Watt, signaled the start of practical, large-scale electrical generation.

Nikola Tesla devised another system for electrical generation: Alternating Current (AC). AC is a type of electric current that reverses direction periodically, usually sixty times per second. George Westinghouse, an American inventor and industrialist, developed Tesla's Alternating Current power system as an option to DC power. The AC system had no problem transferring electricity over hundreds of miles with little power loss. Eventually, AC won the battle and became the standard. AC is the primary type of electrical current used in processing plants. Tesla's name is associated with a unit of measurement for magnetic fields.

The German-American physicist Albert Einstein published his Theory of Relativity which revolutionized physics. His theory also spurred research into the development of atomic energy (splitting atoms to unleash tremendous energy). During World War II, many nations were racing to develop nuclear energy, first by attempting to build nuclear reactors followed by atomic weapons. Although the German and Japanese programs came close, the U.S. was the first to produce atomic weapons.

Concern for the environment and worker safety boomed in the 1960s and 1970s. Governments enacted legislation to provide for cleaner air and water, along with safer working environments.

An energy crisis in the 1970s spurred interest in, and development of, alternative energy sources, such as hydroelectricity, nuclear power plants and solar energy.

Power generation involves producing and distributing electrical energy in large quantities to serve the requirements of industry, businesses, residences, schools and so on. Every process industry requires power to operate.

There are three main segments to the power generation industry, generation, transmission, and distribution. Generating plants produce electricity by converting mechanical or heat into energy using a variety of methods: burning a fuel (coal, oil, or gas), using water, splitting atoms, or harnessing resources such as the sun, wind or heat from the earth.

These sources of power are used to operate turbines and generators which generate electricity. From the generating plant, the electricity travels to a transmission facility that moves the power over high voltage lines to serve a certain region. Distribution facilities then carry the power to end users, such as industrial or residential consumers.

Process technicians must deal with more automation and computerization than past workers in the industry. Although fewer people are now required to work as process technicians due to advances in technology, these workers must have more education and be more highly skilled. A two-year Associates degree is preferred for process technicians.

Process technician jobs are more complex and cross-functional (meaning workers are required to perform a wider variety of tasks) than in the past. Technicians are responsible for entire complex processes, not just simple tasks.

There is an increased emphasis on keeping costs low, by improving safety, performing preventative maintenance, optimizing processes and making efforts to increase efficiency.

Along with technology changes, process technicians must keep up with constantly changing governmental regulations addressing safety and the environment.

Process technicians are required to have strong communication skills (i.e. reading, writing, and verbal) along with solid computer skills and an understanding of math, physics and chemistry.

Futures trend such as increased automation and use of new technology affect process technicians' duties.

CHECKING YOUR KNOWLEDGE

1. Define the following key terms:
   a. Alternating Current (AC)
   b. Atom
   c. Coal
   d. Cogeneration Station
   e. Direct Current (DC)
   f. Electricity
   g. Fission
   h. Generator
   i. Geothermal
   j. Hydroelectric
   k. Nuclear
   l. Solar
   m. Turbine

2. _______________ is a flow of electrons from one point to another along a pathway, called a conductor.

3. Name three sources of power used to generate electricity.

4. Which Greek inventor created a steam powered device over 2,000 years ago?
   a. Hericles
   b. Aristophanes
   c. Casiclees
   d. Heron

5. Put the following events in order, from earliest to latest:
   a. First generator created
   b. Voltaic pile battery created.
   c. Water power used to turn wheat into flour
   d. Atomic bomb tested

6. Who is credit with inventing the first electric generator?
   a. Edison
   b. Faraday
   c. Swan
   d. Franklin

7. *(True or False)* Benjamin Franklin invented electricity.
   a. True
   b. False

8. Which of the following fuel sources is used most often for generating electricity?
   a. Coal
   b. Gas
   c. Nuclear
   d. Water

9. Name one benefit and one drawback of nuclear power.

10. What tasks are associated with troubleshooting and problem solving? Select all that apply.
    a. Apply processes such as root cause analysis
    b. Use statistical tools
    c. Do housekeeping (make sure the work area is clean and organized)
    d. Participate in correct action teams
    e. Perform safety checks

ACTIVITIES

1. If you could travel back in time, which event, period, or person in the history of the power generation industry would you choose to visit and why? Write at least five paragraphs in response.

2. Pick one of the power generation segments (coal/oil/gas, hydroelectric, nuclear, solar, wind, or geothermal) and write at least a one-page research report on benefits, drawbacks, and its future.

3. Consider the responsibilities and expectations of process technicians working in the power generation industry. What aspects would be difficult? What aspects would be rewarding? Discuss this with a classmate or write a paper explaining your answers.

4. Think about the future of the power generation industry. Which skills would you need to learn in order to adapt to the changes and trends? Make a list of these skills.

Chapter 6
Pulp and Paper Industry Overview

OBJECTIVES

Upon completion of this chapter you will be able to:

1. Explain the growth and development of the pulp and paper process industry.
2. Explain the different segments of the pulp and paper industry.
3. Identify the duties, responsibilities, and expectations of the process technician in the pulp and paper industry.
4. Describe changes and future trends in the role of the process technician in the pulp and paper industry.

KEY TERMS

- **Cellulose**—the principal component of the cell walls in plants.
- **Fiber**—a long, thin substance resembling a thread.
- **Fourdriniers**—a papermaking machine, developed by Henry and Sealy Fourdrinier, which produces a continuous web of paper.
- **Mill**—a facility where a raw substance is processed and refined to another form.
- **Pulp**—a cellulose fiber material, created by mechanical and/or chemical means from various materials (e.g. wood, cotton, recycled paper), from which paper and paperboard products are manufactured.

INTRODUCTION

This chapter provides you with a history of the pulp and paper industry and its impact on people, the economy, and the environment. This chapter also describes the tasks and working conditions of process technicians, as well as future trends and how they affect these technicians.

Most paper starts as trees, although recycled paper or other materials can be used. Using a process that is very similar to the one created almost 2,000 years ago, trees are turned into a wide variety of paper products.

Despite the high-tech advancements of the last few decades and the introduction of the phrase "paperless society" to our language, paper continues to play a huge role in everyday life. If, along with paper, you include items made using natural wood chemicals, then the pulp and paper industry creates and distributes thousands of products used daily around the world. Consider just a few of the many different uses for paper:

- **Business and industry**—printer paper, checks, envelopes, documents, business cards, financial statements, catalogs, and boxes.
- **Medical and health**—bandages, medical charts, gowns and masks, sterile filters, prescription pads, and tissues.

- **Construction**—cement bags, fiberboard, insulation, sandpaper, wallpaper, and tar paper.
- **Education**—textbooks, notebooks, class schedules, library cards, folders, tests, and photocopies.
- **Household**—money, books, newspapers and magazines, telephone books, napkins, toilet paper, artwork prints, food packaging, and greeting cards.
- **Recreation**—CD covers, bumper stickers, jigsaw puzzles, party supplies, event tickets, playing cards, and board games.

GROWTH AND DEVELOPMENT OF THE PULP AND PAPER INDUSTRY

Over 2,000 years ago, civilizations began to write down important traditions, ceremonies, religious practices, rules and decrees. Everything from papyrus (writing material made from water reeds), to silk, to clay was used for recording this information.

DID YOU KNOW?

The Chinese invented toilet paper almost 1,200 years ago.

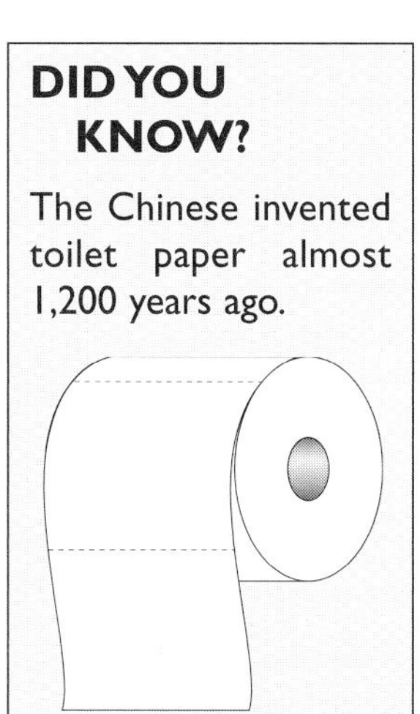

Although many civilizations can claim creating paper or paper-like substances, the first recorded efforts of papermaking involve a Chinese scholar named Ts'ai Lun. Lun ground up bark, linen and hemp mixed with water. He then spread this mixture onto a cloth-covered bamboo frame, and then left the frame in the sun to dry. The finished product: paper. This same basic process can be used today to make homemade paper.

For hundreds of years, the papermaking process remained a secret only known to the Chinese. But starting in the 800s, the process spread to the rest of Asia, Africa and Europe. The process was labor intensive and consumed a considerable amount of raw materials, so paper was made sparingly and was only used for the most important documents.

DID YOU KNOW?

Watermarks, which have been used since the 13th century when the Italians first introduced them, are translucent designs embossed into a piece of paper during its production.

These designs, which are visible when a sheet of paper is held up to the light, are used to identify the paper and the papermaker.

1400s

In these early years, a considerable amount of paper was made from rags and cloth scraps, and all documents were still being written by hand. Because of this, paper continued to be used in a limited way. However, in the mid-1400s a German named Johann Gutenberg created a printing machine. His invention allowed documents to be printed more quickly with less labor.

Soon, books were being printed and distributed widely. Paper became even more important and more uses for it were found. People became better educated due to the increased availability of books and other printed materials, thereby contributing to the Renaissance period in history, often referred to as the "Age of Enlightenment." In addition, science and the arts flourished during this time.

1600s

Around 1690, a paper **mill** (a facility where a raw substance is processed and refined to another form) was established in America near Pennsylvania. Water was essential to the paper making process and for transporting the finished goods.

1700s

In the 1700s, Rene de Reamure observed the paper wasp, a type of insect that consumed wood then spit the mush out to make a nest. Based on his observations, Reamure devised a practical way to create paper, using wood. The basic formula for making paper was set forth: wood fiber, combined with water and energy, would make paper.

In the following decades, many people began to create inventions for turning wood into **pulp,** a cellulose fiber material that could be used to create paper.

1800s

In the early 1800s, Nicholas Louis Robert created the first papermaking machine. Two brothers, Henry and Sealy Fourdrinier, improved up the design and invested large sums of money to develop the machine. These inventions allowed paper to be mass-produced. Today, most modern paper making machines are called **Fourdriniers**.

Various improvements in papermaking came over the next few decades. The use of paper and paper products increased as more consumer products requiring packaging were manufactured and distributed, along with books, newspapers, magazines and more. Papermaking became a science as well as an art.

1900s

Concern for the environment and worker safety boomed in the 1960s and 1970s. Governments enacted legislation to provide for cleaner air, water, and land, along with safer working environments.

From the 1980s on, the industry has struggled with price declines, the need for increased automation, labor issues, and growing worldwide competition.

OVERVIEW OF THE PULP AND PAPER INDUSTRY

Many different materials can be used to make paper, but wood is the primary material. Wood is composed of **fibers** (thin threads), made up of material called **cellulose** and stuck together with a natural adhesive called lignin. The papermaking process involves separating and rearranging those fibers.

Paper can also be made from other materials such as:

- Recovered paper (note: paper can only be recycled so many times before it cannot be processed any more)
- Linen
- Cotton
- Synthetics (man-made materials)

DID YOU KNOW?

U.S. dollar bills are composed of 25% linen and 75% cotton, and that red and blue synthetic fibers of various lengths are distributed evenly throughout the paper.

Prior to World War I these fibers were made of silk.

Trees are harvested from tree farms. Two types of trees are used to make paper:

- **Hardwood**—trees like oak and maple produce short fibers, which makes smooth but weak paper that is suitable for writing.
- **Softwood**—trees like pine and spruce produce long fibers, which makes strong but rough paper that is suitable for packaging (e.g. cardboard).

Papermaking operations can combine hardwood and softwood fibers to make strong, smooth papers. Almost every part of a harvested tree is used for one purpose or another. Along with the wood itself, natural chemicals from a tree, such as resin and oils, are also used to make products. These products include toothpaste, roofing shingles, car wax, crayons, clothing, sports helmets, and film stock.

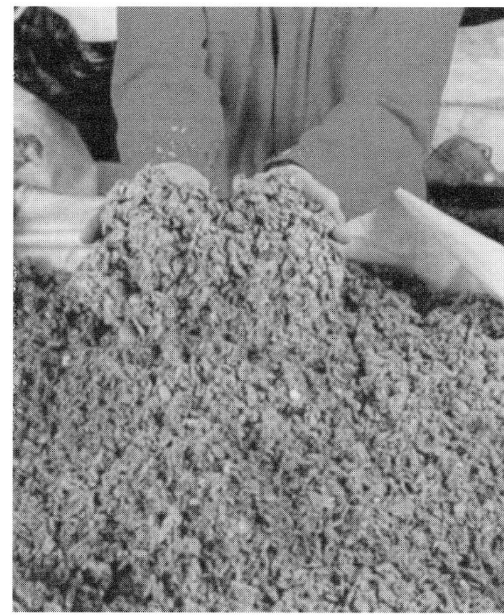

Figure 6-1: Wood pulp
Courtesy of Pacific Northwest National Laboratory

Pulp Mills

Once the trees are harvested, they are sent to the pulp mill. Harvested trees are trimmed to logs and transported to the pulp mill where they are washed and debarked. During the pulping process, the wood fibers must be separated into individual strands. This can be achieved either mechanically, chemically or both:

- If the fibers are separated mechanically, a grinder processes the wood.
- If the fibers are separately chemically, a chipper processes the wood. From there, it is processed in a device called a digester, where it is treated with boiling chemicals. The chemicals are then removed using an extractor.

The type of process used is determined by the strength, appearance, and use of the paper that will be made from the pulp.

The fibers are washed, screen and bleached. The resulting product is a mushy solution of water and fiber called pulp. A considerable amount of water and energy are required to make pulp. Sometimes, fillers, additives and dyes are added to the pulp to make the finished paper glossier, absorbent, water-resistant, or a specific color.

Some pulp mills only make pulp, while others make both paper and pulp.

Paper Mills

The papermaking process produces many different types of paper. The quality of the paper and its characteristics will vary depending on the final application. For example, paper may vary in: strength, weight, brightness, opacity (ability to see through), softness, smoothness, thickness, and more.

When the pulp is ready, it is spread out as a wet mixture, called a slurry, onto a giant screen. This screen passes on a conveyor from what is called the wet end of the process to the dry end. On the wet end, water is removed from the slurry using gravity and vacuums. The remaining fiber bonds into a watery sheet of paper. On the dry end, this sheet is pressed between soft, heated rollers to remove more water.

Additional treatment of the paper, such as coatings, can then be added. Once complete, the resulting dry, uniform paper is placed on giant rolls.

Paper Processing/Converting

Once the paper is placed on rolls, it can be converted into final products. For example, the paper may be:

- Cut
- Folded
- Coated
- Glued
- Screened (printed)
- Embossed (stamped with a raised pattern)

The types of processing or conversion depended on the end product (e.g., whether it is paper napkins, stationery, boxes, or file folders). The final products are then distributed to consumers.

DUTIES, RESPONSIBILITIES, AND EXPECTATIONS OF PROCESS TECHNICIANS

Process technicians must deal with more automation and computerization than past workers in the industry. Although fewer people are now required to work as process technicians due to advances in technology, these workers must have more education and be more highly skilled. A two-year Associates degree is preferred for process technicians.

Figure 6-2: Giant rolls of finished paper
Courtesy of TISEC.

Process technician jobs are more complex and cross functional (meaning workers are required to perform a wider variety of tasks) than in the past. Technicians are responsible for entire complex processes, not just simple tasks.

There is an increased emphasis on keeping costs low, by improving safety, performing preventative maintenance, optimizing processes and making efforts to increase efficiency.

Along with technology changes, process technicians must keep up with constantly changing governmental regulations addressing safety and the environment.

Process technicians are required to have strong communication skills (i.e. reading, writing, and verbal) along with solid computer skills and an understanding of math, physics and chemistry.

Skills

Companies are looking for skills such as:

- Education (preferably a two year Associates degree)
- Technical knowledge and skills
- Interpersonal skills
- Computer skills
- Physical abilities

This section will cover general responsibilities, since there are a wide range of processes involved with pulp and paper and companies have different ways of organizing job duties.

Job Duties

The process technician can be responsible for a wide variety of tasks, including:

- Monitoring and controlling processes
  - Inspect equipment.
  - Start, stop and regulate equipment.
  - View instrumentation readouts.
  - Analyze data.
  - Make process adjustments.
  - Respond to changes, emergencies and abnormal operations.
  - Document activities, issues and changes.
- Assisting with equipment maintenance
  - Change or clean filters or strainers.
  - Lubricate equipment.
  - Monitor and analyze equipment performance.
  - Prepare equipment for repair.
  - Return equipment to service.
- Troubleshooting and problem solving
  - Apply troubleshooting processes such as root cause analysis.
  - Participate in corrective action teams.
  - Use statistical tools.
- Communicating and working with others
  - Write reports.
  - Perform analysis.
  - Write and review procedures.
  - Document incidents.
  - Help train others.
  - Learn new skills and information.
  - Work as part of a team.
- Performing administrative duties
  - Do housekeeping (make sure the work area is clean and organized).
  - Perform safety checks.
  - Perform environmental checks.
- Performing with a focus on safety, health, security, and environment
  - Keeps safety, health, environmental and security regulations in mind at all time.
  - Looks for unsafe acts.
  - Watches for signs of potentially hazardous situations.

Equipment

As a process technician in the pulp and paper industry, you will work with all different types of equipment including:

- Pumps
- Compressors
- Vessels
- Reactors
- Evaporators
- Pipes
- Valves
- Furnaces
- Instrumentation

Later chapters in this textbook explain what this equipment does, how to monitor and maintain it, potential hazards, and other information.

Workplace Conditions and Expectations

The following are some workplace conditions that process technicians can expect:

- A drug and alcohol free environment.
- Being a member of a team.
- A facility that operates 24 hours a day, as part of shift work.
- Using tools and doing some heavy lifting.

Companies expect process technicians to:

- Avoid accidents and errors (work safely, in compliance with government and industry regulation).
- Help with environmental compliance.
- Work smarter and focus on business goals.
- Look for ways to reduce waste and improve efficiency.
- Keep up with industry trends and constantly improve skills.

A range of federal government agencies and their regulations impact the pulp and paper industry, such as OSHA and the EPA. State agencies also regulate certain elements within their jurisdiction.

CHANGES AND FUTURE TRENDS IN THE ROLE OF PROCESS TECHNICIANS

The role of process technicians in the pulp and paper industry is changing. Process technicians should expect to see these following trends:

- Restructured organizations looking to reduce costs and compete more effectively.
- Increased use of computers, automation, and advanced technology.
- New and updated government and industry regulations.
- More involvement in process control and business decision.
- Increased workplace diversity.
- Less traditional methods of supervision.

- Continuous upgrading of skills and knowledge.
- New training methods and technologies, including computer-based training and partnerships between education and industry.

SUMMARY

Most paper starts as trees, although recycled paper or other materials can be used. Using a process that is very similar to what was used almost 2,000 years ago, trees are turned into a wide variety of paper products.

Paper continues to play a huge role in everyday life. If, along with paper, you include items made using natural wood chemicals, then the pulp and paper industry creates and distributes thousands of products used daily around the world.

The first recorded effort of papermaking involves a Chinese scholar called Ts'ai Lun. He ground up bark, linen and hemp mixed with water, which he spread out on a cloth-covered bamboo frame. He then left the frame in the sun to dry.

In the mid-1400s a German named Johann Gutenberg created a printing machine. His invention allowed documents to be printed more quickly with less labor. Around the start of the 1800s, Nicholas Louis Robert created the first papermaking machine. Two brothers, Henry and Sealy Fourdrinier, improved up the design and invested large sums of money to develop the machine. These inventions allowed paper to be mass-produced. Today, most modern papermaking machines are called Fourdriniers.

Concern for the environment and worker safety boomed in the 1960s and 1970s. Governments enacted legislation to provide for cleaner air, water, and land, along with safer working environments. From the 1980s on, the industry has struggled with price declines, the need for increased automation, labor issues, and growing worldwide competition.

Many different materials can be used to make paper, but wood is the primary material. Wood is composed of fibers (thin threads), made up of a material called cellulose and stuck together with a natural adhesive called lignin. The papermaking process involves separating and rearranging those fibers. Wood is turned into pulp, a watery substance created through mechanical and/or chemical methods.

When the pulp is ready, it is spread out as a wet mixture, called a slurry, onto a giant screen. This screen passes on a conveyor from what is called the wet end of the process to the dry end. On the wet end, water is removed from the slurry using gravity and vacuums. The remaining fiber bonds into a watery sheet of paper. On the dry end, this sheet is pressed between soft, heated rollers to remove more water. The paper can then be converted into final products.

Process technicians must deal with more automation and computerization than past workers in the industry. Although fewer people are now required to work as process technicians due to advances in technology, these workers must have more education and be more highly skilled. A two-year Associates degree is preferred for process technicians.

Process technicians are required to have strong communication skills (i.e. reading, writing, and verbal) along with solid computer skills and an understanding of math, physics and chemistry.

CHECKING YOUR KNOWLEDGE

1. Define the following key terms:
   a. Cellulose
   b. Fiber
   c. Fourdriniers
   d. Mill
   e. Pulp

2. Which ancient civilization is generally credited with inventing paper?
   a. The Japanese
   b. The Chinese
   c. The Greeks
   d. The Egyptians

3. Put the following events in order, from earliest to latest:
   a. Printing machine invented
   b. Paper-making process invented
   c. Paper mill built in America
   d. Paper-making machine invented

4. What is the name associated with modern papermaking machines?
   a. Fourdrinier
   b. Justinier
   c. Caitinier
   d. Courtinieir

5. When was the first paper mill built in America?
   a. 1690
   b. 1720
   c. 1836
   d. 1965

6. ___________ is a natural fiber from wood used to make paper.

7. Name three types of material, other than wood, that can be used in the papermaking process.

8. *(True or False)* Natural chemicals from wood are also used to make products.
   a. True
   b. False

9. Name three steps associated with the paper making process.

10. What tasks are associated with troubleshooting and problem solving? Select all that apply.
    a. Apply processes such as root cause analysis
    b. Use statistical tools
    c. Do housekeeping (make sure the work area is clean and organized)
    d. Participate in correct action teams
    e. Perform safety checks

11. What tasks are associated with equipment maintenance? Select all that apply.
    a. Applying processes such as root cause analysis
    b. Lubrication
    c. Housekeeping
    d. Monitoring and analyzing equipment performance
    e. Change filters

ACTIVITIES

1. Conduct a scavenger hunt at home or work, keeping a list of all paper and paper-related products you can find. Limit your search to 1 hour or less. Then, review the list and count the number of products you found.

2. Pick one of the following three topics relevant to the papermaking process, and then research the topic using the Internet, library, or other resources. Write a one-page summary of what you learned.
   a. Early papermaking efforts (Egyptians, Chinese)
   b. The invention of the printing machine
   c. The invention of the papermaking machine.

3. Which skills do you think are the most important for a process technician working in the pulp and paper industry? Make a list and discuss it with a classmate.

Chapter 7

Water and Waste Water Treatment

OBJECTIVES

Upon completion of this chapter you will be able to:

1. Explain the history of water treatment.
2. Explain the importance of water treatment for the community.
3. Describe the various water treatment customers.
4. Explain the water treatment process.
5. Explain the importance of wastewater treatment for the environment.
6. Explain the wastewater treatment process.
7. Identify treatment process technician duties, responsibilities, and expectations.
8. Identify regulations associated with water and wastewater treatment and explain the purpose of each.

KEY TERMS

- **Absorb**—the process of drawing inward.
- **Adsorb**—the process of sticking together.
- **Disinfection**—the process of killing pathogenic organisms.
- **Dissolved Solids**—solids which are held in suspension indefinitely.
- **Filtration**—the process of removing particles from water by passing it through porous media.
- **Pathogen**—a disease-causing microorganism.
- **Settleable Solids**—solids in wastewater that can be removed by slowing the flow in a large basin or tank.
- **Suspended Solids**—solids that can not be removed by slowing the flow.
- **Turbidity**—cloudiness caused by particles suspended in water or some other liquid.

INTRODUCTION

Clean water is essential for life and many industrial processes. Through water treatment facilities, process technicians are able to process and treat water so it is safe to drink and to ensure public health is protected.

To eliminate the spread of pollution and disease, wastewater treatment facilities were built. From the use of early lagoons to modern complex treatments, the spread of pollution and waterborne diseases has been controlled.

In order to work in a water or wastewater treatment facility, process technicians must be able to perform a variety of tasks and be familiar with the rules and regulations pertaining to water and wastewater treatment.

HISTORY OF THE WATER TREATMENT INDUSTRY

The importance of good drinking water was recognized early on in history. However, it took centuries for people to understand that their senses (taste, sight, and smell) were inadequate when it came to judging water quality.

During the eighteenth century, the Europeans devised some of the earliest water treatment systems. Theses systems used **filtration** (removing particles from water by passing it through porous media) to remove the taste and appearance of particles in water. However, filtration did not address the problem of biological contaminants such as typhoid, dysentery, and cholera.

The impact of biological contaminants in water was not known until the second half of the nineteenth century when infectious diseases were first recognized, and the ability to spread or transmit these diseases through water was demonstrated. Following this demonstration, scientists became increasingly more focused on biological contaminants and disease-causing microorganisms (**pathogens**) and their impact on public water supplies.

During their research, scientists discovered that **turbidity** (cloudiness) in water was not only unappealing, but it also indicated a potential health risk. Turbidity can be caused by suspended particles that have the potential to harbour pathogens. Based on this information, new and improved water treatment systems were designed to reduce turbidity and pathogenic organisms.

By the beginning of the twentieth century water treatment had improved. In 1906, chlorine was first introduced as a means of water **disinfection** (a process for killing pathogenic organisms) and thus began the elimination of water borne diseases in drinking water. Since then, scientists and engineers have continued to develop new ways to process water more quickly and effectively, and at a lower cost.

DID YOU KNOW?

Excessive water contamination can occur when rains wash debris from nearby public trails and pastures into water sources.

OVERVIEW OF THE WATER TREATMENT PROCESS

The water treatment process begins with water pumped from wells, rivers, streams, and lakes. This water is sent to a water treatment plant through a series of pumps and valves. Water sources must always be protected against accidental contamination.

Well water is usually low in turbidity and often does not need to be filtered before being disinfected. River water, however, does need filtration.

Once inside the treatment plant, river water is filtered and chemically treated. After disinfection, the filtered water can then distributed to residential, agricultural, and industrial customers.

DID YOU KNOW?

The use of chlorine as a water disinfection agent is considered one of the greatest discoveries of the last millennium.

Water quality analysis and testing is essential when producing water for human consumption. Depending on how the water is used, different standards are needed. For example, treated water for consumption should be:

- Free of corrosive properties
- Free of dissolved substances
- Free of toxic contaminants
- pH controlled (pH level can vary based on the use of the water)
- Proper temperature and pressure range

HISTORY OF THE WASTEWATER TREATMENT INDUSTRY

As water is used, wastewater is formed (e.g., dirty dish water, process water, laundry and biological wastes). In the beginning, waste water was dumped into the street or nearby streams. Over time, stone lined ditches were used to carry wastes to the streams. Eventually, elaborate collection systems were developed to carry wastes to receiving streams. However, in the end the result was the same. Wastes dumped upstream were carried to communities downstream, and diseases were spread.

With the discovery of disease-causing organisms came the diversion of waste water to treatment plants. These facilities treat the water before returning it to the environment (e.g., streams, rivers, and oceans). Increased recycling efforts of today encourage the use of treated wastewater for irrigation purposes such as for golf courses or crops that are not for human consumption.

OVERVIEW OF THE WASTEWATER TREATMENT PROCESS

Collection systems carry residential and industrial wastewater, as well as storm water to the wastewater treatment facility. Large rocks, sticks, and other debris are removed early in the wastewater treatment process to avoid damaging pumps and other equipment. The first stage (known as primary treatment) in the wastewater treatment facility is usually a settling pond or clarifier. The heavier **settleable solids** (solids in wastewater that can be removed by slowing the flow in a large basin or tank) are removed and, in many instances, dewatered. They are then burned or buried in the local landfill.

In secondary treatment the wastewater is further treated to remove **suspended solids** (solids that cannot be removed by slowing the flow) and possibly some **dissolved solids** (solids which are held in suspension indefinitely). Microorganisms like those found on rocks in a stream, are used to **adsorb** (stick together) and **absorb** (draw inward) the dissolved solids for removal. The wastewater continues through the treatment facility and is disinfected with chlorine before being discharged into the receiving stream.

Some water treatment facilities must provide a third stage known as tertiary treatment to further treat the wastewater before discharging it into the receiving stream. The most common tertiary treatment option is filtration, which removes fine suspended solids.

The EPA regulates the National Pollution Discharge Elimination System (NPDES), which issues permits for discharging into receiving streams. Each permit outlines the specific water qualities that must be monitored, and the conditions for discharge.

PROCESS TECHNICIAN DUTIES AND RESPONSIBILITIES

Technicians in water and wastewater treatment plants are responsible for controlling the equipment and processes required to remove or destroy harmful materials, chemical compounds, and microorganisms from water. These technicians are also responsible for controlling pumps, valves, and other equipment used throughout the various treatment processes.

Other duties technicians may be asked to perform include:

- Reading, interpreting, meters and gauges to make sure that plant equipment and processes are working properly.
- Using a variety of instruments to sample and measure water quality.
- Operating chemical feeding devices and adjusting the amount of chemicals in the water.
- Using computers to monitor equipment, store the results of sampling, make process-control decisions, schedule and record maintenance activities, produce reports, and troubleshoot malfunctions.

The specific duties of a water treatment technician depend on the type and size of plant. In a small plant, one technician may be responsible for controlling all of the machinery, performing tests, keeping records, handling complaints, and performing repairs and maintenance. In a larger plant with more employees, technicians may be more specialized and monitor only one process.

From time to time, water treatment technicians must work during emergency situations. These emergency situations may arise from within the facility (e.g., a chlorine gas leak), or be the result of something external (e.g., a heavy rainstorm causing water volume to exceed the plant's treatment capacity). Technicians must be trained to deal with these types of situations, and must be able to work under extreme pressure to correct problems as quickly as possible. Because working conditions may be dangerous, technicians must always exercise caution.

In addition to working in emergency situations, treatment technicians must be able to perform physically demanding work, indoors and outdoors, in various locations. Because of the presence of hazardous conditions, such as slippery walkways, noise, dangerous gases, and open tanks, technicians must always pay close attention to safety procedures.

ENVIRONMENTAL REGULATIONS AND CONSIDERATIONS

Prior to the 1970s, little was done to protect the environment from the hazardous and sometimes lethal effects of pollution. However, in 1970 a major movement was started in an attempt to educate and inform the masses about environmental issues and their impacts. Several pieces of legislation were enacted as a result of that movement. Two of these items were the Clean Water Act of 1972, and the Safe Drinking Water Act of 1974.

The Clean Water Act of 1972 implemented a national system of regulation on the discharge of pollutants, while the Safe Drinking Water Act of 1974 established standards for drinking water. As a result of these two acts, industrial facilities sending their wastes to municipal treatment plants must now meet certain minimum standards to ensure that the wastes have been adequately pretreated so they will not damage the municipal treatment facility.

Municipal water treatment plants must also meet stringent standards. The list of drinking water contaminants regulated by these statutes has continued to grow over time. Because of this, plant technicians must be familiar with the guidelines established by federal, state, and local authorities, and know the impact these regulations have on their plant.

In order to ensure that water treatment technicians are familiar with the various regulations, each technician must receive training on the various aspects of water treatment, and must pass an examination to certify that they are capable of overseeing treatment plant operations. There are different levels of certification, depending on the technician's experience and training.

SUMMARY

The importance of good drinking water was recognized early on in history. However, the need for advanced treatment techniques was not identified until the second half of the nineteenth century when infectious diseases were first recognized, and the ability to spread or transmit these diseases through water was demonstrated.

In the early days, scientists discovered that turbidity (cloudiness caused by suspended particles) in water was not only unappealing, but it also indicated a potential health risks, since the particles that cause turbidity also have the potential to harbor disease causing microorganisms called pathogens.

By the beginning of the twentieth century, water treatment had improved. Techniques like chlorination and sand filtration greatly reduced the rates of waterborne diseases. Since then, scientists and engineers have continued to develop new ways to process water more quickly and effectively, and at a lower cost.

Clean water is essential for life and many industrial processes. Through water treatment facilities, process technicians are able to treat water so it is safe to drink and to ensure public health is protected.

Environmental concerns in the late 20th century lead to the creation of regulations to limit pollution by limiting discharges into receiving waters. Wastewater treatment facilities are the primary method for limiting

In order to work in a treatment facility, process technicians must be able to perform a variety of tasks and be familiar with the rules and regulations used to protect water quality and ensure public health.

CHECKING YOUR KNOWLEDGE

1. Define the following key terms:
   a. Absorb
   b. Adsorb
   c. Disinfection
   d. Dissolved solids
   e. Filtration
   f. Pathogen
   g. Pathogens
   h. Settleable solid
   i. Suspended solids
   j. Turbidity

2. The impact of biological contaminants in water was not known until:
   a. The 17th century
   b. The 18th century
   c. The 19th century
   d. The 20th century

3. Which of the following duties might a water treatment technician be asked to perform?
   a. Reading, interpreting meters and gauges
   b. Using instruments to sample and measure water quality
   c. Operating chemical feeding devices and adjusting chemical levels
   d. Using computers to monitor equipment
   e. All of the above

4. Which environmental statute established standards for drinking water?
    a. The Clean Water Act of 1972
    b. The Water Safety Act of 1974
    c. The Safe Drinking Water Act of 1974
    d. The Water Protection Act of 1973

ACTIVITIES

1. Use the Internet or other library resources to research the water or wastewater treatment process. Write a two-page paper describing what you learned.

2. Contact the local water or wastewater treatment authority and arrange a tour of a water treatment facility. Write a three-page paper describing what you saw and what you learned.

Chapter 8

Food and Beverage Industry Overview

OBJECTIVES

Upon completion of this chapter you will be able to:

1. Explain the growth and development of the food and beverage manufacturing industry.
2. Explain the food manufacturing process and the most common products produced by the industry.
3. Identify the duties, responsibilities, and expectations of the process technician in the food and beverage manufacturing industry.
4. Describe the responsibility of various regulatory agencies and their impact on the food and beverage manufacturing industry.

KEY TERMS

- **Disinfecting**—destroying disease-causing organisms (e.g., through washing, irradiation, or ultraviolet exposure).
- **Manufacturing**—making a product from raw materials by hand or with machinery (e.g., cooking, decorating, grinding, milling, and mixing).
- **Pathogen**—a disease-causing microorganism.
- **Preserving**—prepare for long-term storage (e.g., canning, drying, freezing, salting).

INTRODUCTION

This chapter provides you with a history of the food and beverage manufacturing industry and its impact on people, the economy, and the environment. This chapter also describes common processes used in the industry and the tasks and working conditions of process technicians.

DID YOU KNOW?

Some scientists believe the Salem Witch Trials of 1692 were caused by convulsive ergotism, a disorder caused by the ingestion of grain contaminated with ergot fungus.

Symptoms of convulsive ergotism include muscle spasms, delusions, and hallucinations.

GROWTH AND DEVELOPMENT OF THE FOOD AND BEVERAGE MANUFACTURING INDUSTRY

From the beginning of civilization, people have been concerned about the quality of their foods. So concerned, in fact, that dietary guidelines and restrictions became an integral part of daily culture and religious beliefs.

As science progressed, people began to understand germs and the role they play in food safety. This lead to improved food handling guidelines and storage techniques.

Over the years, many of these techniques and guidelines have become government rules and regulations aimed at protecting public health.

The following sections provide a few examples of the legislation that has been put in place to ensure food safety and quality.

Early 1900s

In 1906 Congress passed the original Food and Drugs Act. This act prohibited interstate commerce (transportation and/or exchange between states in the U.S.) of misbranded and adulterated foods, drinks, and drugs.

The Meat Inspection Act was also passed in 1906. This act was created in reaction to Theodore Roosevelt's investigation of Chicago meat packers. During these investigations, it was discovered that there were unsanitary conditions in meat packing plants, and that poisonous preservatives and dyes were being used in foods intended for human consumption. As a result of this act, cleanliness standards were established for slaughterhouses and processing plants. In addition, all cattle, sheep, horses, swine and goats were required to pass an inspection by the U.S. Food and Drug Administration prior to and after slaughter.

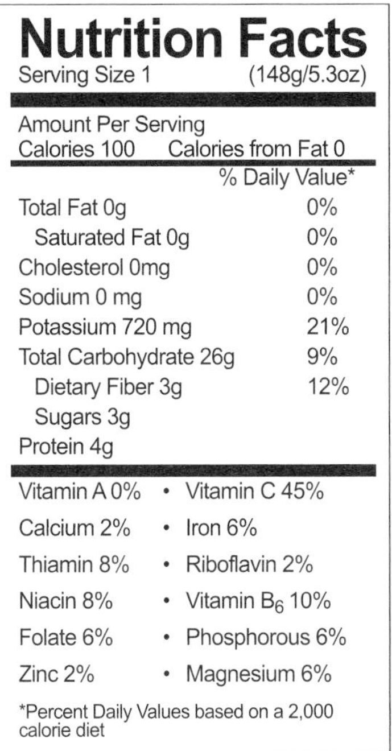

Figure 8-1: Food nutrition label

Mid-1900s

In 1949 the Food and Drug Administration (FDA) published the first industry guidebook for foods.

In 1958 the Food Additives Amendment was enacted. This amendment required manufacturers of new food additives to establish safety. The Delaney proviso prohibited the approval of any food additive shown to induce cancer in humans or animals.

Late 1900s

The Nutrition Labeling and Education Act was enacted in 1990. This act required all packaged foods to bear nutrition labeling, and all health claims and nutrition information to be standardized and consistent with terms defined by the Secretary of Health and Human Services.

OVERVIEW OF THE FOOD AND BEVERAGE MANUFACTURING INDUSTRY

The food and beverage manufacturing industry links farmers to consumers through the processing of fruits, vegetables, grains, meats, dairy products, and other finished goods. Once processed, these finished goods are then delivered to grocers and wholesalers who then supply them to households, restaurants, or institutional food services.

The job roles in the food processing industries can vary widely. For example, technicians may be involved with:

- Slaughtering and processing meat
- Processing milk and other dairy products
- Canning and preserving fruits and vegetables
- Producing grain products such as flour, cereal, and pet foods
- Making bread and other baked goods
- Manufacturing sugar, candy, or other confectionery products
- Brewing or producing beverages such as beer or soft drinks.
- Processing fats and oils into products such as shortening, margarine, and cooking oil
- Preparing packaged seafood

While this list is long, it is not comprehensive. Process technicians can be involved in a wide variety of other tasks.

Food Processing Hazards

A substantial percentage of food-borne illnesses are caused by **pathogens** (disease causing microorganisms) that originated in contaminated environments or in animals prior to slaughter. Process technicians need to be aware of these types of organisms and how they are spread. They also need to know proper handling techniques and methods for preventing the spread of disease.

Food Processing Methods

Process technicians should be familiar with the food processing methods used in their plant. These methods may include techniques for:

- **Disinfecting**—destroying disease-causing organisms (e.g., through washing, irradiation, or ultraviolet exposure).
- **Preserving**—prepare for long-term storage (e.g., canning, drying, freezing, salting).
- **Manufacturing**—making a product from raw materials by hand or with machinery (e.g., cooking, decorating, grinding, milling, and mixing).
- **Drying**—removing moisture from items such as rice (using a conveyor belt moving through a hot air tunnel) or sugar (dried in a turning drum).
- **Agglomerating**—gathering materials into a mass (instant coffee crystals are made using this process).
- **Roasting and toasting**—using heat (from an oven) to process types of food like cereal flakes and coffee.
- **Cooking and frying**—using heat (typically hot oil) to process types of food like chicken strips and potato chips.
- **Mixing**—stirring materials to blend them (such as instant drink mixes).

Figure 8-2: Frozen Food Manufacturing
Courtesy of John Zoiner/Workbook Stock/ Getty Images, Inc.

- **Evaporating**—removing moisture from items such as milk and coffee (to create powered milk or coffee).
- **Applying reverse osmosis**—processing water by forcing it through a membrane through which salts and impurities cannot pass (bottled water is produced this way).
- **Pasteurization**—heating foods to kill organisms, such as bacteria, or make them less likely to cause disease (milk and orange juice are treated this way).
- **Distillation**—removing vapors from liquids using heating and condensing processes (alcoholic beverages are produced this way).

Once food and beverages are processed and ready for packaging, these products are prepared for distribution using materials such as glass, plastic, paper, and metal. Many of these packaging products are created by other process industries (e.g. chemical, pulp and paper). Food can be packaged in a vacuum, low oxygen environment, carbon dioxide environment, or hot/cold environments.

Quality standards apply to all aspects of food and beverage manufacturing, including raw materials, equipment, preparation, packaging materials, and shipping and transportation. These standards are regulated by government agencies, customers and company policy.

Food Processing Systems

Food processing systems can vary from lines set up on a factory floor which are disassembled when the run is finished, to permanent machines that receive a constant product feed and which are run 24 hours a day, 7 days a week.

Food processing can require devices such as augers, conveyers, heaters, fryers, cookers, and freezers. Every food process is different. For example, a soft drink manufacturing plant might use the following equipment:

- Water purification and storage equipment
- Concentrate and sweetener storage containers
- Blending machines
- Sanitary food quality hosing and pipes
- Carbonaters
- Canning and bottling machines
- Packing machines use to pack bottles into crates
- Bottle and can rinsing machines
- Carbonated storage and delivery systems
- Sealing machines
- Labeling machines

Some of these machines may be simple to operate, while others are much more complex.

Consider the following steps in a typical soda bottling process:

1. The water being used to make the beverages must first be treated so the final product has a uniform taste regardless of where it was manufactured.
2. Water, sweetener and concentrate are blended into a syrup.
3. Carbon dioxide (CO_2) is added to the syrup until it becomes carbonated.
4. The soda cans are rinsed and then placed in a filler machine for filling.
5. Once the cans have been filled with soda, a machine called a seamer attaches the metal lid to the can.
6. The cans are then sent to a labeling machine if the container is not already marked with a label.
7. The final product is sent to the retailer.

DUTIES, RESPONSIBILITIES, AND EXPECTATIONS OF PROCESS TECHNICIANS

There are many different types of workers in the food manufacturing industry. According to 2002 Department of Labor (DOL) statistics, more than half of the employees in the food manufacturing industry are production workers. Production jobs require manual dexterity, good hand-eye coordination and, in some industries, strength.

Food and beverage manufacturing facilities use equipment such as boilers, refrigeration equipment, air compressors, fired vessels, heat exchangers, and so on. The use of automation has increased, and workers are needed to operate mixers, blenders, slicers, bottling machines, ovens and other similar machinery. Personnel are also needed to operate and maintain this equipment. Some technicians are responsible for cleaning, maintaining, lubricating, and repairing machines.

Other production workers use their hands or hand tools to do their jobs. For example, cannery workers and meat processors use knives and saws to process meat. Fruit and vegetable processors use their hands to sort, grade, wash, trim, peel or slice. Bakery technicians use mixing machines, ovens, and other types of equipment to mix and create baked goods. Other technicians use their hands to decorate or apply artistic touches to prepared foods.

Skills

Process technicians are required to have strong communication skills (i.e. reading, writing, and verbal), solid computer skills, and an understanding of biology, chemistry and basic food safety.

When searching for potential applicants, companies looking for skills such as:

- Education (preferably a two year Associates degree)
- Technical knowledge and skills
- Interpersonal skills
- Computer skills
- Physical abilities (especially for jobs that require artistic touches or heavy lifting)

Since there are a wide range of processes involved in food manufacturing, and since companies have different ways of organizing job duties, this section will only cover general responsibilities.

Job Duties

Process technicians in the food manufacturing industry may be responsible for a wide variety of tasks, including:

- Monitoring and controlling processes
- Assisting with equipment maintenance
- Troubleshooting and problem solving
- Communicating and working with others
- Performing administrative duties
- Performing with a focus on safety, health, security, and environment

Equipment

Process technicians may also be asked to work with a wide variety of equipment including:

- Pumps
- Compressors
- Tanks
- Blenders
- Grinders
- Mixers
- Ovens
- Slicers
- Conveyors
- Bottling machines
- Labeling machines

Food processing technicians may also work with hand tools and other types of equipment.

Workplace Conditions and Expectations

The workplace conditions will vary from industry to industry. For example, some jobs require minimal physical effort are in climate controlled facilities. Other jobs may occur in non-climate controlled facilities, and may require heavy lifting or other forms of physical exertion.

During the course of their duties, food processing technicians will be required to:

- Use tools and equipment.
- Perform duties that require physical exertion and stamina (does not pertain to all jobs)
- Maintain a drug and alcohol free environment.
- Function as a member of a team.

Companies expect process technicians to:

- Avoid accidents and errors (work safely, in compliance with government and industry regulation).
- Help maintain health and environmental compliance.
- Work smarter and focus on business goals.
- Look for ways to reduce waste and improve efficiency.
- Keep up with industry trends and constantly improve skills.

ENVIRONMENTAL REGULATIONS AND CONSIDERATIONS

The United States has one of the world's safest food supplies, primarily because it is closely monitored, and highly regulated by federal, state, and local authorities. The main authorities involved with food handling and safety include:

- Food and Drug Administration (FDA)
- Centers for Disease Control and Prevention (CDC)
- U.S. Department of Agriculture (USDA)
- U.S. Environmental Protection Agency (EPA)
- National Oceanic and Atmospheric Administration (NOAA)
- Bureau of Alcohol, Tobacco and Firearms (ATF)
- State and local governments

Each of these authorities has a unique set of products that it oversees. Table 8-1 lists these products.

TABLE 8-1: U.S. food regulating agencies and the products they oversee

Agency...	Oversees...
FDA	All domestic and imported food sold in interstate commerce, including shell eggs, but not meat and poultry Bottled water Wine beverages with less than 7 percent alcohol
CDC	All foods
USDA	Domestic and imported meat and poultry and related products, such as meat- or poultry-containing stews, pizzas and frozen foods Processed egg products (generally liquid, frozen or dried pasteurized)
EPA	Drinking water
NOAA	Fish and seafood products
ATF	Alcoholic beverages except wine beverages containing less than 7 percent alcohol
State and local governments	All foods within their jurisdictions

When monitoring and maintaining the safety of various food products, each of these authorities has its own unique set of responsibilities. Table 8-2 lists some of these responsibilities.

TABLE 8-2: The roles U.S. food regulating agencies play in food safety

Agency	Responsibilities
FDA	Enforces food safety laws governing domestic and imported food, except meat and poultry, by: • Inspecting food production establishments and food warehouses and collecting and analyzing samples for physical, chemical and microbial contamination • Reviewing safety of food and color additives before marketing • Reviewing animal drugs for safety to animals that receive them and humans who eat food produced from the animals • Monitoring safety of animal feeds used in food-producing animals • Developing model codes and ordinances, guidelines and interpretations and working with states to implement them in regulating milk and shellfish and retail food establishments, such as restaurants and grocery stores. An example is the model Food Code, a reference for retail outlets and nursing homes and other institutions on how to prepare food to prevent food-borne illness. • Establishing good food manufacturing practices and other production standards, such as plant sanitation, packaging requirements, and Hazard Analysis and Critical Control Point programs • Working with foreign governments to ensure safety of certain imported food products • Requesting manufacturers to recall unsafe food products and monitoring those recalls • Taking appropriate enforcement actions • Conducting research on food safety • Educating industry and consumers on safe food handling practices
CDC	• Investigates with local, state and other federal officials sources of food-borne disease outbreaks • Maintains a nationwide system of food-borne disease surveillance: Designs and puts in place rapid, electronic systems for reporting food-borne infections. Works with other federal and state agencies to monitor rates of and trends in food-borne disease outbreaks. Develops state-of-the-art techniques for rapid identification of food-borne pathogens at the state and local levels.

TABLE 8-2:
Continued

	• Develops and advocates public health policies to prevent food-borne diseases • Conducts research to help prevent food-borne illness • Trains local and state food safety personnel
Food Safety and Inspection Service	Enforces food safety laws governing domestic and imported meat and poultry products by: • Inspecting food animals for diseases before and after slaughter • Inspecting meat and poultry slaughter and processing plants • With USDA's Agricultural Marketing Service, monitoring and inspecting processed egg products • Collecting and analyzing samples of food products for microbial and chemical contaminants and infectious and toxic agents • Establishing production standards for use of food additives and other ingredients in preparing and packaging meat and poultry products, plant sanitation, thermal processing, and other processes • Making sure all foreign meat and poultry processing plants exporting to the United States meet U.S. standards • Seeking voluntary recalls by meat and poultry processors of unsafe products • Sponsoring research on meat and poultry safety • Educating industry and consumers on safe food-handling practices
EPA	Foods made from plants, seafood, meat and poultry • Establishes safe drinking water standards • Regulates toxic substances and wastes to prevent their entry into the environment and food chain • Assists states in monitoring quality of drinking water and finding ways to prevent contamination of drinking water • Determines safety of new pesticides, sets tolerance levels for pesticide residues in foods, and publishes directions on safe use of pesticides
NOAA	• Through its fee-for-service Seafood Inspection Program, inspects and certifies fishing vessels, seafood processing plants, and retail facilities for federal sanitation standards
ATF	• Enforces food safety laws governing production and distribution of alcoholic beverages • Investigates cases of adulterated alcoholic products, sometimes with help from FDA
State and local governments	• Work with FDA and other federal agencies to implement food safety standards for fish, seafood, milk, and other foods produced within state borders • Inspect restaurants, grocery stores, and other retail food establishments, as well as dairy farms and milk processing plants, grain mills, and food manufacturing plants within local jurisdictions • Embargo (stop the sale of) unsafe food products made or distributed within state borders

SUMMARY

Food and beverage quality is an integral to safety, daily culture and even religious beliefs. Various techniques and government guidelines have been enacted to protect public health. Starting in 1906, the U.S. government passed the first of many food and drug-related acts.

The food and beverage manufacturing industry links farmers to consumers through the processing of fruits, vegetables, grains, meats, dairy products, and other finished goods. Once processed, these finished goods are then delivered to grocers and wholesalers who then supply them to households, restaurants, or institutional food services.

A substantial percentage of food-borne illnesses are caused by pathogens (disease causing microorganisms) that originated in contaminated environments or in animals prior to slaughter.

Food processing methods include disinfecting, preserving, manufacturing, drying, agglomerating, roasting/toasting, cooking/frying, mixing, evaporating, pasteurization, distillation and reverse osmosis. Food processing systems, equipment, and methods vary by facility.

Process technicians are required to have strong communication skills (i.e. reading, writing, and verbal), solid computer skills, and an understanding of biology, chemistry and basic food safety. Job duties vary, but can include monitoring and controlling processes, assisting with equipment maintenance, troubleshooting and problem solving, communicating, performing administrative duties, and performing with a focus on safety, health, security and the environment. Workplace conditions vary by the type of facility.

The United States has one of the world's safest food supplies, primarily because it is closely monitored, and highly regulated by federal, state, and local authorities. Various government agencies such as the Food and Drug Administration and the U.S. Department of Agriculture oversee a wide range of food and beverage products.

CHECKING YOUR KNOWLEDGE

1. Define the following key terms:
   a. Disinfecting
   b. Manufacturing
   c. Pathogen
   d. Preserving

2. When did the U.S. Congress pass the original Food and Drugs Act?
   a. 1906
   b. 1954
   c. 1965
   d. 1990

3. A substantial percentage of food-borne illnesses are caused by __________.

4. Disinfecting is the process of:
   a. preparing for long-term storage
   b. making a product from raw materials
   c. removing moisture
   d. destroying disease-causing organisms

5. Name the process that involves forcing water through a membrane to remove impurities:
   a. Evaporation
   b. Reverse osmosis
   c. Agglomeration
   d. Distillation

6. Name three tasks that process technicians

7. *(True or False)* Computer skills are important for process technicians.

8. Name three government agencies that regulate the food and beverage industry.

9. Which government agency oversees fish and seafood products?
   a. ATF
   b. NORAD
   c. NOAA
   d. NCAA

10. Which government agency oversees domestic and imported meat and poultry products?
    a. EPA
    b. USDA
    c. NOAA
    d. NRC

ACTIVITIES

1. Using library resources or the Internet, research the history of food manufacturing. Identify five items or events that you believe had a significant impact on food quality today. Write a two-page paper that lists and describes each of these items.
2. Select two government agencies that impact the food and beverage manufacturing industry. Using the Internet or other resources, write a one-paper paper on what food and/or beverage-related regulations these agencies oversee and how they affect the industry and consumers.

Chapter 9

Pharmaceutical Industry Overview

OBJECTIVES

Upon completion of this chapter you will be able to:

1. Explain the history of the pharmaceutical industry.
2. Provide an overview of the drug manufacturing process.
3. Identify process technician duties, responsibilities, and expectations.
4. Identify future trends in the pharmaceutical industry.

KEY TERMS

- **Antibiotics**—substances derived from mold or bacteria that inhibit the growth of other microorganisms (e.g., bacteria or fungi).
- **Apothecary**—a person who studies the art and science of preparing medicines; in modern times we call these individuals pharmacists.
- **Biologicals**—products (e.g., vaccines) derived from living organisms that detect, stimulate or enhance immunity to infection.
- **Compounding**—mixing two or more substances or ingredients to achieve a desired physical form.
- **Drugs**—substances used as medicines or narcotics.
- **Pharmaceuticals**—are man-made and natural drugs used to treat diseases, disorders, and illnesses.

INTRODUCTION

Pharmaceuticals (drugs) are man-made or naturally derived chemical substances with medicinal properties that may be used to treat diseases, disorders, and illnesses. Modern drug companies produce a variety of products, including finished drugs; biological products, bulk chemicals and botanicals that are used in making finished drugs. These companies may also produce substances that are used for diagnostic purposes such as the chemicals used for blood glucose monitoring and in pregnancy test kits.

Clearly, modern drugs save lives and can improve the well being of countless patients. They can improve health and quality of life, extend lifespan and reduce healthcare costs by keeping people out of healthcare facilities like hospitals, emergency rooms, and nursing homes.

The process for developing new drugs is both a time consuming and expensive. From discovery to market the process may take as long as 10 years and involve countless failures before a new safe and effective medicine is approved.

HISTORY OF THE INDUSTRY

Civilizations have been using pharmaceuticals since before the dawn of history. Early man, for example, used leaves, mud, and other naturally occurring substances to treat ailments. However, their knowledge of pharmaceuticals was very limited and their methods were crude. Discoveries were purely based on trial and error.

2000 B.C.

An **apothecary** is a person who studies the art and science of preparing medicines (in modern times we call these individuals pharmacists). The earliest apothecary records date back to ancient Babylonia in 2600 B.C. These medical texts, which were recorded on clay tablets, listed the symptoms of the illness, the prescribed treatment, and the method for creating the remedy.

Around 2000 B.C., Chinese emperor Shen Nung also began investigating the medicinal properties of hundreds of products, specifically medicinal plants and herbs. During his investigation, Nung identified more than 365 **drugs** (substances used as medicines or narcotics).

1000 B.C.

While Egyptian medicine dates back to 2900 B.C., many of the pharmaceutical records were not created until much later. One of the best-known and most important pharmaceutical records is the "Papyrus Ebers," a collection of more than 800 prescriptions and 700 drugs.

200 A.D.

One of the most revered names in the professions of Pharmacy and Medicine is a Roman scientist named Galen. Galen practiced and taught medicine and pharmacy. His principles for **compounding** (mixing two or more ingredients to achieve a desired physical form) and preparing medicines ruled the western world for 1,500 years. Many of Galen's original procedures for compounding are still in use today. In fact, in Europe drug products are often referred to as galenicals.

DID YOU KNOW?

Galen created the original formula for cold cream, a cream used cosmetically to soften and clean the skin.

The cold cream formula we use today is very similar to Galen's original formula.

1600s–1700s

Pharmaceutical manufacturing as an industry began in the 1600s, but didn't get underway until the mid 1700s. Early manufacturing processes began in Germany and Switzerland, mostly as an outgrowth of the chemical industry, and then spread to England, France, and the United States.

1800s

Quality control for drugs, and standards for education and manufacturing were not put in place until the 1800s.

In the 1820s the first "United States Pharmacopoeia" was created. This book was the first book of drug standards to gain national acceptance. The name pharmacopoeia means "the art of the drug compounder." Inside this book are directions for identifying and preparing compound medicines.

The end of the 1800s marked the introduction of **biologicals**, products (e.g., vaccines) derived from living organisms that detect, stimulate or enhance immunity to infection. When Behring and Roux announced the effectiveness of diphtheria antitoxin in 1894, scientists in the United States and Europe rushed to put this new discovery into production. As a result, the lives of thousands of children were saved.

In addition to biologicals, chemically manufactured compounds also came into existence. The inspiration to begin chemically manufacturing medicinal compounds began in the late 1800s with the creation of antipyrine, a substance used to reduce fever.

1900s–Present

Interest in pharmaceutical research began in the late 1930s and early 1940s when it became clear that organic chemistry could be used to synthesize complex chemical structures that could be used to make drugs. The field has continued to grow ever since. Today, our detailed understanding of diseases and our understanding of the human genome promise rapid advances that were once unthinkable.

In the mid-1900s, the antibiotic era began. **Antibiotics** are substances derived from mold or bacteria that inhibit the growth of other microorganisms (e.g., bacteria or fungi). While antibiotics were not new (Louis Pasteur first observed them in the late 1800s), the mass production of them was.

Sparked by pressures from World War II, pharmaceutical manufacturers were tasked with rapidly adapting their processes so they could mass-produce the antibiotic penicillin. This mass production saved many lives, allowed for a wider distribution of the drug, and cut the cost of production to 1/1000th of the original cost.

OVERVIEW OF THE DRUG MANUFACTURING PROCESS

Modern drug manufacturing establishments produce a variety of products. The process for developing new medicines is lengthy and expensive. Before a drug can be marketed to the public, it must go through many stages of testing. The U.S. Food and Drug Administration (FDA) was charged with regulating the approval of new medicines in 1906. Since then a very well regulated and well-defined process has evolved. Through this process, new medicines are tested to establish safety and efficacy. The stages or phases of this testing are listed in Table 9-1

TABLE 9-1: Stages of the drug manufacturing process

Stage	Description
1. Preclinical studies	Studies conducted on non-human subjects that provide information pertaining to how the body processes the drug and any potential side effects.
2. Clinical evaluation	Documents submitted to the FDA requesting approval to the test the new product in a select group of people.
3. Clinical trials	FDA approved studies conducted on human, volunteer test subjects; used to determine the safety and effectiveness of the drug.
4. Regulatory filing	Documents submitted to the FDA requesting marketing approval.
5. FDA approval and post-approval monitoring	Approved medication is prescribed, and patients are monitored for any adverse reactions.
6. Manufacturing	Drugs are produced; manufacturing facilities receive FDA inspection; samples are submitted to the FDA for testing; approved drugs are manufactured and distributed to consumers.

The drug manufacturing process requires a significant amount of time and financial investment. As the candidate drugs moves through each of the stages, the potential number of compounds decreases. According to some estimates, for every 5,000 compounds that enter preclinical testing, only five will continue on to clinical trials in humans, and only one will be approved for marketing in the United States.

Quality

Pharmaceutical manufacturing is a highly regulated business operating under a strict set of guidelines published by the FDA. These guidelines are often referred to as GMPs or Good Manufacturing Practices. GMPs were mandated by Congress in 1938 after a number of deaths occurred which were attributable to poisonous ingredients being mistakenly used in pharmaceutical manufacturing processes. As a result of these deaths, quality control and quality assurance organizations became a required and vital part in the pharmaceutical industry. Many production workers are assigned to quality control and quality assurance functions on a full-time basis.

DID YOU KNOW?

According to the Tufts Center for the Study of Drug Development, the cost of developing a new drug averages about $897 million over 10 to 15 years.

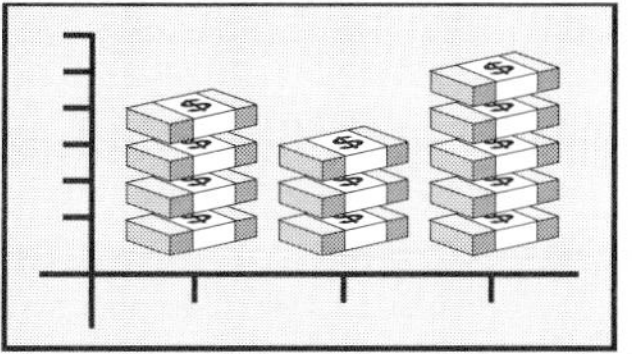

Throughout the production process, inspectors, testers, sorters, samplers, and weighers ensure consistency and quality (e.g. tablet testers inspect tablets for hardness, chipping, and weight to assure conformity with specifications).

PROCESS TECHNICIAN DUTIES AND RESPONSIBILITIES

Process technicians in the pharmaceutical industry perform a wide variety of duties which include operating drug-producing equipment and inspecting products.

Transportation and material-moving employees package and transport finished drugs to other facilities.

Workers on the production side may rotate through different processes (e.g., assemblers and fabricators), while others may specialize in one part of the production process (e.g., compacting machine setters, filling machine operators, and milling machine tenders).

Many production facilities are highly automated. In some processes, milling and micronizing machines (machines which pulverize substances into extremely fine particles) are used to reduce the size of bulk chemicals. These finished chemicals are then combined and processed further in mixing machines. Once mixed, these formulations are then mechanically made into capsules, pressed into tablets, or made into solutions.

In order to create these products, process technicians must be able to operate a variety of equipment (e.g., pumps, valves, conveyors, and milling machines), must have strong technical and computer skills, and must be able to keep up with constantly changing governmental rules and regulations. A two-year Associates degree is also preferred.

Skills

The skills companies look for most when searching for potential applicants include:

- Education (preferably an Associates degree)
- Technical knowledge and skills
- Interpersonal skills
- Computer skills
- Physical abilities

Process technicians must also be able to operate with the principles of cost savings, efficiency, and safety in mind, since there is a strong emphasize on these items within industry.

Job Duties

Process technicians may be asked to perform a wide variety of job duties. The duties may include:

- Monitoring and controlling processes and equipment
- Sampling process streams and analyzing data

- Inspecting, operating and maintaining equipment
- Troubleshooting processes and equipment
- Making process adjustments
- Responding to changes, emergencies and abnormal operations
- Documenting activities, issues and changes
- Communicating and working with others as part of a team
- Learning new skills and information
- Helping train others
- Performing administrative and housekeeping duties (e.g., making sure the work area is clean and organized)
- Performing safety and environmental checks and looking for unsafe acts or potentially hazardous situations
- Keeping safety, health, environment and security regulations in mind at all time
- Handling chemical substances
- Following chemical procedures
- Adhere to the Good Manufacturing Practices regulations that govern the manufacture of drugs and drug products

Equipment

The pharmaceutical manufacturing industry uses a wide variety of equipment, including:

- Pumps
- Compressors
- Milling and micronizing machines
- Valves
- Capsule filling, sealing, and stamping machines
- Conveyors
- Distillation columns
- Centrifuges
- Vessels
- Bottle filling, labeling, and packaging machines

Some of these types of equipment are explained in more detail in the equipment chapter of this textbook.

Work Place Conditions and Expectations

The working conditions in pharmaceutical plants are usually better than those in most other chemical manufacturing facilities. A strong emphasis is placed on keeping equipment and work areas clean because of the potential for contamination (GMP regulations). GMP training is required. Work areas are usually quiet, well lit, air conditioned, and access controlled. In addition, ventilation systems are used to protect workers from dust, fumes, and odors. With the exception of work performed by material handlers and maintenance workers, most jobs require little physical effort.

Only about 5 percent of the workers in the pharmaceutical and medicine manufacturing industry are union members or are covered by a union contract. This is compared with about 15 percent of workers throughout private industry.

Some employees work shift work in plants that operate 7 days a week, 365 days per year. Most operations occur indoors.

The pharmaceutical industry places a heavy emphasis on continuing education for employees. In addition, many companies provide classroom training in safety, environmental and quality control, and technological advances.

Many companies encourage production workers to take courses related to their jobs at local schools and technical institutes or to enroll in correspondence courses. Often continuing college education is sponsored and encouraged. College courses in chemistry, biology, mathematics, and related areas are particularly valuable for highly skilled production workers who operate sophisticated equipment.

FUTURE TRENDS

Pharmaceutical research and development ranks among the fastest growing manufacturing industries, so demand for this industry's products is expected to remain strong, as demand for new and effective medicines is ever increasing. The industry is also relatively insensitive to changes in economic conditions, so work is likely to be stable, even during periods of high unemployment.

SUMMARY

Pharmaceuticals have been used since early times, when knowledge of pharmaceuticals was very limited and the methods of identification were very crude. While the area of pharmaceutical research improved throughout the centuries, it didn't become standardized until the 1800s.

Modern drug manufacturing establishments produce a variety of pharmaceutical products. The process for developing new drugs is a both time consuming and expensive.

During the production of pharmaceutical products, quality is paramount.

Process technicians in the pharmaceutical industry must possess strong technical and interpersonal skills, technical knowledge, computer skills, and physical abilities (depending on the job). They must also be able to work safely and efficiently, and be able to keep up with government rules and regulations that are always subject to change.

The working conditions within the pharmaceutical industry are better than most. Technicians work primarily indoors in air-conditioned facilities that are clean, quiet, and well lit.

Pharmaceutical and medicinal product manufacturing ranks among the fastest growing industries. Demand for this industry's products is expected to remain strong. The industry is also relatively insensitive to changes in economic conditions, so work is likely to be stable, even during periods of high unemployment.

CHECKING YOUR KNOWLEDGE

1. Define the following key terms
   a. Pharmaceuticals
   b. Apothecary
   c. Compounding
   d. Biologicals
   e. Antibiotics

2. In what year were the earliest apothecary records on clay tablets?
   a. 2000 B.C.
   b. 2600 B.C.
   c. 2900 B.C.
   d. 1000 B.C.

3. The introduction of the "United States Pharmacopoeia" in the 1820s was important because:
   a. It marked the introduction of biologicals (vaccines)
   b. It encouraged scientists to conduct more pharmaceutical research
   c. It was the first book of drug standards to gain national acceptance

4. Which world event encouraged pharmaceutical manufacturers to rapidly adapt their processes so they could mass-produce antibiotics?
   a. World War I
   b. World War II
   c. The Cold War
   d. The industrial revolution

5. It is estimated that for every __________ compounds that enter pre-clinical testing, only five will continued on to clinical trials in humans.
   a. 100
   b. 1,000
   c. 2,000
   d. 5,000

ACTIVITIES

1. List and describe each of the stages of the drug manufacturing process.

2. Use the internet or other library resources to research the history of drug manufacturing. Write a 2-3 page paper describing what you learned.

3. Use the internet or other library resources to research current drug manufacturing processes. Write a 2-3 page paper describing what you learned.

Chapter 10
Basic Physics

OBJECTIVES

Upon completion of this chapter you will be able to:

1. Define the application of physics in the petrochemical process industries.
2. Define matter and the states in which it exists.
3. Use physical property characteristics to describe various states of matter (liquid, gas, and solids).
4. Define and provide examples of the following terms:
   - Mass
   - Density
   - Elasticity
   - Viscosity
   - Buoyancy
   - Specific Gravity
   - Flow
   - Evaporation
   - Pressure
   - Velocity
   - Friction
   - Temperature
   - British Thermal Unit
5. Describe the three (3) methods of heat (BTU) transfer:
   - Convection
   - Conduction
   - Radiation
6. Describe how Boyle's Law explains the relationship between pressure and volume of gases.
7. Describe how Charles' Law explains the relationship between temperature and volume of gases.
8. Describe how Dalton's Law explains the relationship between pressure and gas.
9. Describe how the General (Ideal) Gas Law explains the relationship between temperature, pressure, and volume of gas.
10. Describe how Bernoulli's Law relates energy conversions in a flowing fluid.

KEY TERMS

- **Absolute Pressure (psia)**—gauge pressure plus atmospheric pressure; pressure referenced to a total vacuum (zero PSIA).
- **API Gravity**—the American Petroleum Institute (API) standard used to measure the density of hydrocarbons.

- **Atmospheric Pressure**—the pressure at the surface of the earth (14.7 PSIA at sea level).
- **Baume Gravity**—the industrial manufacturing measurement standard used to measure the gravity of non-hydrocarbon materials.
- **Bernoulli's Law**—a physics principle that states as the speed of a fluid increases, the pressure inside the fluid, or exerted by it, decreases.
- **Boiling Point**—the temperature at which liquid physically changes to a gas at a given pressure.
- **Boyle's Law**—a physics principle that states, at a constant temperature, as the pressure of a gas increases, the volume of the gas decreases.
- **British Thermal Unit (BTU)**—the amount of heat energy required to raise the temperature of one pound of water one degree Fahrenheit.
- **Buoyancy**—the principle that a solid object will float if its density is less than the fluid in which it is suspended; the upward force exerted by the fluid on the submerged or floating solid is equal to the weight of the fluid displaced by the solid object.
- **Calorie**—the amount of heat energy required to raise the temperature of one gram of water by one degree Celsius.
- **Charles' Law**—a physics principle that states, at constant pressure, as the temperature of a gas increases, the volume of the gas also increases.
- **Conduction**—the transfer of heat through matter via vibrational motion.
- **Convection**—the transfer of heat through the circulation or movement of a liquid or gas.
- **Dalton's Law**—a physics principle that states the total pressure of a mixture of gases is equal to the sum of the individual partial pressures.
- **Density**—the ratio of an object's mass to its volume.
- **Elasticity**—an object's tendency to return to its original shape after it has been stretched or compressed.
- **Flow**—the movement of fluids.
- **Fluid**—substances, usually liquids or vapors, that can be made to flow.
- **Friction**—the resistance encountered when one material slides against another.
- **Gas (vapor)**—substance with a definite mass but no definite shape, whose molecules move freely in any direction and completely fill any container it occupies, and which can be compressed to fit into a smaller container.
- **Gauge Pressure (psig)**—pressure measured with respect to the Earth's surface at sea level (zero PSIG).
- **General Gas Law (Combined Gas Law)**—relationship between pressure, volume and temperature in a closed container; pressure and temperature must be in absolute scale ($P_1V_1/T_1 = P_2V_2/T_2$).
- **Heat**—the transfer of energy from one object to another as a result of a temperature difference between the two objects.

- **Heat Tracing**—a coil of heated wire or tubing that is adhered to or wrapped around a pipe in order to increase the temperature of the process fluid, reduce fluid viscosity, and facilitate flow.
- **Hydrocarbon**—organic compounds, that contain only carbon and hydrogen, which are most often found occurring in petroleum, natural gas and coal.
- **Hydrometer**—an instrument designed to measure the specific gravity of a liquid.
- **Ideal Gas Law**—the mathematical expression PV = nRT which expresses the simplest relationship between temperature, pressure, volume and moles of a gas.
- **Latent Heat**—heat that does not result in a temperature change but causes a phase change.
- **Latent Heat of Condensation**—the amount of heat energy given off when a vapor is converted to a liquid without a change in temperature.
- **Latent Heat of Fusion**—the amount of heat energy required to change a solid to a liquid without a change in temperature.
- **Latent Heat of Vaporization**—the amount of heat energy required to change a liquid to a vapor without a change in temperature.
- **Liquid**—substances with a definite volume, but no fixed shape, that demonstrate a readiness to flow with little or no tendency to disperse, and are limited in the amount in which they can be compressed.
- **Lubrication**—a friction-reducing film placed between moving surfaces in order to reduce drag and wear.
- **Mass**—the amount of matter in a body or object measured by its resistance to a change in motion.
- **Matter**—anything that takes up space and has mass.
- **Phase Change**—when a substance changes from one physical state to another, such as when ice melts to form water.
- **Plasma**—a gas that contains positive and negative ions.
- **Pressure**—The force exerted on a surface divided by its area.
- **Radiation**—the transfer of heat energy through electromagnetic waves.
- **Sensible Heat**—heat transfer that results in a temperature change.
- **Solids**—substances, with a definite volume and a fixed shape, that are neither liquid, nor gas, and that maintain their shape independent of the shape of the container.
- **Specific Gravity**—the ratio of the density of a liquid or solid to the density of pure water, or the density of a gas to the density of air at standard temperature and pressure (STP).
- **Specific Heat**—the amount of heat required to raise the temperature of one gram of a substance one degree Celsius, or one pound of a substance one degree Fahrenheit.

- **Temperature**—The degree of hotness or coldness that can be measured by a thermometer and a definite scale.
- **Vacuum Pressure (psiv)**—any pressure below atmospheric pressure.
- **Vapor Pressure**—a measure of a substance's volatility and its tendency to form a vapor.
- **Velocity**—the distance traveled over time or change in position over time.
- **Viscosity**—the measure of a fluid's resistance to flow.
- **Weight**—a measure of the force of gravity on an object.

INTRODUCTION

The concepts of basic physics are important to process technicians. By learning these concepts, process technicians increase their ability to visualize what is occurring in a process, improve their ability to troubleshoot, and reduce the likelihood they will make costly or catastrophic mistakes.

APPLICATIONS OF PHYSICS IN THE PROCESS INDUSTRIES

During normal, every day operations, process technicians are required to open valves, check pressures, monitor fluid and gas flows, monitor furnace operations and much more. In order to perform these tasks effectively, a technician must have a firm understanding of physics.

Consider the task of rerouting fluid flow using valves like the ones show in Figure 10-1. If a technician does not have a firm understanding of fluid flow principles, he or she might not realize that you must open a new path (valve A) before closing off the original path (valve B). Failure to perform these steps in this order could cause an increase in back flow pressure that could damage or "dead head" the pump.

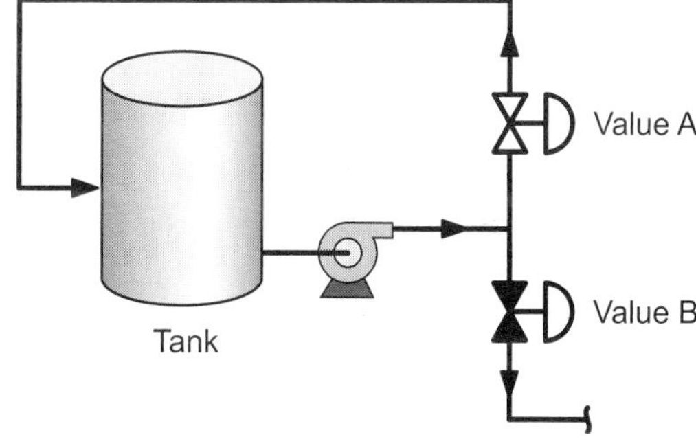

Figure 10-1: Technicians should never "dead-head" a pump

STATES OF MATTER

The term matter is used frequently in physics. **Matter** is anything that takes up space and has inertia and mass. In other words, matter is the substance things are made of. An atom is the smallest indivisible unit of matter.

Matter comes in four states: solid, liquid, gas and plasma. Each of these states has its own unique properties and characteristics.

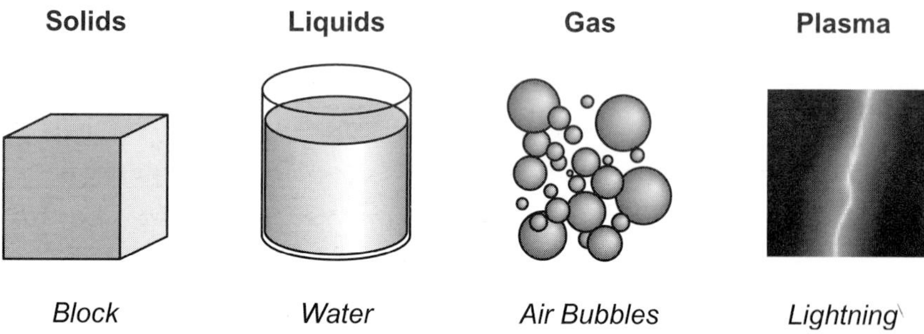

Figure 10-2: Matter can be a solid, liquid, gas or plasma

Solids are substances with a definite volume and a fixed shape. They are neither liquid nor gas and they maintain their shape independent of the shape of the container.

Liquids are substances with a definite volume but no fixed shape. They demonstrate a readiness to flow with little or no tendency to disperse, and are limited in the amount in which they can be compressed (unlike gases, which are highly compressible).

Gases are substances with definite mass but no definite shape. The molecules in a gas move freely in any direction, completely fill any container they occupy, and can be compressed to fit into a smaller container. Gases can mix freely with each other, and can be liquefied through compression or temperature reduction.

Plasma is a gas that contains positive and negative ions. Because plasma is not commonly found in the process industries, it will not be discussed in great detail in this chapter.

As the states of matter (specifically, solids, liquids, and gases) are exposed to different environmental conditions they may be induced to change form (e.g., ice melting to into a puddle of water). This process is called a **phase change**. During a phase change the substance (solid, liquid or gas) maintains the same chemical composition, but changes physical properties (e.g., ice is a solid that melts to form liquid, yet both are made up of the water).

Process technicians need to understand the concept of phase changes, since some of the substances they work with may undergo these changes. Failure to understand phase changes and their impacts could cause process problems or pose serious safety risks. The table below lists and describes the six phases of change as well as potential problems or safety risks associated with each.

TABLE 10-1: Six Phase Changes

Description	Term	Movement of Heat	Example of Change	Examples of Problems or Risks
Solid to liquid	Melting	Heat is absorbed by the solid as it melts	Ice melting to form a puddle of water	Melting ice could form a puddle of liquid that poses a slipping hazard.
Liquid to solid	Freezing	Heat leaves the liquid as it freezes	Water freezing to form ice	Leaving an unheated vessel (e.g., a tank or pipe) full of water during freezing temperatures could cause equipment damage or vessel rupture as ice builds up and expands inside.
Liquid to vapor	Vaporization (includes boiling and evaporation)	Heat absorbed by the liquid as it vaporizes	Water boiling to make steam or a puddle evaporating after a rain shower	Liquids exposed to extreme heat could form vapor that could over pressurize an enclosed vessel or pose a burn risk.
Vapor to liquid	Condensation	Heat leaves the vapor as it condenses	Water vapor (humidity) in the air condensing as water droplets on a cold drinking glass	Sensitive electronic equipment left outside overnight could be damaged by humidity or condensation.
Solid to vapor	Sublimation	Heat absorbed by the solid as it sublimates	"Dry ice" (compressed carbon dioxide) changing from a solid to a fog-like vapor	Carbon dioxide (CO_2) "fog" displaces oxygen; in a confined space this could pose a suffocation risk.
Vapor to solid	Deposition	Heat leaves the vapor as it solidifies	Water vapor condensing to form frost on the surface of a pipe.	Exposing uncovered skin to frost could result in a burn that requires medical attention.

Mass

Mass is the amount of matter in a body or object measured by its resistance to a change in motion. It is also a measure of an object's resistance to acceleration.

Mass should not be confused with weight. **Weight** is a measure of the force of gravity on an object. To illustrate, think of an apple. An apple would weigh more on Jupiter than it does on Earth because Jupiter's gravity is stronger. However, the mass is the same, regardless of where the apple is located.

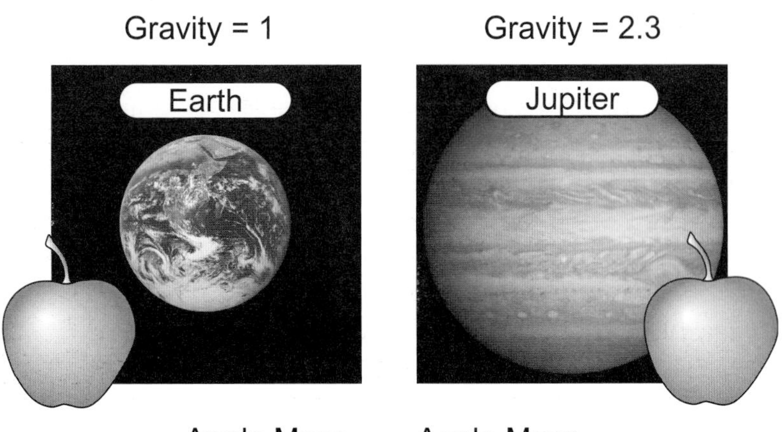

Figure 10-3: Gravity impacts weight

Density

Density is a scientific way to determine or compare the "heaviness" of an object or objects. To determine density one must compare the ratio of the object's mass to its volume.

Consider a canister of marshmallows like the one shown in Figure 10-4.

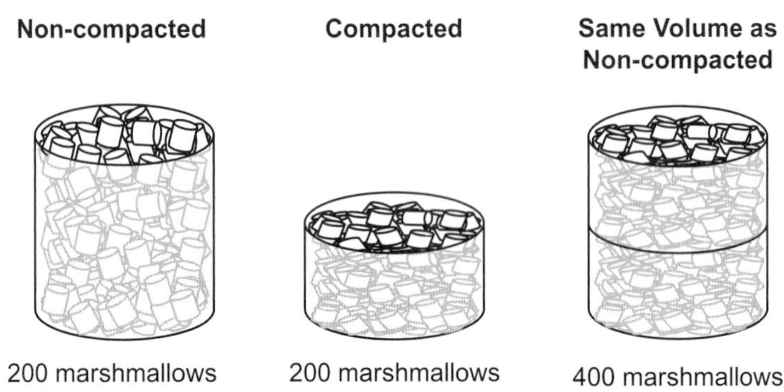

Figure 10-4: Marshmallows become denser when air is removed

In this example, a canister of marshmallows at normal atmospheric pressure weighs 16 ounces. If you were to compress or compact all of those marshmallows so they only filled half the container, they would still weigh 16 ounces, but would only occupy half the volume. In other words, the weight of the marshmallows remains the same, but the density increases. If you then come back and add 16 more ounces of compressed marshmallows to the container you find the volume stays the same, but the mass increases from 16 ounces to 32 ounces and the marshmallows are more compact (dense). This is the concept of density.

Elasticity

Elasticity is an object's tendency to return to its original shape after it has been stretched or compressed. The elastic limit is the greatest amount of stress a material is capable of sustaining without permanent deformation.

To illustrate the concept of elasticity, think of a rubber band. If you stretch a rubber band you see it deform and become thinner and longer. However, once the band is released, it returns to its original shape and size. This is elasticity.

Viscosity

Liquids can be either free-flowing or viscous. The term **viscosity** refers to the fluid's resistance to flow. To demonstrate the concept of viscosity, imagine a spoon full of water and a spoon full of honey (see Figure 10-5).

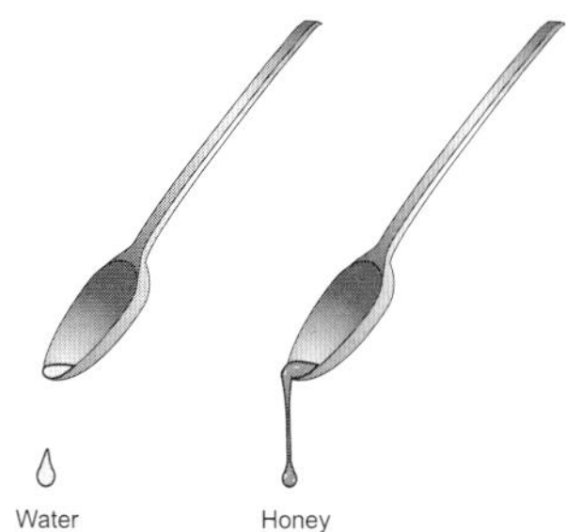

Figure 10-5: Honey is more viscous than water

If you fill a spoon with water and then tilt it you notice the water drips rapidly off the spoon. This is because water is a free-flowing liquid. If you try the same experiment with honey, however, you find the honey drips off the spoon much more slowly. This is because honey is more viscous than water and is, therefore, less ready to flow.

Specific Gravity

Specific Gravity is the ratio of the density of a liquid or solid to the density of pure water, or the density of a gas to the density of air at standard temperature and pressure (STP). In other words, it is the "heaviness" of a substance.

Pure water has a density of 1 gram per cubic centimeter (g/cc). Materials which are lighter than water (specific gravity less than 1.0) will float. Those that are heavier than water (specific gravity greater than 1.0) sink.

Hydrocarbons are organic compounds, found in petroleum, natural gas and coal, which contain only carbon and hydrogen. Process technicians in the chemical and refining areas need to be aware that most hydrocarbons have a specific gravity below 1.0.

Knowing the specific gravity of a substance is important. By knowing the specific gravity a technician can better understand the characteristics of a substance and how it will react in the presence of other substances with differing specific gravities. This can be very important in both day-to-day operations and emergency situations.

For example, light, flammable liquids such as gasoline may spread and, if ignited, burn on the surface of a body of water. If a remediation or fire fighting crew is unaware of this they may take actions that make the situation worse rather than better (e.g., dousing the fire with water, causing it to spread).

In non-emergency situations process technicians need to understand the concept of specific gravity since this measurement is often used when testing samples.

For example, in a heat exchanger, many process fluids are routed through a set of cooling tubes surrounded by a shell full of flowing cool water. In order to make sure the fluid is moving properly and verify that the tubes are not leaking, process technicians must regularly sample the cooling water and test its specific gravity using a **hydrometer** (an instrument designed especially for measuring the specific gravity of a liquid). If the hydrometer reading is normal, the technician knows the system is working properly. However, if the specific gravity indicates the presence of hydrocarbons, the technician will know there is a leak and can take corrective action to isolate the exchanger.

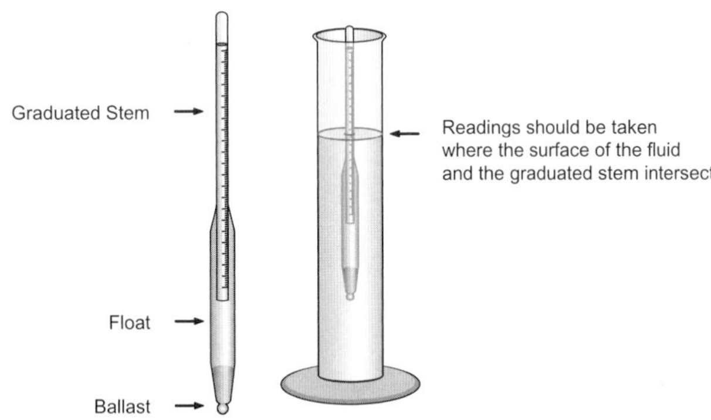

Figure 10-6: Hydrometer example

Hydrometers are also used in the food and beverage industries to test the specific gravity of beer, wine and other food products. Figure 10-6 shows an example of a hydrometer.

In the process industries the two most common gravity measurement types are Baume gravity and API gravity.

Baume gravity is the industrial manufacturing measurement standard used to measure non-hydrocarbon heaviness, while **API gravity** is the American Petroleum Institute (API) standard used to measure the density of hydrocarbons. In order to determine API gravity, technicians must use a specially designed hydrometer marked in API units (the higher the API reading, the lower the fluid's gravity).

Buoyancy

Buoyancy refers to the principle that a solid object will float if its density is less than the fluid in which it is suspended, and if the upward force exerted by the fluid on the submerged or floating solid is equal to the weight of the fluid displaced by the solid object.

The higher the specific gravity of the fluid, the more buoyant the object is. For example, distilled water has a specific gravity of 1.000 while natural seawater has a specific gravity somewhere around 1.025. For this reason, objects such as boats and scuba divers float better in salt water than they do in fresh water.

Flow

Flow is the movement of fluids. When we use the term flow we refer to any uninterrupted movement of a liquid or gas.

There are many factors that can restrict or enhance the flow of a substance. These factors may include mechanical restriction (e.g., valves), environmental factors (e.g., extreme heat or extreme cold), or the physical characteristics of the substance itself (e.g., viscosity).

Process technicians need to understand the concept of flow, especially as it applies to process monitoring, since a large part of a process technician's job involves monitoring process variables, troubleshooting, and opening and closing valves.

Valves, which will be discussed in more detail in later chapters, can have a definite impact on flow rates of fluids. Depending on the type of valve used, process technicians can regulate the flow of a substance so it is completely on, completely off, or somewhere in between.

DID YOU KNOW?

Objects are more buoyant in salt water than fresh water because salt water has a higher specific gravity?

Because of this, scuba divers must wear more weight when diving in salt water to maintain the same neutral buoyancy (a position in water where you go neither up nor down).

So how does temperature impact flow? There is a direct relationship between temperature and molecular movement. As temperatures decreases, so does molecular movement. Thus, if a process occurs in the presence of extremely cold temperatures, either due to the weather outside or an endothermic (heat absorbing) chemical reaction, flow rate may be impacted (e.g., consider the viscosity and flow rate of molasses in January). If the impact of this decreased flow is significant enough, plant technicians may have to apply **heat tracing** (a coil of heated wire or tubing that is adhered to or wrapped around a pipe) in order to increase the temperature of the process fluid, reduce fluid viscosity, and facilitate flow.

In addition to the standard valve and temperature impacts, there are other factors that may impact flow as well. However, many of these are more

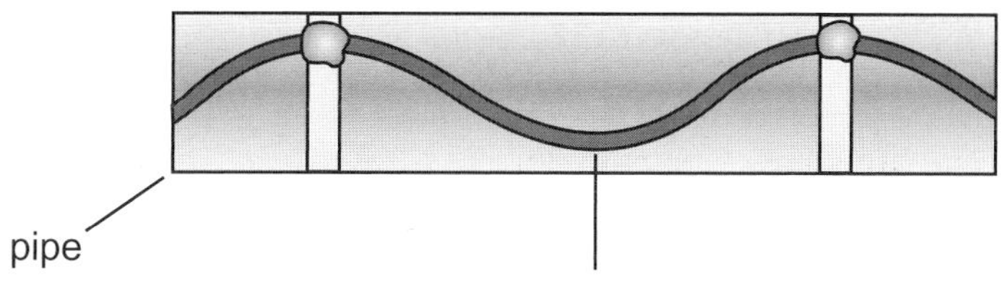

Figure 10-7: Heat tracing applied to process piping

directly related to malfunctions that would be detected during troubleshooting (e.g., a plugged tube might stop the flow of a substance, while a ruptured tube might only change the flow rate depending on the size of the ruptured area, the pressure of the flow, and the meter location).

Velocity

Velocity is the distance traveled over time or change in position over time.

In the process industries, the concept of velocity is most often used in relationship to equipment such as pumps and compressors, which are discussed in later chapters. As velocity increases, so does the amount of flow.

The simplest relationship between weight flow rate and fluid velocity is as follows:

$w = \rho * v * A$

Where:

w = weight flow rate of fluid in lb/sec

ρ = fluid density in lb/ft^3

v = average fluid velocity in ft/sec

A = the cross-sectional area of the pipe carrying the fluid

This equation shows, that for a constant weight flow rate, the velocity of fluid will increase if the cross-sectional area (hence diameter [d]) of the pipe decreases. Because area is proportional to d^2, velocity increases rapidly as pipe diameter decreases.

In systems the temperature can affect the density as discussed later and also alter the fluid velocity even if weight flow rate and pipe diameter do not change.

Friction

Friction is the resistance encountered when one material slides against another. Friction can be both desirable and undesirable. The amount of friction generated is dependent on how smooth the contact surfaces are and how much force is applied when the two surfaces are pressed together. Friction can produce heat and potential wear to two sliding surfaces.

In order to reduce friction, lubrication is often employed. **Lubrication** is the introduction of a friction-reducing film between moving surfaces in order to reduce drag and wear. Lubricants may be fluid, solid, or plastic substances.

It is important to understand that *sliding* an object back and forth over a surface produces more friction, requires more energy, and causes more wear than *rolling* an object (e.g., a tire or ball bearing). Because of this, lubrication, ball bearings and other friction reducing techniques are often used in rotating equipment such as pumps and compressors. Failure to maintain proper lubrication during operations may cause excessive wear or heat that could damage or ruin equipment.

While friction caused by the movement of two solid objects seems to be the type most often thought of, it is also possible to have friction with the movement of fluids. Fluid friction occurs when the molecules of a gas or liquid are in motion. Unlike solid friction, fluid friction varies with velocity and area. It is important to understand the concept of fluid friction, since there are some chemical substances that will ignite or explode as a result of friction that is built up when flow is changed too rapidly.

Temperature

Temperature is a measure of the thermal energy of a substance (hotness or coldness) that can be determined using a thermometer. Process technicians use a variety of temperature scales to measure temperature. These include Kelvin, Celsius (formerly called Centigrade), Fahrenheit and Rankine. Figure 10-8 compares and contrasts each of these four temperature measurement scales.

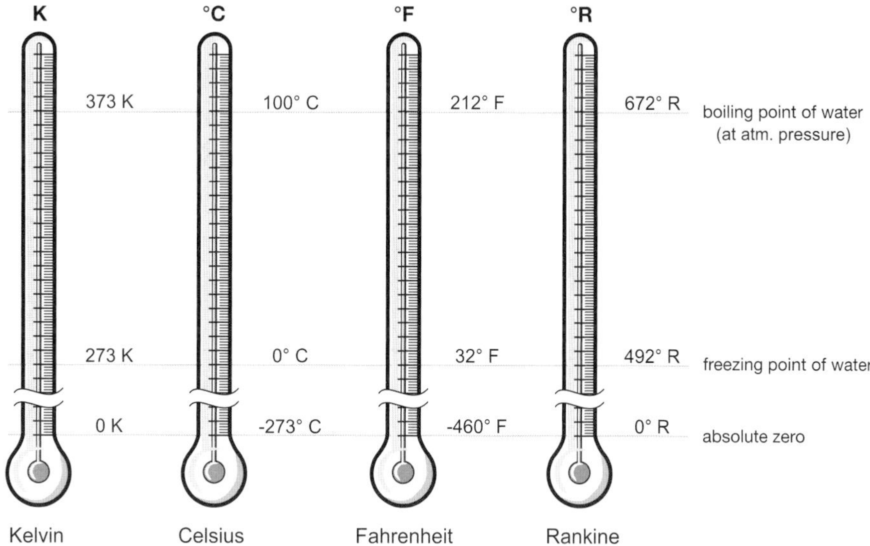

Figure 10-8: Kelvin, Celsius, Fahrenheit and Rankine temperature scales

It is worth noting that both K and °R are known as absolute temperature scales based on a true zero temperature in which molecular motion is believed to cease. When dealing with **ideal gases** later in this chapter we will have to use absolute temperatures to calculate changes in vapor pressures and temperatures.

As you can see, there is a great deal of variability between the four temperature measuring scales. For example, the boiling point of water in Celsius is 100°. In Rankine, however, it is 672°. For this reason, process technicians need to know which scale they are working with when they report their measurements.

Table 10-2 lists several formulas process technicians will find useful when performing temperature conversions.

TABLE 10-2:
Temperature Conversion Formulas

Conversion Type	Formula
Celsius to Kelvin (approximate)	K = °C + 273
Celsius to Fahrenheit	°F = 1.8 * °C + 32
Fahrenheit to Rankine (approximate)	°R = °F + 460
Fahrenheit to Celsius	°C = (°F – 32) / 1.8

When studying temperature, it is also important to know that temperature and heat are not the same thing. Temperature is a measure of the thermal energy of a substance (hotness or coldness) that can be determined using a thermometer. **Heat** is the transfer of energy from one object to another as a result of a temperature difference between the two objects.

According to the principles of physics, heat energy always moves from hot to cold and it cannot be created or destroyed; it can only be transferred from one object or substance to another.

DID YOU KNOW?

The term Calorie (with a capital "C"), found on food labels and in diet books, is really 1000 of the calories referred to in this chapter.

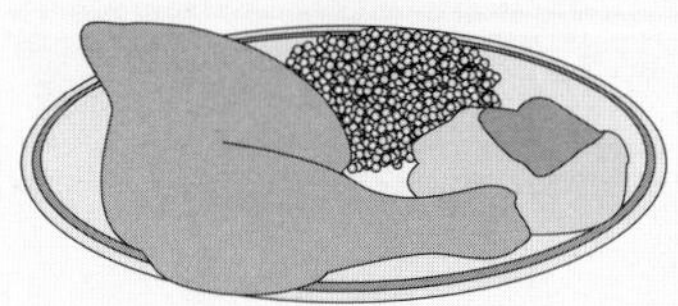

Heat is typically measured in British Thermal Units or in calories. A **British Thermal Unit (BTU)** is the amount of heat energy required to raise the temperature of one pound of water by one degree Fahrenheit. A **calorie** is the amount of heat energy required to raise the temperature of one gram of water by one degree Celsius. One BTU is equivalent to 252 calories.

Heat comes in many different forms: specific heat, sensible heat, latent heat, latent heat of fusion, latent heat of crystallization, latent heat of vaporization and latent heat of condensation. Table 10-3 lists and describes each of these different forms of heat.

Type of Heat	Description
Specific heat	The amount of heat required to raise the temperature of one gram of a substance by one degree Celsius, or one pound of a substance by one degree Fahrenheit. *Example: The specific heat of water is 1 cal/g °C or 1 BTU/lb °F. The specific heat of copper is .10 cal/g °C or .10 BTU/lb °F.*
Sensible heat	Heat transfer that results in a temperature change. *Example: The heat required to bring a pot of water up to boiling.*
Latent heat	Heat that does not result in a temperature change but causes a phase change. *Example: The heat required to keep a pot of water boiling with the temperature remaining constant.*
Latent heat of fusion	The amount of heat energy required to change a solid to a liquid without a change in temperature. *Example: The heat required to change ice to water at a constant temperature.*
Latent heat of vaporization	The amount of heat energy required to change a liquid to a vapor without a change in temperature. *Example: The heat required to change water into steam at a constant temperature.*
Latent heat of condensation	The amount of heat given off when a vapor is converted to a liquid without a change in temperature. *Example: The heat removed to change steam into water at a constant temperature.*

TABLE 10-3: Different Types of Heat

Figure 10-9 illustrates the difference between sensible heat and latent heat when an ice cube at 0°F is converted to steam at 300°F.

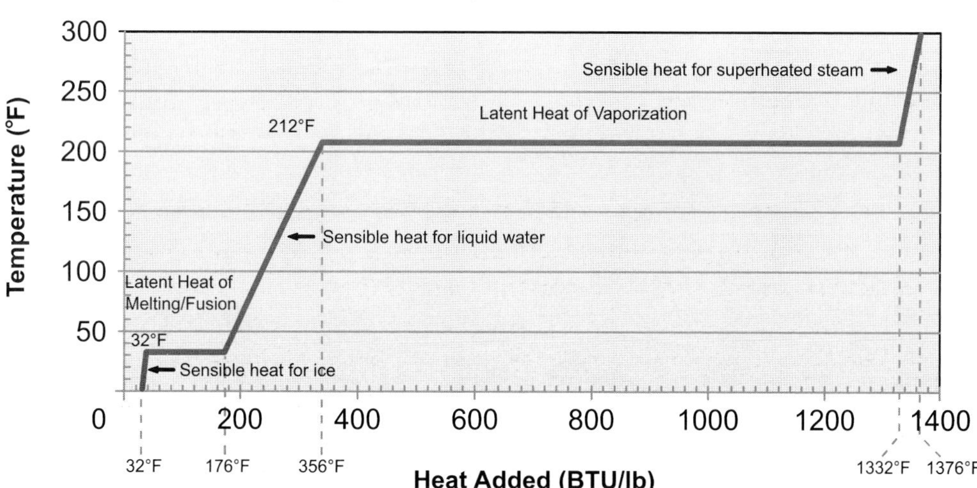

Figure 10-9: Comparison of sensible heat and latent heat

Heat energy is commonly transferred by three methods: conduction, convection, and radiation. **Conduction** is the transfer of heat through matter via vibrational motion (e.g., transferring heat energy from a frying pan to an egg or from heat tracing attached to a pipe to a process fluid).

Convection is the transfer of heat through the circulation or movement of a liquid or gas (e.g., warm air being circulated in a furnace). **Radiation** is the transfer of heat energy through electromagnetic waves (e.g., warmth emitted from the sun or an open flame).

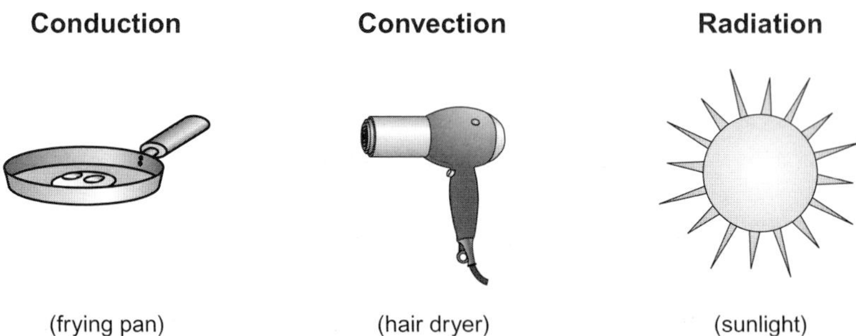

Figure 10-10: Heat may be transferred through conduction, convection or radiation

Pressure

Pressure is the amount of force a substance or object exerts over a particular area. The formula for pressure is defined as force per unit area and is usually expressed in pounds per square inch (psi).

It is important to understand the concept of pressure, especially with regard to fluids, since pressure is a fundamental part of process operations.

Within a plant a technician may encounter pressure in components such as hydraulic lines, storage tanks, fluid lines and more. By understanding the concept of pressure, technicians improve their ability to troubleshoot and reduce the likelihood of equipment damage or serious injury.

ATMOSPHERIC PRESSURE

Air pushes on every surface it touches. Air pressure is the weight of the air from the Earth's atmosphere down to the surface. At sea level, air exerts a pressure of 14.7 pounds per square inch (psia). This is called **atmospheric pressure**. One atmosphere is equal to 14.7 psia or 760 millimeters of mercury (mmHg) absolute. Any pressure less than atmospheric pressure (14.7 psia or 760 mmHg absolute) is referred to as **vacuum**. Vacuum, which will be discussed later, is an important part of many industrial processes.

To illustrate the concept of atmospheric pressure, think of a piece of paper that is 10 inches square. At sea level, the atmosphere exerts a pressure of 14.7 pounds per square inch (psi). In a 10" x 10" piece of paper, there is a total of 100 square inches. So, at sea level, the atmosphere is actually exerting 1,470 pounds of pressure on the top surface of the paper! "So why doesn't the paper collapse under all of that pressure?" you might ask. The answer is simple. The same pressure is also being exerted on the opposite side of the paper, so the two pressures counteract one another.

VAPOR PRESSURE AND BOILING POINT

Vapor pressure is a measure of a substance's volatility and its tendency to form a vapor. Vapor pressure can be directly linked to the strength of the molecular bonds of a substance.

There is an indirect relationship between molecular bond strength and vapor pressure. The stronger the molecular bond, the lower the vapor pressure. The weaker the molecular bond, the higher the vapor pressure. As heat is added to a liquid, molecular motion and vapor pressure increase.

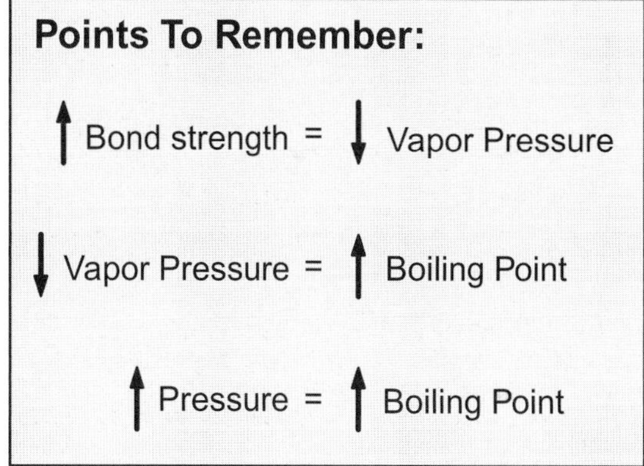

Points To Remember:
↑ Bond strength = ↓ Vapor Pressure
↓ Vapor Pressure = ↑ Boiling Point
↑ Pressure = ↑ Boiling Point

There is also an indirect relationship between vapor pressure and boiling point. If a substance has a low vapor pressure, then its boiling point will be high and vice versa.

Boiling Point is the temperature at which a liquid boils at a specified pressure.

When the boiling point of a liquid is reached, the vapor pressure of the liquid becomes slightly greater than the pressure being exerted on the liquid by the surrounding atmosphere. As a result, bubbles form inside the liquid and rapid vaporization begins.

The boiling point of pure water at atmospheric pressure (14.7 psi) is 100° C or 212° F. However, liquids do not have to reach their boiling point in order for the process of evaporation to occur. Think of a puddle after a rain shower. Over time, the puddle will evaporate and disappear, especially if the sun comes out, warms the air, and increases the molecular activity of the water.

PRESSURE IMPACT ON BOILING

System pressure has a direct impact on the boiling point of a substance. As pressure increases boiling point increases because the molecules are forced closer together. In addition, increases in pressure may also cause vapors to be forced back into solution.

As pressures decrease, as in the case of a vacuum, boiling points decrease as well. By lowering the boiling point using a vacuum, plants can reduce energy costs, and reduce heat-induced molecular damage or product degradation.

PRESSURE GAUGE MEASUREMENTS

In order to determine the amount of pressure present in a particular process, process technicians must be able to read and interpret various types of pressure gauges and measurements. The three main types of pressure measurements a technician will encounter are: gauge, absolute and vacuum.

- **Gauge pressure (psig)** is pressure measured with respect to the Earth's surface at sea level (zero PSIG).
- **Absolute pressure (psia)** is gauge pressure plus atmospheric pressure (pressure referenced to a total vacuum, which is zero PSIA).
- **Vacuum pressure (psiv)** is any pressure below atmospheric pressure. Vacuum is typically measured in inches of mercury (in. Hg) or millimeters of mercury (mmHg).

Figure 10-11 below shows examples of gauge, absolute and vacuum pressure gauges.

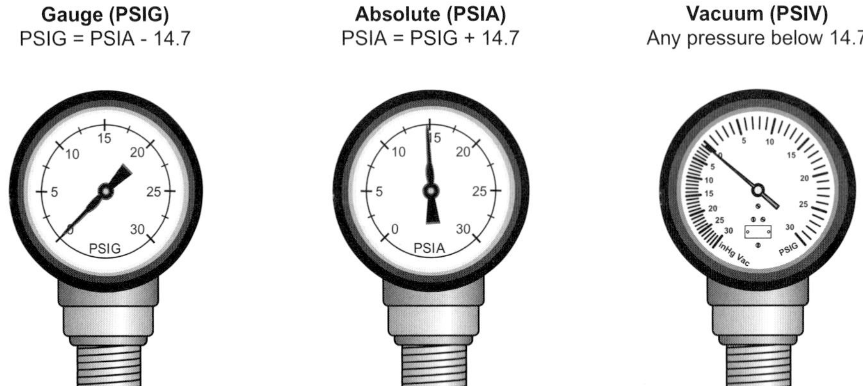

Figure 10-11: Pressure gauges may present gauge, absolute, or vacuum pressure readings

Gas Laws

Gas laws are scientific principles that describe the properties of gases and how they will react under different circumstances. The gas laws that process technicians encounter most are Boyle's Law, Charles' Law, Dalton's Law, the General Gas Law, the Ideal Gas Law, and Bernoulli's Law (Principle).

When performing gas law calculations, it is important to remember that these types of calculations require the use of absolute temperature and pressure units.

BOYLE'S LAW (PRESSURE–VOLUME LAW)

Robert Boyle was an English scientist who, in the late 1600s, discovered there is an inverse relationship between the pressure of a gas and its volume if the sample and temperature are held constant. During his experiments Boyle learned that if he took a gas sample and doubled the absolute pressure, the volume decreased by 50 percent. If he decreased the pressure by 50 percent, the volume doubled. Based on his observations, Boyle created Boyle's Law. Boyle's Law states that the pressure of an ideal gas is inversely proportional to its volume, if the temperature and amount of gas is held constant.

In layman's terms, this means that there is an opposite (inverse) relationship between pressure and volume. As the pressure on a quantity of gas increases, the volume of the gas decreases. In other words, if you double the pressure, you decrease the volume by half.

The formula for Boyle's Law is $\mathbf{P_1 V_1 = P_2 V_2}$

Where:

P = Pressure

V = Volume

(Note: Another formula for Boyle's law is: **PV=K)**

> **BOYLE'S LAW**
>
> States that the pressure of an ideal gas is inversely proportional to its volume, if the temperature and amount of gas is held constant.
>
> The formula for Boyle's Law is:
>
> $$\mathbf{P_1 V_1 = P_2 V_2}$$
>
> *(P = Pressure, V = Volume)*
>
> Note: One atmosphere is equal to 14.7psi

CHARLES' LAW (TEMPERATURE–VOLUME LAW)

Jacques Alexander Cesar Charles was a French physicist who, in the early 1800's (a time when there was much excitement over hot-air balloons) became interested in the effect of temperature on a volume of gas. In order to study the effects of temperature, Charles created an apparatus that allowed him to maintain a fixed mass of gas at a constant pressure. What he determined from his experiments was that as the temperature increased, the volume of gas also increased.

> **CHARLES' LAW**
>
> States that the volume of a gas varies directly with the temperature when the pressure is held constant.
>
> The formula for Charles' Law is:
>
> $$\mathbf{K = V/T}$$
>
> *(K = Temperature in Kelvin, V = Volume, T = Temperature)*

DALTON'S LAW (LAW OF PARTIAL PRESSURES)

John Dalton was an English scientist who, in the early 1800's, discovered the relationship between gases when two or more were mixed together. What Dalton determined was that the total pressure of a mixture of gases is actually equal to the sum of the individual partial pressures.

Partial pressure is the pressure a particular gas, contained in a mixture of gases, would have if it were alone in the container. In other words, it is the pressure that a particular gas would exert if all of the other gases were removed.

> **DALTON'S LAW**
>
> States the total pressure of a mixture of gases is equal to the sum of the individual partial pressures.
>
> The formula for Dalton's Law is:
>
> $$P_{total} = P_a + P_b + P_c \ldots P_n$$
>
> *(P = Pressure)*

The subscripts used in Dalton's formula (e.g., P_a) are used symbolically in the formula to represent different gases. However, in normal applications, the symbolic subscript would be replaced with the actual formula for the gas being referenced. For example, the partial pressure of oxygen (O_2) would be written as P_{O2} or PO_2.

GENERAL (IDEAL) GAS LAW

Earlier in this chapter we explained that Boyle's Law applies to situations in which the temperature remains constant, and that Charles' Law pertains to situations in which pressure remains constant. However, in many settings it is not possible to control pressure or temperature, especially if the gas storage containers (e.g., tanks or bottles) are exposed to harsh environmental conditions such as extreme heat and extreme cold. Therefore, in order to compensate for these pressure and temperature variances, Boyle's Law and Charles' Law have been combined to form the General (Ideal) Gas Law.

The formula for the Ideal Gas Law is: $$\frac{P_1V_1}{T_1} = \frac{P_2V_2}{T_2}$$

Where:

P = Pressure
V = Volume
T = Temperature

General Gas Law allows you to calculate what will happen if you change conditions (either pressure, temperature or volume) given an existing set of conditions. For example, a gas is stored in a container (a fixed volume, that can't change) at a given temperature (T_1) and pressure (P_1). If you increase the temperature (T_2) on the container, you can calculate how much the pressure will change (P_2) using the formula above. V_1 and V_2 would be the same because gas cylinders or storage tanks are rigid and the gas would not be able to change volume.

These principles are important when maintaining control of pressures in plant equipment.

GENERAL (IDEAL) GAS LAW

A law that combines the principles of Boyle's Law and Charles' Law.

The formula for the General Gas Law is:

$$\frac{P_1V_1}{T_1} = \frac{P_2V_2}{T_2}$$

(P = Pressure, V = Volume, T = Temperature)

BERNOULLI'S LAW (BERNOULLI'S PRINCIPLE)

Daniel Bernoulli was a Dutch-born Swiss mathematician who, in the early 1700's, began formulating theories of fluid flow and aerodynamics (remember, in scientific terms, gases and liquids are both considered fluids). Through his experiments Bernoulli created a principle which states that as the speed of a fluid increases, the pressure inside or exerted by the fluid decreases. To better illustrate this concepts, refer to Figure 10-12.

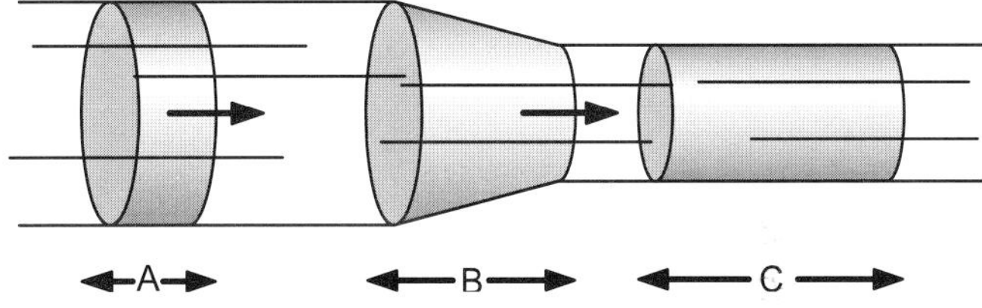

Figure 10-12: As speed increases, pressure decreases

Figure 10-2 shows a section of water traveling through a pipe that is gradually narrowing. When the section of water gets to point C it elongates and occupies a much wider space than it does at point A. Also, it has to take the same amount of time to travel its own length at either point. Therefore, the water has to travel faster at point C, because it has to travel farther in the same amount of time. In order to allow an increase in speed the forces must change, meaning the pressure has to drop at point C. There for the pressure at C is lower than the pressure at A.

This can also be shown in Figure 10-13. In this figure, the water in the vertical tubes stands highest in the widest part of the pipe, showing that the pressure is greatest where the water flows slowest.

To illustrate this principle in more practical terms, think about what happens to your shower curtain when you first turn on the water. As the water/air

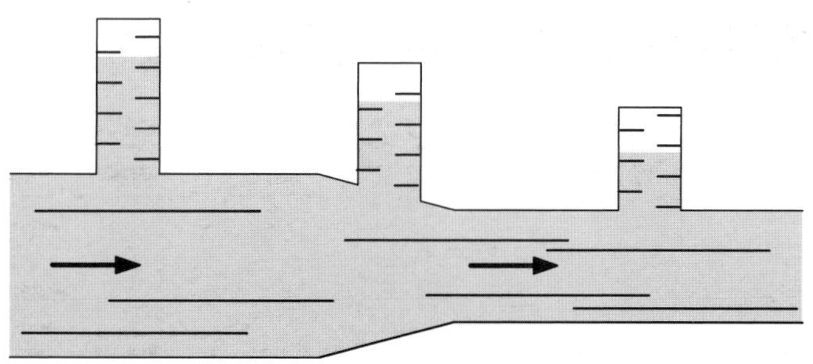

Figure 10-13: Another example of the speed/pressure relationship

velocity inside the curtain increases, the pressure inside the shower decreases relative to the still air outside the shower. This pressure differential is what causes the shower curtain to be sucked into the tub.

Another practical example of Bernoulli's principle pertains to hurricanes. If you have ever seen the aftermath of a hurricane or tornado on television you have noticed that often times many windows are broken. Why is this?

While it is possible that some of the windows were broken by flying debris, many of them were actually broken as a result of an air pressure differential. During a storm, the outside air speed is significantly higher than the speed of the air inside. This increased air speed causes a dramatic drop in pressure on the outside of the house. As a result, the extreme pressure differential causes the windows to be forced outward until they explode. This is why many people open the windows to equalize the pressure and prevent breakage.

BERNOULLI'S LAW

States that in a flowing liquid or gas, the pressure is least where the speed is greatest.

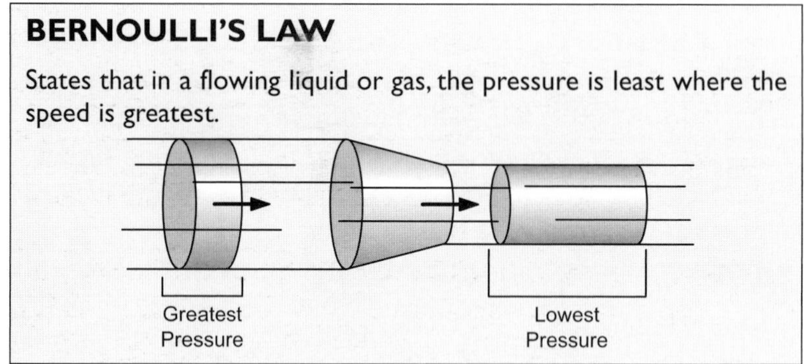

SUMMARY

Process technicians encounter physics concepts everyday as they open and close valves, check process variables like pressure, temperature and flow, and work with different states of matter such as solids, liquids and gases.

Each of state of matter (solid, liquid or gas) has its own unique characteristics and the potential to undergo a phase change, during which it changes from one form to another (e.g., ice, which is a solid, can melt and become a liquid).

Other concepts technicians need to be familiar with include mass and weight, density, elasticity, viscosity, specific gravity, buoyancy, flow, velocity, friction, temperature and pressure.

Mass, weight and density help process technicians determine how solid or heavy an object is.

Elasticity is how much stress (stretching or compressing) a substance can sustain without becoming permanently deformed.

Viscosity describes a fluid's readiness to move or flow. The more viscous a substance is, the less willing it is to flow. Honey is more viscous than water.

Specific gravity is the ratio of the density of a substance compared to that of pure water (or, in the case of gases, the density of the gas compared to air at standard temperature and pressure). In other words, it is the "heaviness" of a substance.

As specific gravity increases, so does the buoyancy of objects floating in the substance. Salt water has a higher specific gravity than fresh water. For this reason, scuba divers must wear more weight when diving in salt water. Hydrocarbons, however, tend to have a specific gravity lower than that of water. That is why many of them float on the surface of a body of water.

Velocity is the distance an object travels over time. As velocity increases, so does friction.

Friction is the resistance encountered when one material slides against another. Friction can occur in both liquids and solids. As friction increases, so does temperature and equipment wear. For this reason, lubrication is often employed to reduce friction.

Temperature is the amount of thermal energy (hotness or coldness) a substance has that can be determined by a thermometer.

Heat is the transfer of energy from one object to another object as a result of the temperature difference between the two objects. Heat can be measured in British Thermal Units (BTUs) or calories. There are many different types of heat, including: specific heat, sensible heat, latent heat, latent heat of fusion, latent heat of vaporization, and latent heat of condensation.

Another concept technicians need to be familiar with is pressure. Pressure is the amount of force a substance or object exerts over a particular area, and it is usually measured in pounds per square inch (PSI).

There is a direct relationship between pressure and boiling point. As pressure increases, so does boiling point, the temperature at which a liquid boils at a specified pressure.

When working with temperature and pressure, it is important to understand the various laws that describe the characteristics of gases and how they will react under different circumstances.

These laws include Boyle's law (pressure-volume law), Charles' law (temperature-volume law), Dalton's law (law of partial pressures), the General/Ideal gas law (a combination of Boyle's law and Charles' law), and Bernoulli's law (the relationship between speed and pressure).

Boyle's Law states that the pressure of an ideal gas is inversely proportional to its volume, if the temperature and amount of gas is held constant.

Charles' law (temperature-volume law) states that the volume of a gas varies directly with the temperature when the pressure is held constant.

Dalton's law (law of partial pressures) states the total pressure of a mixture of gases is equal to the sum of the individual partial pressures.

The General/Ideal gas law (a combination of Boyle's law and Charles' law), combines the principles of Boyle's Law and Charles' Law.

Bernoulli's law (the relationship between speed and pressure) states that in a flowing liquid or gas, the pressure is least where the speed is greatest.

Failure to understand the laws and the principles of physics can result in process problems or safety issues which could lead to costly or catastrophic events.

CHECKING YOUR KNOWLEDGE

1. Define the following key terms:
   a. Mass
   b. Density
   c. Elasticity
   d. Viscosity
   e. Buoyancy
   f. Specific Gravity
   g. Flow
   h. Evaporation
   i. Pressure
   j. Velocity
   k. Friction
   l. Temperature

2. The phase change in which a solid (e.g., dry ice) changes to a vapor is:
   a. Condensation
   b. Sublimation
   c. Transformation
   d. Vaporization

3. The transfer of heat from heat tracing to a pipe is an example of:
   a. Conduction
   b. Convection
   c. Radiation

4. *(True or False)* A Hydrocarbon with a specific gravity less than the specific gravity of water will sink to the bottom if the two substances are mixed together.
   a. True
   b. False

5. Using the thermometers shown in Figure 10-8, list the freezing point of water for each of the following temperature scales.
   a. Kelvin
   b. Celsius
   c. Fahrenheit
   d. Rankine

6. Using the formulas found in Table 10-2, make the following temperature conversions (round to the nearest tenth if necessary).
   a. 79°C is equal to ________ K
   b. 68°C is equal to ________ °F
   c. 200°F is equal to ________ °R
   d. 182°F is equal to ________ °C
   e. 102°F is equal to ________ K

7. At what temperature are the Celsius and Fahrenheit readings the same?

8. The amount of heat required to raise one pound of water one degree Fahrenheit is a:
   a. Calorie
   b. British Thermal Unit (BTU)
   c. Rankin

9. Atmospheric pressure is 14.7 pounds per square inch (psi). What do we call any pressure lower than atmospheric pressure.
   a. Vapor pressure
   b. Gauge pressure
   c. Vacuum

10. As atmospheric pressure increases, boiling point:
    a. Increases
    b. Decreases
    c. Stays the same

11. List and describe the four states of matter.

12. Describe what is meant by the term phase change.

13. Explain the difference between specific heat, sensible heat, and latent heat.

14. Explain the differences between convection, conduction and radiation.

15. According to Boyle's Law, what is the relationship between the absolute pressure of a gas and its volume?

16. According to Charles' Law, what is the relationship between the absolute temperature of a gas and its volume?

17. According to Dalton's Law, what is the total pressure of a mixed gas equal to?

18. Boyle's Law and Charles' Law both pertain to situations in which temperature or pressure remains constant. However, in many settings it is not possible to keep these variables constant. What gas law was created to compensate for these pressure and temperature differences?

19. According to Bernoulli's Law, what is the relationship between the speed of a fluid and the pressure it exerts?

20. Which of the five gas laws discussed in this chapter best explains why windows blow out during a hurricane?

21. Use the formula for Boyle's Law to determine the answer to the following gas problem:
    a. If 100 milliliters (mL) of oxygen gas is compressed from 10 atmospheres (atm) of pressure to 50 atm of pressure, what is the new volume at a constant temperature?

22. Use the formula for Charles' Law to determine the answer to the following gas problem:
    a. If we have 2 cubic feet of gas at 100°F and we heat it up to 450°F, what happens to the volume.

23. Use the formula for Dalton's Law to determine the answer to the following gas problem:
    a. The total pressure of a sample is 760 millimeters of mercury (mm Hg). In clean, dry air at sea level and 0°C, the partial pressure of nitrogen (N2) is 601 mm Hg. If oxygen (O2) is the only other component, what is the partial pressure of oxygen?

24. Use the formula for the General (Ideal) Gas Law to determine the answer to the following gas problem:
    a. Five cubic feet of a gas at 90 psig and 75°F are compressed to a volume of 2 cubic feet and then heated to a temperature of 250°F. What is the new pressure?

25. According to Bernoulli's Law, which point on the diagram below will have the greatest pressure, point a or point b?

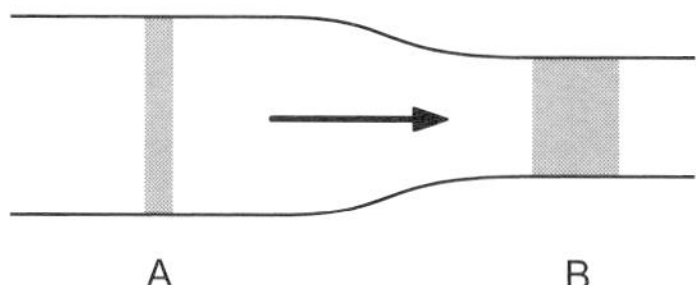

ACTIVITIES

1. Investigate how adding salt to a glass of water can increase the buoyancy of an egg by doing the following:
   a. Obtain a glass of water, a raw egg, a container of table salt, a teaspoon measure, and a stirring spoon.
   b. Form a hypothesis as to how much salt you think it will take to float an egg in a glass of water.
   c. Place the egg in the glass of water (no salt added) and observe whether or not it floats (If the egg floats, go to Step E. If the egg does not float, go to Step D).
   d. Remove the egg from the glass, add a teaspoon of salt to the water, stir until the salt is dissolve, and then place the egg back in the water to see if it floats (If the egg floats, go to Step E. If the egg did not float, repeat Step D).
   e. Record your results (i.e. record how much water was in the cup, how many teaspoons of salt it took to make the egg float and whether or not your hypothesis was correct).

2. Complete the following phase change experiment and record your results.
   a. Fill a small plastic container with 1 cup of water. Place the container in the freezer and then monitor it to see how long it takes for the water to freeze solid. Record your results.
   b. Once the water has frozen solid, remove the ice from the cup (note: you may have to run the outside of the cup under warm water for a few seconds to make the ice turn loose). Place the ice in a glass bowl at room temperature, and then monitor to see how long it takes for the ice to melt completely. Record your results.
   c. Once the ice has melted, pour the water from the ice into a sauce pan. Place the saucepan on the stove and turn the heat on high. Monitor the water to see how long it takes for the water to start forming steam. Record your results.

      Safety Note: Once the steam has begun to form, turn the fire off. Do NOT allow an empty pan to continue heating on the stove, as this poses a safety risk and will damage the pan.

Chapter 11

Basic Chemistry

OBJECTIVES

Upon completion of this chapter you will be able to:

1. Define the application of chemistry in the petrochemical Process industries.
2. Describe the relationship between molecules, atoms, protons, neutrons, and electrons.
3. Define the difference between organic and inorganic chemistry.
4. Explain the difference between chemical properties and physical properties.
5. Define and provide examples of the following terms:
   - Hydrocarbon
   - Boiling Point
   - Chemical Reaction
   - Acidic
   - Alkaline
   - Exothermic
   - Endothermic
   - Compounds
   - Mixtures
   - Solutions
   - Homogenous
   - Equilibrium
   - Catalyst
6. Describe the difference between an acid and a base (caustic).
7. Describe the method of measure for acids-bases (what is pH?).

KEY TERMS

- **Acid**—substances with a pH less than 7.0 that release hydrogen (H+) ions when mixed with water.
- **Alkaline**—a term used to refer to a base (e.g., substances with pH greater than 7).
- **Atomic Number**—the number of protons found in the nucleus of an atom.
- **Atomic Weight**—the sum of protons and neutrons in the nucleus of an atom.
- **Atoms**—the smallest particles of an element that still retain the properties and characteristics of that element.
- **Base**—a substance with a pH greater than 7.0 that releases hydroxyl (OH-) anions when dissolved in water.

- **Boiling Point**—the temperature at which liquid physically changes to a gas at a given pressure.
- **Caustic**—a term used to describe a substance capable of destroying or eating away human tissues or other materials by chemical action; also a process industries term used to refer to a strong base.
- **Catalyst**—a substance used to change the rate of a chemical reaction without being consumed into the reaction.
- **Catalytic Cracking**—the process of adding heat and a catalyst to facilitate a chemical reaction.
- **Chemical Change**—a reaction in which the properties of a substance do change and a new substance is produced.
- **Chemical Formula**—a short-hand, symbolic expression that represents the elements in a substance and the number of atoms present in each molecule (e.g., water, H20, is two hydrogen atoms and one oxygen atom bonded together).
- **Chemical Property**—a property of elements or compounds that is associated with a chemical reaction.
- **Chemical Reaction**—a chemical change or rearrangement of chemical bonds to form a new product.
- **Chemical Symbol**—one or two letter abbreviations for elements on the periodic table.
- **Chemistry**—the science that describes matter, its chemical and physical properties, the chemical and physical changes it undergoes, and the energy changes that accompany those processes.
- **Compound**—a pure and homogeneous substance that contains atoms of different elements in definite proportions, and that usually has properties unlike those of its constituent elements.
- **Electrons**—negatively charged particles that orbit the nucleus of an atom.
- **Elements**—substances composed of like atoms that cannot be broken down further without changing its properties.
- **Endothermic**—a chemical reaction that requires the addition or absorption of energy.
- **Equilibrium**—a point in a chemical reaction in which the rate of the products forming from reactants is equal to the rate of reactants forming from the products.
- **Exothermic**—a chemical reaction that releases energy.
- **Heterogeneous**—matter with properties that are not the same throughout.
- **Homogeneous**—matter that is evenly distributed or consisting of similar parts or elements.
- **Hydrocarbon**—organic compounds that contain only carbon and hydrogen which are most often found occurring in petroleum, natural gas and coal.

- **Inorganic Chemistry**—the study of substances that do not contain carbon.
- **Insoluble**—describes a substance that does not dissolve in a solvent.
- **Ions**—charged particles.
- **Mixture**—occurs when two substances are mixed together but do not react chemically.
- **Molecular Weight**—a unit of measure for a substance equal to the sum of the atomic weights of the elements that are present in a substance.
- **Molecule**—two or more atoms held together by chemical bonds.
- **Neutron**—a neutrally charged particle found in the nucleus of an atom.
- **Organic Chemistry**—the study of carbon-containing compounds.
- **Periodic Table**—a chart of all known elements listed in order of increasing atomic number and grouped by similar characteristics.
- **pH**—a measure of the amount of hydrogen ions in a solution that can react and indicates whether a substance is an acid or a base.
- **Physical Change**—an event in which the physical properties of a substance (e.g., how it looks, smells or feels) may change, the change may be reversible, and a new substance is not produced.
- **Physical Property**—the properties of an element or compound that is observable and does not pertain to a chemical reaction.
- **Products**—the substances that are produced during a chemical reaction.
- **Proton**—a positively charged particle found in the nucleus of an atom.
- **Reactants**—the starting substances in a chemical reaction.
- **Soluble**—describes a substance that will dissolve in a solvent.
- **Solute**—the substance being dissolved in a solvent.
- **Solution**—a homogeneous mixture of two or more substances.
- **Solvent**—the substance present in a solution in the largest amount.

INTRODUCTION

Chemistry is a very important part of the process industries. Through chemistry, scientists and process technicians are able to understand various elements and compounds, their proportions, and how they interact with one another in the presence of heat, cold, catalysts, and other variables. By understanding these principles, the process industries are able to produce better products and safer processes.

APPLICATIONS OF CHEMISTRY IN THE PROCESS INDUSTRIES

Chemistry is the science that describes matter, its chemical and physical properties, the chemical and physical changes it undergoes, and the energy changes that accompany those processes. This includes the study of elements, the compounds they form, and the reactions they undergo. It is the study of substances, what they are made of and how they react.

The field of chemistry is divided into many branches including physical chemistry, organic chemistry and inorganic chemistry.

Physical chemistry is the branch of chemistry that studies the relationships between the physical properties of substances and their chemical compositions and transformations. **Organic chemistry** is the study of carbon-containing compounds. **Inorganic chemistry** is the study of substances that do not contain carbon. Process technicians focus primarily on organic or inorganic chemistry.

Technicians in the process industries may work with a variety of substances including organics such as hydrocarbons. **Hydrocarbons** are organic compounds that contain only carbon and hydrogen. They are naturally found in petroleum, natural gas and coal.

ELEMENTS AND COMPOUNDS

Matter is anything that has mass and takes up space. All matter is composed of building blocks called elements, individually or in combination.

Elements are substances composed of atoms that cannot be broken down further. There are currently 118 known elements, 92 of which occur naturally. Hydrogen, oxygen, nitrogen and phosphorous are all examples of naturally occurring elements.

Each element has physical and chemical characteristics that distinguish it from other elements. For example, oxygen is a colorless, odorless gas at room temperature, while carbon is a black solid at room temperature.

The Periodic Table

In order to keep up with all of the elements and their properties, scientists have created a reference chart called the periodic table. The **periodic table** contains all known elements listed in order of increasing atomic number, and grouped by similar characteristics (i.e. how they react with other elements, their chemical and physical properties). This table allows elements to be classified into categories of metals, metalloids and non-metals.

1 ← Atomic Number
Hydrogen ← Name
H ← Symbol
1.008 ← Atomic Weight

Metals
Metalloids
Non-metals

Period	1 1A	2 2A	3 3B	4 4B	5 5B	6 6B	7 7B	8	9 8B	10	11 1B	12 2B	13 3A	14 4A	15 5A	16 6A	17 7A	18 8A
1	1 Hydrogen H 1.008																	2 Helium He 4.003
2	3 Lithium Li 6.94	4 Beryllium Be 9.01											5 Boron B 10.81	6 Carbon C 12.01	7 Nitrogen N 14.01	8 Oxygen O 16.00	9 Fluorine F 19.00	10 Neon Ne 20.18
3	11 Sodium Na 22.99	12 Magnesium Mg 24.31											13 Aluminum Al 26.98	14 Silicon Si 28.09	15 Phosphorus P 30.97	16 Sulfur S 32.07	17 Chlorine Cl 35.45	18 Argon Ar 39.95
4	19 Potassium K 39.10	20 Calcium Ca 40.08	21 Scandium Sc 44.96	22 Titanium Ti 47.87	23 Vanadium V 50.94	24 Chromium Cr 52.00	25 Manganese Mn 54.94	26 Iron Fe 55.85	27 Cobalt Co 58.93	28 Nickel Ni 58.69	29 Copper Cu 63.55	30 Zinc Zn 65.39	31 Gallium Ga 69.72	32 Germanium Ge 72.61	33 Arsenic As 74.92	34 Selenium Se 78.96	35 Bromine Br 79.90	36 Krypton Kr 83.80
5	37 Rubidium Rb 85.47	38 Strontium Sr 87.62	39 Yttrium Y 88.91	40 Zirconium Zr 94.22	41 Niobium Nb 22.99	42 Molybdenum Mo 95.94	43 Technetium Tc (98)	44 Ruthenium Ru 101.07	45 Rhodium Rh 102.91	46 Palladium Pd 106.42	47 Silver Ag 107.87	48 Cadmium Cd 112.41	49 Indium In 114.82	50 Tin Sn 118.71	51 Antimony Sb 121.76	52 Tellurium Te 127.60	53 Iodine I 126.90	54 Xenon Xe 131.29
6	55 Cesium Cs 132.91	56 Barium Ba 137.33	71 Lutetium Lu 174.97	72 Hafnium Hf 178.49	73 Tantalum TA 180.95	74 Tungsten W 183.84	75 Rhenium Re 186.21	76 Osmium Os 190.23	77 Iridium Ir 192.22	78 Platinum Pt 195.08	79 Gold Au 196.97	80 Mercury Hg 200.59	81 Thallium Tl 204.38	82 Lead Pb 207.2	83 Bismuth Bi 208.98	84 Polonium Po (209)	85 Astatine At (210)	86 Radon Rn (222)
7	87 Francium Fr (223)	88 Radium Ra (226)	103 Lawrencium Lr (262)	104 Rutherfordium Rf (261)	105 Dubnium Db (262)	106 Seaborgium Sg (263)	107 Bohrium Bh (264)	108 Hassium Hs (265)	109 Meitnerium Mt (268)	110 Ununnilium Uun (269)	111 Unununium Uuu (272)	112 Ununbium Uub (227)		114 Ununquadium Uuq (285)		116 Ununhexium Uuh (289)		118 Ununoctium Uuo (293)

Lanthanides	57 Lanthanum La 138.91	58 Cerium Ce 140.12	59 Praseodymium Pr 140.91	60 Neodymium Nd 144.24	61 Promethium Pm (145)	62 Samarium Sm 150.36	63 Europium Eu 151.96	64 Gadolinium Gd 157.25	65 Terbium Tb 158.93	66 Dysprosium Dy 162.50	67 Holmium Ho 167.93	68 Erbium Er 167.26	69 Thulium Tm 168.93	70 Ytterbium Yb 173.04
Actinides	89 Actinium Ac (227)	90 Thorium Th 232.04	91 Protactinium Pa 231.04	92 Uranium U 238.03	93 Neptunium Np (237)	94 Plutonium Pu (244)	95 Americium Am (243)	96 Curium Cm (247)	97 Berkelium Bk (247)	98 Californium Cf (251)	99 Einsteinium Es (252)	100 Fermium Fm (257)	101 Mendelevium Md (258)	102 Nobelium No (259)

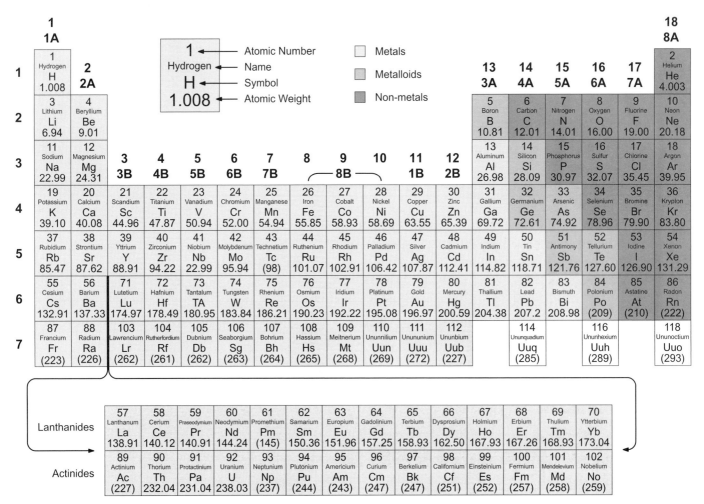

Figure 11-1: The periodic table

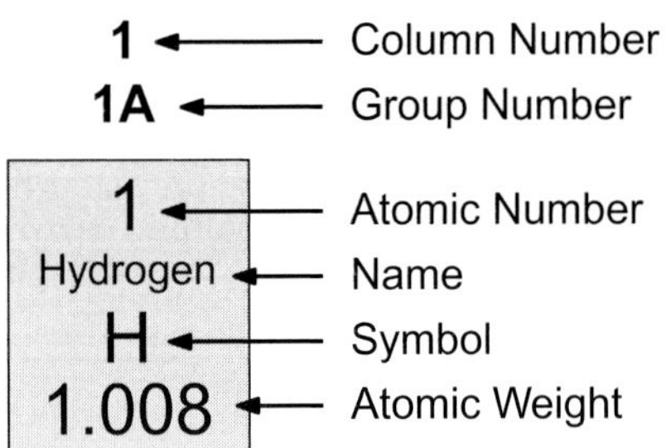

Figure 11-2: Sample element from a periodic table

Each row in a periodic table is referred to as a **period**. Each column is referred to as a **group**. Within each group are elements that contain similar properties. Each group is given a number (e.g., 8A).

For example, column 18 (Group 8A) on the far right of the periodic table contains a group of elements known as "noble gases." Noble gases, also called inert gases, include helium, neon, argon, krypton, xenon, and radon. All of these are rare gases that exhibit great chemical stability and extremely low reaction rates. By knowing the major properties of a chemical family, it is possible for process technicians to predict the behavior of other elements in that family or group.

In the periodic table there are eight groups or "families" of elements. Within each group, each element behaves similarly to all of the other elements in the group, although the level or intensity of the behavior may vary. As we transition from one side of the periodic table to the other, there is a noticeable change in chemical behavior and physical properties.

To properly read a periodic table, a process technician must be familiar with the different components of the table. The diagram in Figure 11-2 demonstrates each of these components.

At the top of the table you will see a column number and the group number. The column number is simply a reference to make it easier to find a particular set of elements. In all, there are 18 columns on the periodic table. Subdivided within these columns are 8 groups.

Each group represents a unique set of physical and/or chemical properties that each member of that group possesses. Beneath the column and group numbers is an atomic number. The **atomic number** is the number of protons found in the nucleus of an atom (protons will be explained in more detail later on in this chapter).

After the atomic number is the name of the element. Beneath the element's name is a one or two letter abbreviation called a **chemical symbol**. This abbreviation is a shorthand way to refer to the element when writing chemical equations.

Finally, beneath the element's symbol is the atomic weight of the element. The **atomic weight** is the approximate sum of the number of protons and neutrons found in the nucleus of an atom (also to be explained in more detail later on in this chapter).

DID YOU KNOW?

Chemical symbols containing two letters only have the first letter capitalized.

Example:

Element

11
Sodium
Na
22.99

Characteristics of Atoms

All elements are composed of atoms. **Atoms** are the smallest particles of an element that still retain the properties and characteristics of that element. A **molecule** is a set of two or more atoms held together by chemical bonds. A molecule is the smallest unit of a compound that displays the properties of the compound.

Since atoms are too small to see with the naked eye, scientists use drawings and models to represent the atoms and their components. Figure 11-3 and Figure 11-4 show examples of one of these drawings.

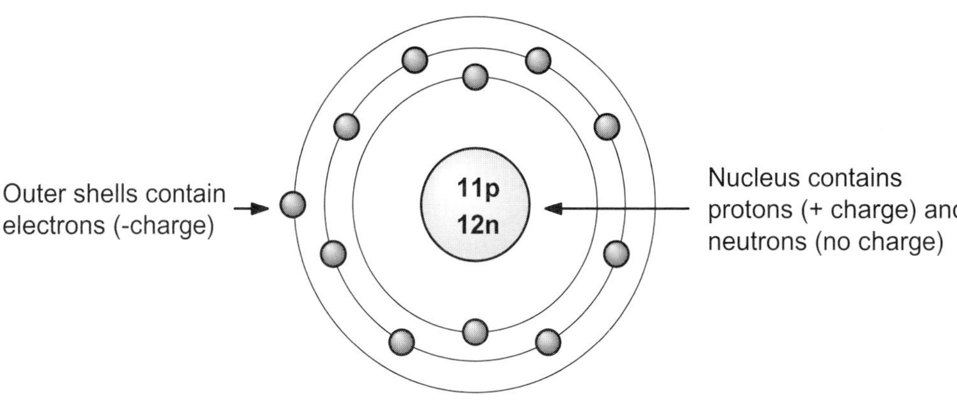

Figure 11-3: Two-dimensional diagram of a sodium atom

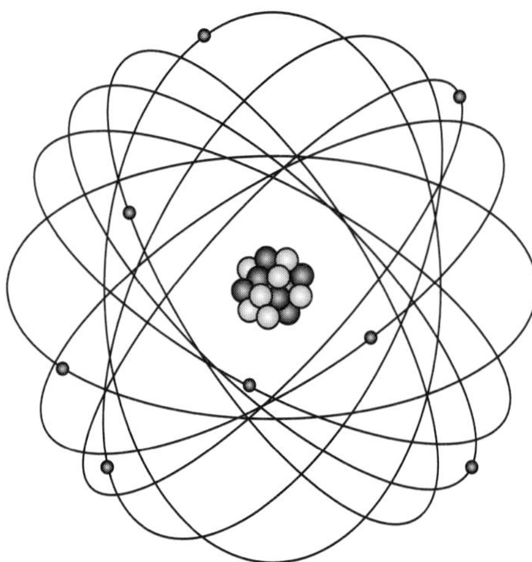

Figure 11-4: Three dimensional representation of a sodium atom

Figure 11-4 shows a three dimensional diagram of the atom shown in Figure 11-3.

In these drawing you will see that an atom has a nucleus at its center. Inside the nucleus are positively charged particles called **protons** and neutrally charged particles called **neutrons**. Surrounding the nucleus are shells which contains negatively charged particles called **electrons**. Every atom has an equal number of protons and electrons. This allows the atom to remain electrically balanced (neutral) when they are not reacted.

Within an atom the positive electrical charge of a proton is equal to the negative charge of an electron. The mass of these particles is different, however. A proton has about the same mass as a neutron. Electrons, however, are much lighter. It takes about 1840 electrons to equal the mass of one proton. Note: for the sodium atom the atomic number is 12 and the approximate atomic weight is 12 + 11 or 23.

Characteristics of Compounds

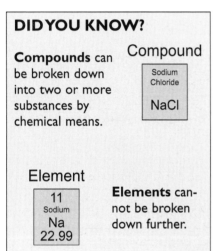

DIDYOU KNOW?

Compound

Sodium Chloride

NaCl

Compounds can be broken down into two or more substances by chemical means.

Element

11
Sodium
Na
22.99

Elements cannot be broken down further.

With only 112 known elements, how there can be so many different substances. The reason for this is simple. Elements combine together chemically to form compounds.

A **compound** is a pure and **homogeneous** (consisting of similar parts or elements) substance that contains atoms or ions of different elements in definite proportions, and that usually has properties unlike those of its constituent elements. Compounds can be broken down into two or more elements by chemical means.

Scientists use chemical formulas to show the kinds of atoms in a compound and their proportions. The table below lists some examples of common chemical formulas and their constituents.

FIGURE 11-5: Common chemical formulas

Substance	Chemical Formula	Elements the Compound is Made From
Table Salt	NaCl	Sodium (Na) and Chlorine (Cl)
Water	H_2O	Hydrogen (H) and Oxygen (O)
Hydrogen Peroxide	H_2O_2	Hydrogen (H) and Oxygen (O)
Oxygen Gas	O_2	Oxygen (O)
Carbon Dioxide Gas	CO_2	Carbon (C) and Oxygen (O)
Ethanol	C_2H_5OH	Carbon (C), Hydrogen (H) and Oxygen (O)
Acetic Acid	$HC_2H_3O_2$	Carbon (C), Hydrogen (H) and Oxygen (O)
Glucose	$C_6H_{12}O_6$	Carbon (C), Hydrogen (H) and Oxygen (O)

If you look closely at the formulas in Figure 11-5 you will see that several of the substances contain the same components. For example, ethanol, acetic acid and glucose all contain carbon, hydrogen and oxygen. What is different, however, are the quantities of each element and how they are attached.

To better illustrate this concept, consider water and hydrogen peroxide.

Water contains two hydrogen atoms and one oxygen atom. Hydrogen peroxide contains two hydrogen atoms and two oxygen atoms. Because of this extra oxygen, hydrogen peroxide is a good oxidizer, an effective antiseptic and a good bleaching agent.

Water Molecule (H_2O)

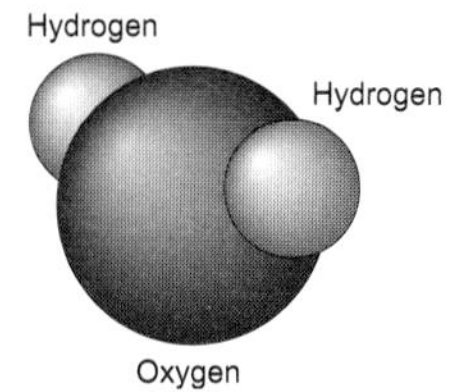

Hydrogen Peroxide Molecule (H_2O_2)

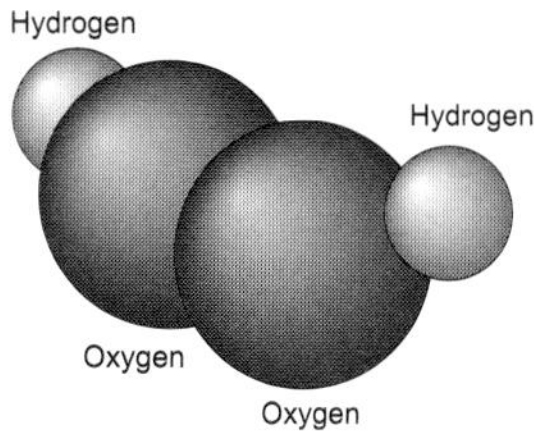

CHEMICAL VS. PHYSICAL PROPERTIES

In order to understand chemical reactions and the properties of matter, process technicians must be able to determine if the characteristics of a substance and any changes associated with it are physical or chemical.

Chemical properties are properties of elements or compounds that are associated with a chemical reaction. **Physical properties** are the properties of elements or compounds that are observable and do not pertain to a chemical reaction.

During a **physical change** the physical properties of a substance, such as how it looks, smells or feels, may change. The change will be reversible, and a new substance is *not* produced. In a **chemical change**, however, the physical properties of the substance produced will change, the change is readily reversible, and a new substance *is* produced. Chemical changes usually involve changes in color or odor, the production of gas bubbles, the release or absorption of heat, or the creation of a new substance with different properties.

The table below provides some examples of physical and chemical changes.

TABLE 11-1: Examples of physical and chemical changes

Physical Change	Chemical Change
Cutting firewood	Burning firewood to make carbon and heat
Leaves falling from a tree	Composting leaves to form soil
Mining bauxite from the ground	Making aluminum from bauxite

One physical property that is important to process technicians in the refining industry is boiling point. **Boiling point** is the temperature at which liquid physically changes to a gas at a given pressure. Because gasoline, kerosene, and other organic compounds have different boiling points, scientists are able to separate these substances out using heat and special equipment.

Chemical Reactions

Elements react with one another to form compounds. This change, or a rearrangement of bonds to form a new product, is called a **chemical reaction.** Compounds can be represented symbolically by a chemical formula.

A **chemical formula** is a short-hand, symbolic expression that represents the elements in a substance and the number of atoms present in each element.

In a chemical reaction there are two main components: reactants and the products.

Reactants are the starting substances in a chemical reaction. **Products** are the substances that are produced during a chemical reaction. To better illustrate these components, look at the following reaction:

Sodium Metal (Na) + Chlorine Gas (Cl) ⟶ Sodium Chloride (NaCl)

Reactants | **Products**

On the left-hand side of the equation you will see the reactants. On the right-hand side of the equation you will see the products.

In this particular example, the reactants include 1 molecule of sodium (Na) and one molecule of chlorine (Cl). These molecules reactant to form 1 molecule of sodium chloride (NaCl), which is table salt.

This process is complex and outside the scope of this textbook. However, these types of reactions are discussed in more detail in the general chemistry courses that are required for most process technology degree programs.

What you do need to know is that conditions must be right for a chemical reaction to occur. During a chemical reaction the chemical bonds in reactants are broken and re-formed so that the end product is different from the original reactants. When the chemical reaction reaches a point where the rate of products forming from reactants is equal to the rate of reactants forming from the products, the reaction is in a state of **equilibrium**.

CATALYSTS

In some chemical reactions, substances called **catalysts** are introduced to speed up the reaction. When a catalyst is added, the speed of the chemical reaction is increased but the catalyst itself is not consumed in the reaction.

Catalysts help facilitate chemical reactions immensely. In the refining industry catalysts and heat are used to break large hydrocarbon molecules down into smaller molecules. This process of adding heat and catalyst to facilitate a chemical reaction is called **catalytic cracking**. Because the catalyst, which starts out in solid form and flows through the process with the process fluids, is not incorporated into the reaction, it can be reclaimed at the end of the cracking process, "regenerated," and used again (providing there is still some surface area left on the catalyst particles themselves).

ENDOTHERMIC VS. EXOTHERMIC REACTIONS

Another characteristic of chemical reactions is their tendency to release energy or require the addition of energy. Chemical reactions that release energy are called **exothermic** reactions. Reactions that require the addition or absorption of energy (often as heat) are called **endothermic** reactions.

Mixtures and Solutions

The previous section discussed chemical reactions and how reactants are combined together to form a product that is chemically different from the original reactants. But what is produced if the combined substances do not react chemically? A mixture.

A **mixture** occurs when two substances are combined but they do not react chemically. If you have ever put sand and water together in a container and mixed them, you are familiar with the concept of mixtures.

In a mixture, each substance retains its own identity. The components of a mixture can easily be identified as two separate substances and can be separated by physical means (e.g., filtering through a screen).

A **solution** is a homogeneous mixture of two or more substances. The individual particles are uniformly distributed throughout the substance.

Solutions have two parts: a solute and a solvent. A **solvent** is the substance present in a solution in the largest amount. A **solute** is the substance being dissolved in the solvent. An example of a solute and a solvent would be salt and water. If you dissolve salt (the solute) in water (the solvent) you form a solution of salt and water. However, during this process the salt and the water do not actually react chemically. Thus, if the water were to evaporate or boil away, the salt would remain intact

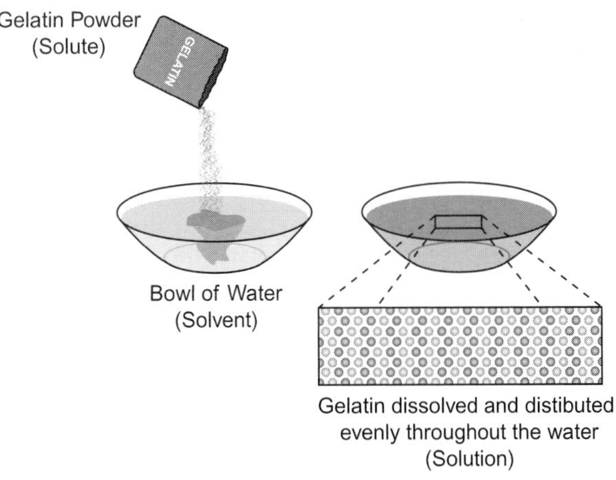

Figure 11-6: Solute molecules evenly distributed among solvent molecules in a solution

and chemically unchanged. Figure 11-6 shows an example of how solutes and solvents interact in a solution.

If a solute will dissolve in a solvent it is considered **soluble**. If a solute will not dissolve in a solvent it is considered **insoluble**. Oil is insoluble in water.

Another way homogeneous liquid mixtures can be separated out is through the process of distillation. In distillation, the components of the liquid mixture are separated out by boiling point. The component or "fraction" with the lowest boiling point is collected first.

ACIDS AND BASES

Acids and bases are two common compounds that react in water. **Acids** are corrosive substances, with a pH less than 7.0, that release hydrogen (H+) ions when mixed with water. Acids neutralize bases and conduct electricity. If tested with litmus paper, acids turn litmus paper red. The term **acidic** is used to refer to materials that are acids. Consumable acids, like lemon or grapefruit juice, have a sour taste.

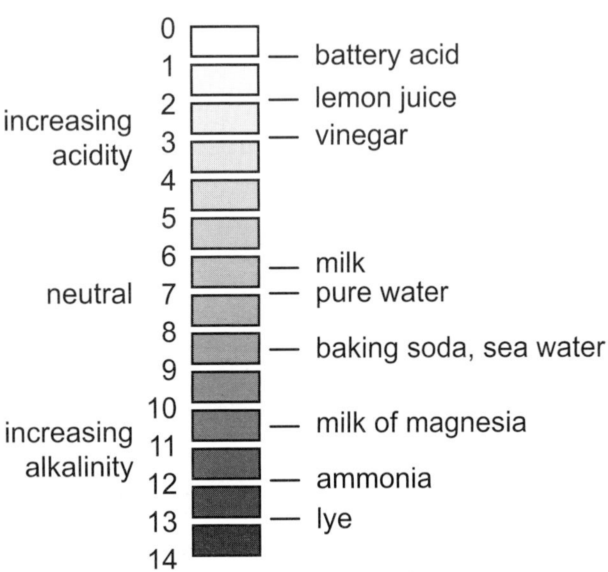

Figure 11-7: Example of a pH scale

Bases are corrosive substances with a pH greater than 7.0 that usually release hydroxyl (OH-) anions when dissolved in water. Bases react with acids to form salts and water. If tested with litmus paper, they turn litmus paper blue. A strong base is called **alkaline** or **caustic**. The term **basic** is used to refer to materials that are bases. Bath soap is an example of a base.

pH is a reference to the amount of hydrogen ions in a solution. The scale for measuring pH ranges from 0 to 14, with 0 being a strong acid, 14 being a strong base, and 7 being neutral (neither acid or base). Figure 11-7 shows an example of a pH scale.

It is important for process technicians to understand that the pH scale is based on powers of 10. This means that a substance with a pH of 3 is ten times more acidic than a substance with a pH of 4, and a thousand times more acidic that a substance with a pH of 6.

SUMMARY

Chemistry is an important part of the process industries. It is the study of elements, the compounds they form, and the reactions they undergo. A person who studies chemistry is called a **chemist**.

The field of chemistry is divided into many fields including organic and inorganic. **Organic chemistry** is the study of carbon containing compounds called hydrocarbons. **Inorganic chemistry** is the study of compounds that don't contain carbons.

Chemistry helps process technicians understand various elements and compounds and how they interact with one another. By understanding these principles, the process industries are able to produce better products and safer processes.

All matter is composed of building blocks called **elements**. Elements are recorded on a reference chart called a periodic table.

Atoms are the smallest particle of an element that still retains the properties and characteristics of that element. Each atom contains positively charged particles called **protons**, neutrally charged particles called **neutrons**, and negatively charged particles called **electrons**.

A **molecule** is two or more atoms held together by chemical bonds. A molecule is the smallest unit of a compound that still retains the properties of the compound.

Compounds are homogeneous (consisting of similar parts or elements) substances that can be chemically broken down into two or more substances. Scientists use chemical formulas to represent these substances.

In any chemical reaction there is a **reactant** (the beginning substance) and a **product** (the ending substance). **Catalysts** may be used to facilitate or speed up a chemical reaction, but are not chemically incorporated into the reaction.

When a chemical reaction reaches a point where the products forming from reactants are equal to the rate of reactants forming from the products, the reaction is in a state of **equilibrium.**

Chemical reactions that release energy (often in the form of heat) are called **exothermic** reactions. Reactions that absorb energy are called **endothermic** reactions.

Mixtures occur when two substances combine but do not react chemically. **Solutions** are a class of mixtures in which the molecules are uniformly distributed (homogeneous). In any given solution there is a **solute** (the substance being dissolved) and a **solvent** (an agent in which the solute is being dissolved). If a solute will not dissolve in a solvent (e.g., oil and water) it is said to be **insoluble**.

Substances can be an **acid**, **base** or **neutral**. The scale we use to determine this is called the **pH scale**. The pH scale ranges from 0 to 14 with zero being a strong acid, fourteen being a strong base, and seven being neutral.

CHECKING YOUR KNOWLEDGE

1. Define the following terms:
   a. Hydrocarbon
   b. Boiling Point
   c. Chemical Reaction
   d. Acidic
   e. Alkaline
   f. Exothermic
   g. Endothermic
   h. Compound
   i. Mixture
   j. Molecule
   k. Solution
   l. Homogenous
   m. Equilibrium
   n. Catalyst
   o. Atom
   p. Proton
   q. Neutron
   r. Electron

2. Explain the difference between organic and inorganic chemistry.

3. Explain the difference between chemical properties and physical properties.

4. Tell whether the following change is physical or chemical (remember: in a chemical change a new substance is formed).
   a. Ice melting
   b. Gasoline burning
   c. Iron nails rusting
   d. Dissolving sugar into a pitcher of lemonade
   e. Vinegar and baking soda reacting with one another to form a foamy substance
   f. Water boiling
   g. Butter melting
   h. Aluminum foil being cut in half

5. If a mixture is uniformly distributed it is said to be:
   a. Homogeneous
   b. Heterogeneous

6. Which of the following is the smallest unit of an element?
   a. Atom
   b. Molecule
   c. Compound
   d. Mixture

7. Every element has a set number of protons, neutrons, and electrons. If the element silicon has 14 protons, how many electrons will it have? Will these electrons be located in the nucleus or the shells?

8. Define the term pH.

9. Given several pH test strips or litmus paper, determine the pH of the following household items and tell if each one is an acid, base, or neutral?

   a. Bleach
   b. Hand lotion
   c. Carbonated beverage or soda
   d. Window cleaner with ammonia
   e. Vinegar
   f. Tap water
   g. Bath soap
   h. Rubbing alcohol
   i. Mouthwash or toothpaste
   j. Shampoo

10. Explain why a scientist or a process technician might use the process of distillation.

11. Using the period table shown in Figure 11-1, locate the information for calcium, lead and chlorine, and then use that information to complete the chart below.

	Calcium	Lead	Chlorine
Column number (e.g., 1, 2, 3)			
Group number (e.g., 1A, 2A)			
Atomic number			
Symbol			
Atomic weight			

12. If each dot represents a molecule, which of the following diagrams best represents a chemical change (remember: in a chemical change a new substance is formed).

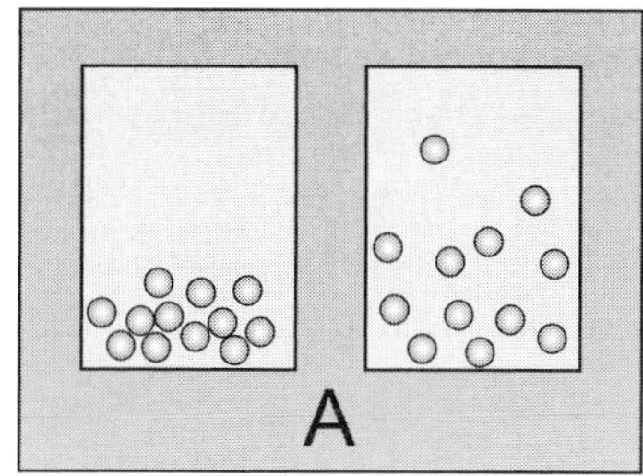

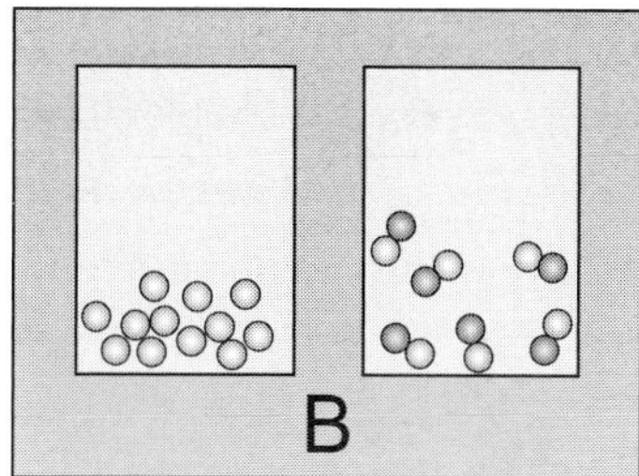

13. On the following equation, draw a circle around the reactants and a square around the products.

NH_3	+	HCl	$\rightarrow$	NH_4Cl
(Ammonia)		(Hydrochloric acid)		(Ammonium chloride)

If you have a packet of hot chocolate mix and you pour it into a cup of boiling water, which of these items is the solute?

ACTIVITIES

1. Studying the effect of an acid on metal.

   Perform the following steps and identify the change that occurs.

   a. Obtain a paper towel, a glass dish, a shiny penny, and a small bottle of vinegar.

   b. Place the paper towel in the dish and then pour vinegar over it until it is completely saturated.

   c. Place the shiny penny on the paper towel and leave it for 24 hours.

   d. Describe the change(s) that occurred and identify if the change(s) are chemical or physical.

2. Studying the effects of water and vinegar on baking soda.

   Perform the following steps and identify the change that occurs.

   a. Obtain 2 small glass dishes, 2 tsp. baking soda, 1 Tbsp. water, and 1 Tbsp. vinegar.

   b. Place 1 tsp. of baking soda in each dish.

   c. Pour 1 Tbsp. of water into dish #1 and observe what happens.

   d. Pour 1 Tbsp. of vinegar into dish #2 and observe what happens.

   e. Record your results and identify whether the reaction that occurred was physical or chemical.

Chapter 12

Safety, Health, Environment and Security

OBJECTIVES

Upon completion of this chapter you will be able to:

1. Discuss the safety, health, and environmental hazards found in the process industries.
2. Explain and describe the responsibility of the following regulatory agencies:
   - EPA
   - OSHA
   - DOT
   - NRC
3. Describe the intent and application of the primary regulations impacting the process industries:
   - OSHA 1910.119—Process Safety Management (PSM)
   - OSHA 1910.132—Personal Protective Equipment (PPE)
   - OSHA 1910.1200—Hazard Communication (HAZCOM)
   - OSHA 1910.120—Hazardous Waste Operations and Emergency Response (HAZWOPER)
   - DOT CFR 49.173.1—Hazardous Materials—General Requirements for Shipments and Packaging
   - EPA CFR 264.16—Resource Conservation and Recovery Act (RCRA)
4. Describe the role of the process technician in protecting the safety and health of the company, employees, and community while achieving successful compliance with regulations.
5. Describe the personal attitudes and behaviors that can help to prevent workplace accidents and incidents.
6. Describe the components of the fire triangle and the fire tetrahedron.
7. Identify the consequences of non-compliance with regulations:
   - Legal
   - Moral and Ethical
   - Safety, Health, and Environmental
8. Explain the managerial and engineering controls used in the industry to minimize hazards and maximize worker and system protection in the workplace.
9. Describe (demonstrate, if possible) the correct use of personal protective equipment (PPE).
10. Describe the intent of the OSHA—Voluntary Protection Program (VPP).
11. Describe the application of the International Organization of Standards (ISO) 14000 as it relates to the process industries.
12. Explain physical and cyber security requirements in the process industries.

KEY TERMS

- **Administrative Controls**—the implementation programs (e.g., policies and procedures) and activities to address a hazard.
- **Air Pollution**—the contamination of the atmosphere, especially by industrial waste gases, fuel exhausts, smoke or particulate matter (finely divided solids).
- **Attitude**—a state of mind or feeling with regard to some issue or event.
- **Behavior**—an observable action or reaction of a person under certain circumstances.
- **Biological Hazard**—any hazard that comes from a living, or once living, organism such as viruses, mosquitoes, or snakes, which can cause a health problem.
- **Chain Reaction**—a series of reactions in which each reaction is initiated by the energy produced in the preceding reaction.
- **Chemical Hazard**—any hazard that comes from a solid, liquid or gas element, compound, or mixture that could cause health problems or pollution.
- **Cyber Security**—security measures intended to protect information and information technology from unauthorized access or use.
- **DOT**—U.S. Department of Transportation; a U.S. government agency with a mission of developing and coordinating policies to provide efficient and economical national transportation system, taking into account need, the environment and national defense.
- **Engineering Controls**—controls that use technological and engineering improvements to isolate, diminish, or remove a hazard from the workplace.
- **EPA**—Environmental Protection Agency; a Federal agency charged with authority to make and enforce the national environmental policy.
- **Ergonomic Hazard**—Hazards that can create physical and psychological stresses because of forceful or repetitive work, improper work techniques, or poorly designed tools and workspaces.
- **Fire Triangle/Tire Tetrahedron**—the three elements (fuel, oxygen and heat) required for a fire to start and sustain itself; a fire tetrahedron adds a fourth element: a chemical chain reaction.
- **Fuel**—any material that burns; can be a solid, liquid or gas.
- **Hazardous Agent**—the substance, method, or action by which damage or destruction can happen to personnel, equipment, or the environment.
- **Heat**—added energy that causes an increase in the temperature of a material (sensible heat) or a phase change (latent heat); the energy required by the fuel to generate enough vapors for the fuel to ignite.

- **ISO 9000**—an international standard that provides a framework for quality management by addressing the processes of producing and delivering products and services.
- **ISO 14000**—an international standard that addresses how to incorporate environmental aspects into operations and product standards.
- **Material Safety Data Sheet (MSDS)**—A document that provides key safety, health and environmental information about a material.
- **NRC**—Nuclear Regulatory Commission; a U.S. government agency that protects public health and safety through regulation of nuclear power and the civilian use of nuclear materials.
- **OSHA**—Occupational Safety and Health Administration (OSHA); a U.S. government agency created to establish and enforce workplace safety and health standards, conduct workplace inspections and propose penalties for noncompliance, and investigate serious workplace incidents.
- **Personal Protective Equipment (PPE)**—specialized gear that provides a barrier between hazards and the body and its extremities.
- **Physical Hazard**—any hazards that comes from environmental factors such as excessive levels of noise, temperature, pressure, vibration, radiation, electricity or mechanical hazards (note: this is not the OSHA definition of physical hazard).
- **Physical Security**—security measures intended to protect specific assets such as pipelines, control centers, tank farms, and other vital areas.
- **Soil Pollution**—the accidental or intentional discharge of any harmful substance into the soil.
- **Voluntary Protection Program (VPP)**—an OSHA program designed to recognize and promote effective safety and health management.
- **Water Pollution**—the introduction, into a body of water or the water table, any EPA listed potential pollutant that affects the chemical, physical or biological integrity of that water.

INTRODUCTION

In the process industries, workers routinely work with hazardous agents, or environmental factors that can cause injury, illness or death. Some of these hazardous agents can also impact the environment in the short and long term.

Government regulations are in place to protect worker health and safety, the community and the environment. Industries comply with regulations by using engineering and administrative controls, and PPE.

Companies use physical and cyber security measures to protect assets and workers from internal and external threats. Physical security focuses on protecting facilities and components such as pipelines, control centers and other vital areas from damage or theft. Cyber security protects information assets and computing systems.

Process technicians must be trained and be able to recognize hazardous agents and security threats, and understand the impact on themselves and the plant or facility where they work.

This chapter provides an overview of various hazardous agents that process technicians might encounter in the workplace, the government agencies and regulations that address Safety, Health, Environment and Security (SHE), controls for hazards, Personal Protective Equipment, the cost of non-compliance, some voluntary programs that promote workplace safety and measures to protect against physical and cyber security threats.

SAFETY, HEALTH, AND ENVIRONMENTAL HAZARDS FOUND IN THE PROCESS INDUSTRIES

Hazardous agents are the substances, methods, or actions by which damage or destruction can happen to personnel, equipment, or the environment.

Different government agencies, industry groups and individuals have created various ways of classifying and describing hazardous agents. Many companies and their safety professionals use the following classification system to categorize hazardous agents, dividing these agents into five major types: chemical, physical, ergonomic, biological, or physical security & cyber security.

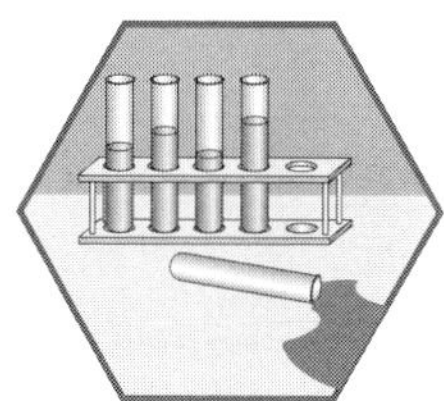

Chemical hazard—any hazard that comes from a solid, liquid or gas element, compound, or mixture that could cause health problems or pollution.

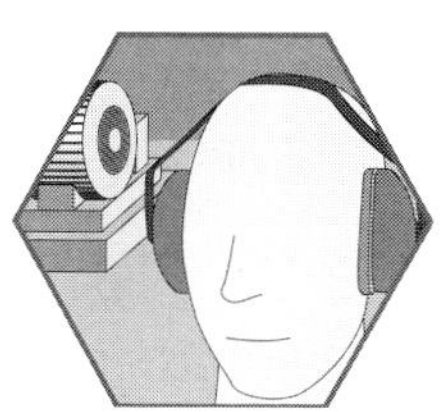

Physical hazard—any hazards that comes from environmental factors such as excessive levels of noise, temperature, pressure, vibration, radiation, electricity or mechanical hazards (note: this is not the OSHA definition of physical hazard).

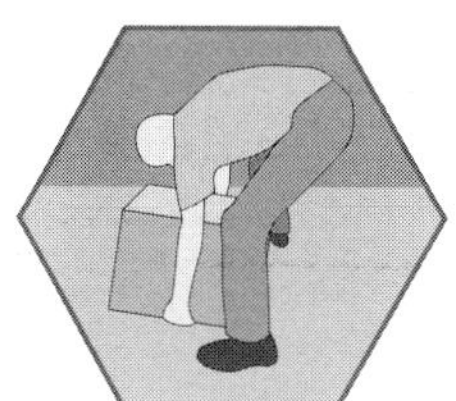

Ergonomic hazard—Hazards that can create physical and psychological stresses because of forceful or repetitive work, improper work techniques, or poorly designed tools and workspaces.

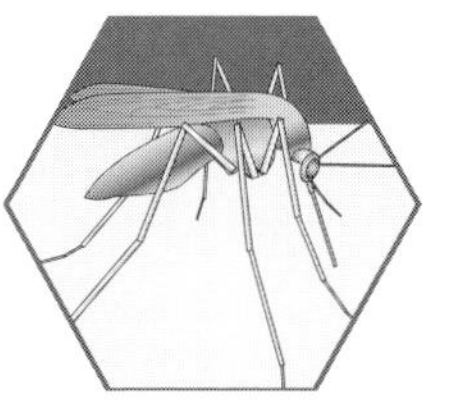

Biological hazard—any hazard that comes from a living, or once living, organism such as viruses, mosquitoes, or snakes, which can cause a health problem.

Security hazard—a hazard or threat from a person or group seeking to intentionally harm people, computer resources, or other vital assets.

- **Physical security**—security measures intended to prevent physical threats from a person or group seeking to intentionally harm other people or vital assets.
- **Cyber security**—security measures intended to protect electronic assets from illegal access and sabotage.

Environmental hazards fall into one of three broad categories: air pollution, water pollution, and soil pollution.

Air pollution

The contamination of the atmosphere, especially by industrial waste gases, fuel exhausts, smoke or particulate matter (finely divided solids).

Water pollution

The introduction of any EPA listed potential pollutant that affects the chemical, physical or biological integrity of water.

Soil pollution

The accidental or intentional discharge of any harmful substance into the soil.

REGULATORY AGENCIES AND THEIR RESPONSIBILITIES

In order to protect workers, the public, and the environment from environmental and safety hazards, the U.S. government has created several different agencies. Among these agencies are the Environmental Protection Agency (EPA), the Occupational Safety and Health

Administration (OSHA), the Department of Transportation (DOT), and the Nuclear Regulatory Commission (NRC).

Environmental Protection Agency (EPA)

On January 1, 1970, President Richard Nixon signed the National Environmental Policy Act (NEPA). NEPA was enacted to set national policy regarding the protection of the environment, to promote efforts to prevent or eliminate pollution of the environment, to advocate knowledge of ecological systems and natural resources, and to establish a Council on Environmental Quality (CEQ) to oversee the aforementioned policy.

President Nixon soon realized that the CEQ, as structured, did not have the resources or the power to fulfill its mission. In Reorganization Order No. 3, issued on July 9, 1970, the President stated: "It also has become increasingly clear that only by reorganizing our federal efforts can we develop that knowledge, and effectively ensure the protection, development and enhancement of the total environment itself."

Through Reorganization Order No. 3, the **Environmental Protection Agency (EPA)** and the **National Oceanic and Atmospheric Administration (NOAA)** were formed by transferring control of many environmentally related functions from other governmental offices and agencies to the EPA and NOAA.

The EPA's mission is "to protect human health and the environment". The EPA works for a cleaner, healthier environment for Americans. NOAA seeks to "observe, predict and protect our environment."

Occupational Safety and Health Administration (OSHA)

On December 29, 1970, President Richard Nixon signed the Occupational Safety and Health Act of 1970. The purpose of the OSH Act was, and continues to be, "to assure so far as possible every working man and woman in the nation safe and healthful working conditions and to preserve our human resources."

The OSH Act established several agencies to oversee the protection of the American worker. These included:

- **The Occupational Safety and Health Administration (OSHA)**—created to establish and enforce workplace safety and health standards, conduct workplace inspections and propose penalties for noncompliance, and investigate serious workplace incidents.
- **The Occupational Safety and Health Review Commission (OSHRC)**—formed to conduct hearings when employers who were cited for violation of OSHA standards and contested their penalties.

- **The National Institute for Occupational Safety and Health (NIOSH)**—established to conduct research on workplace safety and health problems, specifically injuries and illnesses that may be attributed to exposure to toxic substances.

Department of Transportation (DOT)

On October 15, 1966, President Lyndon Johnson signed Public Law 89-670, which established the **Department of Transportation (DOT)**. DOT's mission is "to develop and coordinate policies that will provide an efficient and economical national transportation system, with due regard for need, the environment and the national defense."

DID YOU KNOW?

The Department of Transportation (DOT), the governmental institution responsible for regulating our highways,

is also responsible for regulating the transportation of natural gas, petroleum, and other hazardous materials through pipelines.

On September 23, 1977, Secretary of Transportation Brock Adams established the Research and Special Programs Administration (RSPA) within the Department of Transportation consolidating various diverse functions that dealt with intermodal activities.

Eventually, the RSPA came to oversee the Office of Pipeline Safety (OPS) and the Office of Hazardous Materials Safety (OHMS), two entities with considerable jurisdiction over the petrochemical industry.

Nuclear Regulatory Commission (NRC)

Congress established the **Atomic Energy Commission (AEC)** in the Atomic Energy Act of 1946. The AEC's mission was regulation of the nuclear industry. Eight years later, Congress replaced that act with the Atomic Energy Act of 1954, which enabled the development of commercial nuclear power. The AEC's mission became twofold: encouraging the use of nuclear power and regulating its safety.

DID YOU KNOW?

Many scientists believe that the Chernobyl nuclear disaster, which happened in the Ukraine in 1986, occurred because the plant was improperly designed and plant operators ignored important safety measures.

As a result, large amounts of radioactive materials were emitted into the environment. This has led to serious health problems and/or death for many of those who were exposed.

In the 1960s, critics charged that the AEC's regulations were not rigorous enough in several important areas, including radiation protection standards, reactor safety, plant location, and environmental protection. The AEC was disbanded in 1974 under the Energy Reorganization Act. This act created the **Nuclear Regulatory Commission (NRC)**, which started operations in 1975. Today, the NRC's regulatory activities focus on reactor safety oversight, materials safety oversight, materials licensing and management of both high- and low-level radioactive waste.

The NRC also regulates instruments in the process industries that use radioactive materials, such as testing devices (e.g. gas chromatographs) and inspection equipment (e.g. x-ray machines).

REGULATIONS IMPACTING THE PROCESS INDUSTRIES

The U.S. government has enacted numerous regulations to minimize workplace hazards. These regulations are administered through various federal agencies such as OSHA, the EPA, and the DOT. Some regulations are generic in scope and affect a variety of industries. Other regulations were created to specifically regulate a certain industry and even certain hazardous substances.

OSHA administers many of the government regulations that significantly impact the day-to-day operations of the process industries. Four of the most important regulations are described in this section: Process Safety Management (PSM), Personal Protective Equipment (PPE), Hazard Communication (HAZCOM) and Hazardous Waste Operations and Emergency Response (HAZWOPER).

Two other major regulations administered by other agencies, the EPA and DOT, are also described in this section. These regulations address hazardous materials and their shipment.

Additional regulations that are not covered in this textbook are discussed in a separate textbook, *Safety, Health and Environment.*

OSHA 1910.119—Process Safety Management (PSM)

The OSHA Process Safety Management of Highly Hazardous Materials (PSM)—29 CFR 1910.119 standard seeks to prevent or minimize the consequences of catastrophic releases of toxic, reactive, flammable, or explosive chemicals.

This standard establishes 14 elements aimed at improving worker safety:

- Employee Involvement
- Process Safety Information
- Process Hazard Analysis
- Operating Procedures
- Training
- Contractors
- Pre-Startup Safety Review
- Mechanical Integrity
- Hot Work Permit System
- Management of Change
- Incident Investigation
- Emergency Planning and Response

- Compliance Audits
- Trade Secrets

OSHA 1910.132—Personal Protective Equipment (PPE)

The OSHA Personal Protective Equipment (PPE)—29 CFR 1910.132 standard aims to prevent worker exposure to potentially hazardous substances through the use of equipment that establishes a barrier between the hazardous substance and the individual's eyes, face, head, respiratory system, and extremities.

This standard requires employers to assess workplace hazards to:

- Determine if Personal Protective Equipment (PPE) is necessary.
- Provide required PPE to their employees.
- Train employees in the proper use and care of the PPE.
- Ensure that employees use the PPE appropriately.

OSHA 1910.1200—Hazard Communication (HAZCOM)

The OSHA Hazard Communication (HAZCOM)—29 CFR 1910.1200 standard seeks to ensure that the hazards of all produced or imported chemicals are evaluated and that information relating to the hazards is provided to employers and employees.

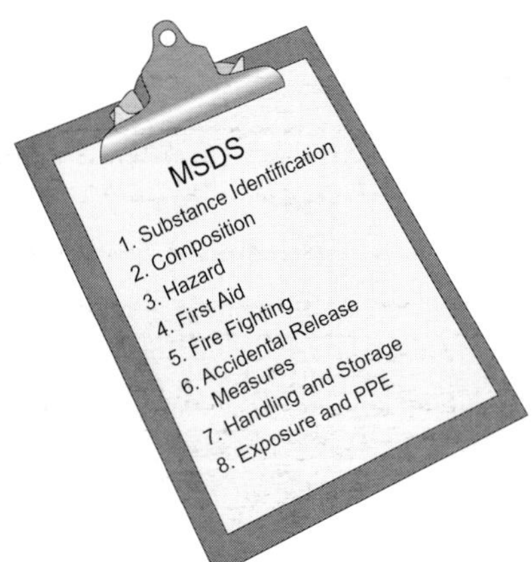

Figure 12-1: Material Safety Data (MSDS) Sheet

This standard requires this transmittal of information through comprehensive hazard communication programs. Information must include container labeling and other forms of warning, Material Safety Data Sheets (MSDSs) and employee training.

MSDS sheets provide key safety, health and environmental information about a material. This information includes physical properties, proper storage and handling, toxicological data, established exposure limits, fire fighting information, and other useful data. This information is provided in a standardized format. MSDS information must be made available for any material manufactured, used, stored, or repackaged by an organization.

OSHA 1910.120—Hazardous Waste Operations and Emergency Response (HAZWOPER)

The OSHA Hazardous Waste Operations and Emergency Response (HAZWOPER)—29 CFR 1910.120 standard outlines the establishment of safety and health programs and level of training for employees in hazardous waste operations and emergency response. Employers must identify, evaluate, and control safety and health hazards in operations involving hazardous waste or emergency response.

OSHA 1910.1000 Air Contaminants

The OSHA Air Contaminants standard (29 CFR 1910.1000) establishes the Permissible Exposure Limits (PELs) for a variety of toxic and hazardous substances. A PEL describes the amount of an airborne toxic or hazardous substance to which an employee can be exposed to over a specified amount time. OSHA 1910.1000 through 1910.1500 list specific toxic or hazardous substances and the assigned PEL for each.

DOT CFR 49.173.1—Hazardous Materials—General Requirements for Shipments and Packaging

The DOT Hazardous Materials—General Requirements for Shipments and Packaging—49 CFR 173.1 standard establishes requirements for preparing hazardous materials to ship by air, highway, rail or water.

This standard establishes requirements for preparing hazardous materials to be shipped by air, highway, rail, water, or any combination of these. It also covers the inspection, testing/retesting responsibilities for persons who retest, recondition, maintain, repair and rebuild containers used or intended for transporting hazardous materials.

EPA CFR 264.16—Resource Conservation and Recovery Act (RCRA)

The EPA Resource Conservation and Recovery Act (RCRA)—40 CFR 264.16 standard promotes "cradle-to-grave" management of hazardous wastes.

This standard classifies and defines requirements for hazardous waste generation, transportation and treatment, storage, and disposal facilities. Additionally, it requires industries to identify, quantify, and characterize their hazardous wastes prior to disposal. It holds the generator of the hazardous waste responsible for management from the point of inception to the final disposal of materials.

EPA Clean Air and Clean Water Acts

The 1990 Clean Air Act "sets limits on how much of a pollutant can be in the air anywhere in the United States". The act ensures that all Americans are covered using the same basic health and environmental protections. Each state must carry out its own implementation plan to meet the standards of the act. For example, it would be up to a state air pollution agency to grant permits to power plants or chemical facilities, fines companies for violating the air pollution limit, etc.

In 1972, the Federal Water Pollution Control Act Amendments were enacted, reflecting growing public concern for controlling water pollution.

When amended in 1977, this law became commonly known as the Clean Water Act. This Act regulates the discharges of pollutants into the waters in the U.S. The act gives the EPA the authority to implement pollution control programs (e.g. setting wastewater standards for industry). The act also sets water quality standards for all contaminants in surface waters, making it illegal to discharge any pollutant from a source into navigable waters unless a permit was obtained.

The Role of the Process Technician in Regulations Compliance

Process technicians play a vital role in a company's efforts to comply with government regulations and other safety, health and environmental policies and procedures. Some different ways that process technicians can help comply with regulations are shown in Table 12-1.

TABLE 12-1: Ways process technicians can help comply with regulations

Process technicians can help comply with regulations by:	
■ Familiarizing themselves with applicable government regulations. ■ Following all plant policies and procedures, since many of these are written to ensure compliance with regulations. ■ Attending all mandatory training to stay current with applicable regulations. ■ Learning to recognize hazards and reporting/handling them appropriately.	■ Performing all job tasks in a timely and accurate way while following safe work practices. ■ Maintaining a safe work environment by performing good housekeeping functions, as required by your job. ■ Having a safe attitude and exhibiting safe behavior.

Table 12-2 provides a list of general safety tips process technicians should follow regardless of the plant they are work in.

TABLE 12-2: General safety tips

Process technicians should always:	
■ Recognize all alarms and know the corresponding response procedures. ■ Smoke only in designated areas. ■ Stay focused and alert (e.g., get adequate rest, eat properly, and refrain from abusing drugs and alcohol). ■ Report injuries and incidents immediately to appropriate personnel. ■ Obey traffic regulations in the plant and never park in fire lanes. ■ Use the proper tool for the job.	■ Review all safety procedures. ■ Understand and properly use the equipment with which you work. ■ Watch for hazardous conditions and report or correct them. ■ Be prepared and keep a clear head in emergency situations. ■ Stay in your assigned area; if you must go to another area, make sure to tell appropriate personnel. ■ Know how to use safety equipment and protective gear.

ATTITUDES AND BEHAVIORS THAT HELP TO PREVENT ACCIDENTS

Safety studies show that, historically, human errors play a significant factor in almost every accident at a plant.

One study (Figure 12-2) examined almost 200 accidents in various chemical plants (Nimmo 1995), reporting the most frequent causes as:

1. Insufficient knowledge (34%)
2. Procedural errors (24%)
3. Operator errors (16%)

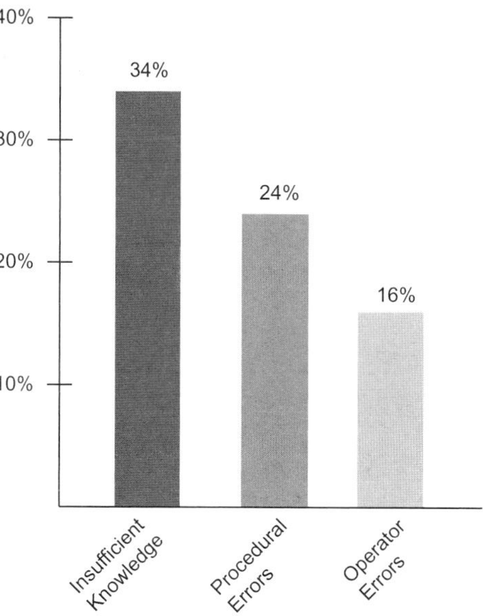

Figure 12-2: The top three causes of accidents in chemical plants

Another study in the petrochemical and refining industry (Figure 12-3) found the following accident causes:

1. Equipment and design failures (41%)
2. Operator and maintenance errors (41%)
3. Inadequate or improper procedures (11%)
4. Inadequate or improper inspection (5%)
5. Miscellaneous causes (2%)

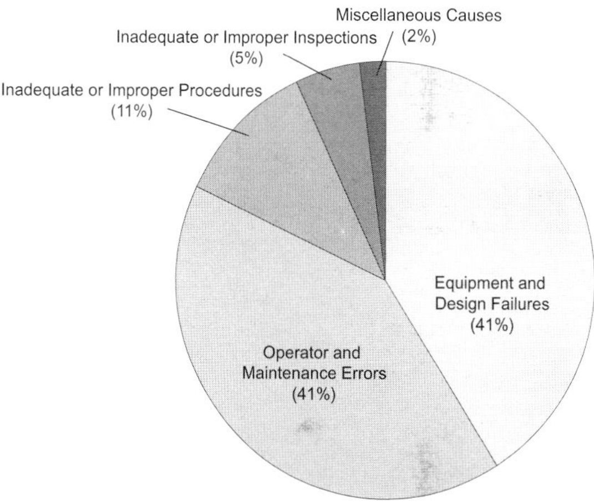

Figure 12-3: The top five causes of accidents in the petrochemical and refining industries

Although human error did not account for all of these accidents and incidents, it did account for many of them.

Personal attitudes and behaviors towards safety can play a significant part in preventing accidents or incidents.

An **attitude** is defined as a state of mind or feeling with regard to some issue or event. Process technicians who maintain a safety mind-set, always thinking about safety, experience fewer (if any) accidents than process technicians who are not safety-oriented.

A **behavior** can be defined as an action or reaction of a person under certain circumstances. Process technicians must respond immediately and appropriately to potential hazards, do a job right the first time, and perform housekeeping duties in a timely manner.

Process technicians must understand and follow not only governmental regulations on safety, health and the environment, but also plant policies and procedures, general safety principles and common sense. He or she must obey Safety, Health, Environment and Security (SHE) rules and report unsafe conditions or unsafe behaviors of co-workers.

Many employers try to determine during a job interview if a candidate will exhibit a safe attitude and behave safely on the job.

COMPONENTS OF THE FIRE TRIANGLE AND THE FIRE TETRAHEDRON

Hydrocarbons, chemicals and many other materials used in the process industries are extremely flammable and/or combustible. Because of this, one of the greatest potential hazards to process technicians is fire and/or explosions.

Figure 12-4: Fire triangle

Fire is a chemical reaction. Fire starts when a substance (fuel), in the presence of air (oxygen), is heated to an ignition point (heat) resulting in combustion. Fire must have all of these elements (fuel, oxygen, heat) present to start. Removing one of these elements or will extinguish a fire.

These three elements are referred to as a **fire triangle**:

1. **Fuel**—Any combustible material; can be a solid, liquid or gas.
2. **Oxygen**—Air is composed of 21% oxygen; generally, fire only needs 16% oxygen to ignite.
3. **Heat**—The energy or heat required by a fuel to produce ignition.

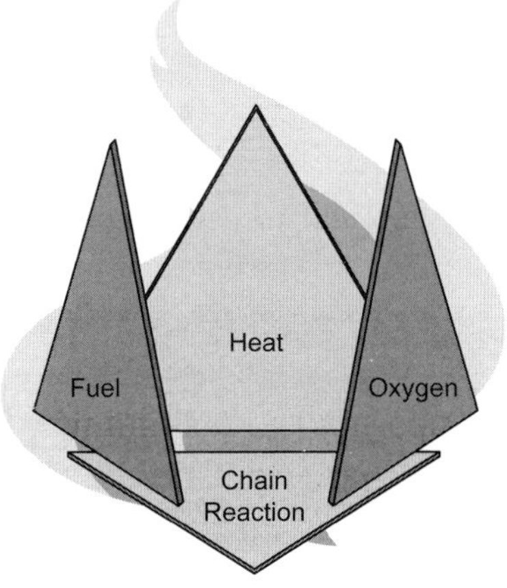

Figure 12-5: Fire tetrahedron

Fire Tetrahedron

The fire triangle represents the elements necessary to create a fire. Once a fire has started, the fire tetrahedron represents the elements necessary to sustain combustion.

The fire tetrahedron consists of the components of the fire triangle and another component—the chain reaction. A **chain reaction** is a series of reactions in which each reaction is initiated by the energy produced in the preceding reaction. (e.g., toppling dominoes; when the first domino is knocked over it causes the second domino in the series to topple. This, in turn, makes the third domino topple and so on). This type of reaction occurs when fuel, oxygen and heat come together in proper amounts under certain conditions. Chain reactions are what cause fires to build on themselves and spread. In order to stop a fire, one of the four components of the fire tetrahedron must be removed.

Extinguishing agents (such as halon gas) stop a fire, not by removing fuel heat, or oxygen, but by preventing the chain reaction from occurring.

Classes of Fire

Fires are classified according to four groups:

- **Class A**: Combustible materials such as wood, paper and plastic.
- **Class B**: Grease or combustible and flammable gases or liquids.
- **Class C**: Fire involving live electrical equipment.
- **Class D**: Combustible metals (e.g., aluminum, sodium, potassium and magnesium).
- **Class K:** Cooking oil, fat, grease or other kitchen fires; intended to supplement a fire suppression system (like in a commercial kitchen).

Figure 12-6: The five classes of fire

Process technicians must understand the elements required to start a fire, how to prevent fires and control them, and the combustible/flammable properties of the materials they are working with.

CONSEQUENCES OF NON-COMPLIANCE WITH REGULATIONS

If a process technician fails to comply with regulations, this can cause many consequences: legal, moral and ethical, or safety, health and environment. These consequences can be imposed as a result of a minor accident, a major accident, or from an on-site inspection by a government agency representative.

Legal

Legal consequences fall into one of two major types:

- Fines and/or citations levied by federal, state, or local regulatory agencies (and possibly even criminal charges).
- Lawsuits filed by affected parties, such as injured workers or local citizens.

Moral and Ethical

Moral and ethical consequences can manifest as:

- Burden of contributing to injuries or deaths.
- Responsibility for causing damaged equipment, lost production and associated costs.
- Guilt for not complying with regulations, policies and procedures.

Safety, Health and Environment

Numerous safety, health and environmental consequences can result from non-compliance. These include:

- Exposed or injured workers
- Exposed or injured citizens
- Air pollution
- Water pollution
- Soil pollution

DID YOU KNOW?

Noise-Induced Hearing Loss (NIHL) can be caused by a single loud impulse noise (e.g., an explosion) or by loud, continuous noise over time (e.g., noise generated in a woodworking shop).

Hearing protection should always be worn when working in environments with sounds louder than 80 decibels (normal conversation is around 60 decibels).

Other sounds that can cause NIHL include motorcycles, firecrackers, and firearms, all of which range from 120 to 140 decibels.

ENGINEERING CONTROLS, ADMINISTRATIVE CONTROLS AND PPE

The process industries use three methods to minimize or eliminate worker exposure to hazards. These methods, listed from highest priority down, are:

- **Engineering controls**—controls that use technological and engineering improvements to isolate, diminish, or remove a hazard from the workplace. Examples of some engineering controls are:
  - Using a non-hazardous material in a process that will work just as well as a hazardous material.
  - Placing a sound reducing housing around a pump to muffle the noise it makes.
  - Adding guards to rotating equipment.
- **Administrative controls**—if an engineering control cannot be used to address a hazard, an administrative control is used. Administrative controls involve implementing programs and activities to address a hazard.

  Programs consist of written documentation such as policies and procedures. Activities involve putting a program into action.

  Administrative control is also called a work practice control or managerial control. Examples of administrative controls are:
  - Writing a procedure to describe the safe handling of a hazardous material.
  - Limiting the amount of time a worker is exposed to loud noises.

- Training a worker on how to safely perform a potentially dangerous activity.
- Documenting how workers should select and properly wear Personal Protective Equipment suited to a specific task.

- **Personal Protective Equipment (PPE)**—When engineering and administrative controls are not adequate enough to protect workers, PPE is used. PPE is specialized gear that provides a barrier between hazards and the body and its extremities. Examples of PPE are:
  - Hearing protection
  - Hard hats
  - Flame retardant clothing (FRC)
  - Gloves and shoes

  The next section covers PPE in more detail.

THE CORRECT USE OF PERSONAL PROTECTIVE EQUIPMENT (PPE)

OSHA requires employers to use personal protective equipment (PPE) to reduce employee exposure to hazards when engineering and administrative controls are not feasible or effective. Employers are required to determine all exposures to hazards in their workplace and determine if PPE should be used to protect their workers.

Different types of PPE are used to protect process technicians from head to toe in a variety of situations and hazards. Figure 12-7 shows some examples of PPE.

Types of PPE include:

- Head protection
- Face protection
- Eye protection
- Ear protection
- Respiratory protection
- Body protection
- Hand protection
- Foot protection

The process technician should also be familiar with the location and operation of eye washes and safety showers in the operating area. Eye washes and safety showers, if used quickly and properly, can greatly reduce the severity of a chemical exposure.

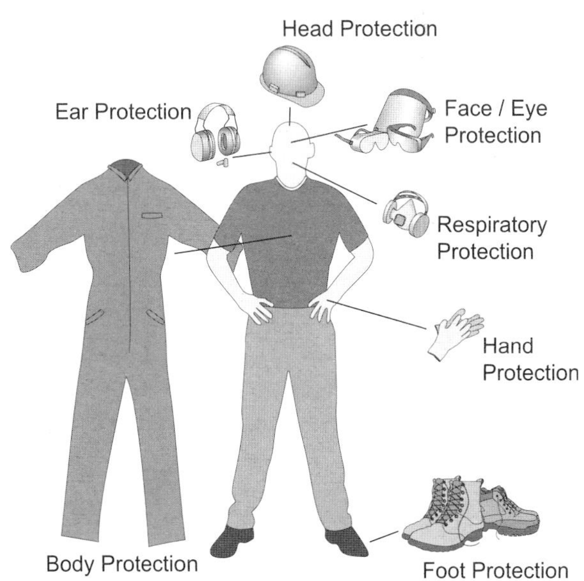

Figure 12-7: Examples of Personal Protective Equipment (PPE)

Head Protection

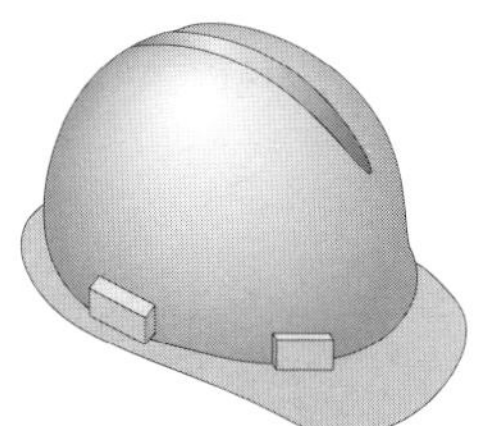

Figure 15-8: Hard Hat

Head protection is used when a person's head is in danger of being bumped or struck by falling or flying objects.

Safety helmets (also referred to "hard hats") must be impact resistant and meet the American National Standard Institute standard (ANSI) for protective headwear.

Bump hats or caps do not meet the ANSI standard, as they are intended only to protect against bumping an obstruction.

Face Protection

Face protection is used to protect the face (and often the head and neck) against impact, chemical or hot metal splashes, heat, radiation and other hazards.

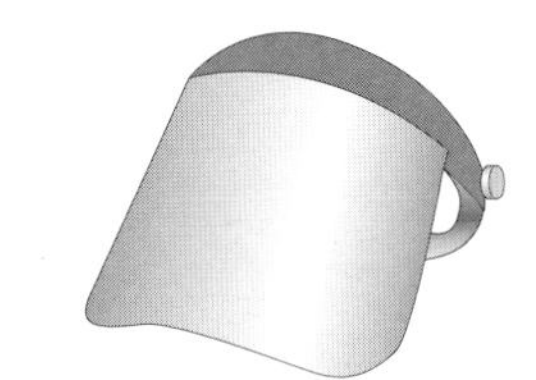

Figure 12-9: Plastic face shield

Plastic face shields protect the face and eyes during activities such as sawing, buffing, sanding, grinding, or handling chemicals.

Acid-proof hoods protect the head, face, and neck against splashes from corrosive chemicals.

Welding helmets protect against splashes of molten metal and radiation.

Eye Protection

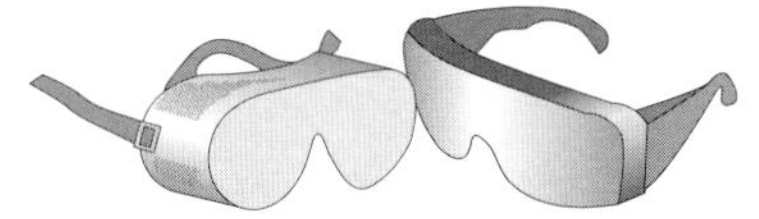

Figure 12-10: Goggles and safety glasses

Eye protection is used to protect the eyes from flying objects, splashes of corrosive liquid or molten metals, dust, and harmful radiation.

Cover goggles are used when there is a danger of flying objects or splashing. Regular safety glasses with side shields are used during all other process-related activities.

Ear Protection

Figure 12-11: Earplugs and earmuffs

Ear protection is used when excessive noise is present in the workplace.

Ear muffs generally have a higher noise reduction rating than ear plugs. Earplugs and earmuffs can be used together when extreme noise is present (i.e., double hearing protection).

Respiratory Protection

Respiratory protection is used when airborne contaminants are present.

Filter and Cartridge Respirators protect against nuisance dusts and hazardous chemicals. You must choose the correct filter or cartridge based on the chemical(s) to which you will be exposed. Respirators are not suitable for high concentrations of contaminants. Also, they do not supply oxygen so you cannot use them in oxygen-deficient atmospheres.

Figure 12-12: Respirator

Air supplying respirators deliver breathing air through a hose connected to the wearer's face piece. The air can be supplied through a tank attached to the back of the wearer (called a Self Contained Breathing Apparatus, or SCBA) or from a tank that stays in one place with a long hose connecting the tank to the face piece (called a hose line respirator).

Body Protection

Body protection is used to protect the body. This type of protection comes in various forms depending on the level of chemical protection required:

- Aprons protect the worker from chemical splashes, while harnesses with lifelines provide workers with fall protection.
- Protective clothing comes in various forms.
- Reflective clothing protects against radiant heat.
- Flame-resistant clothing protects against sparks and open flames.

Figure 12-13: Flame Retardant Coveralls (FRCs)

Hand Protection

Hand protection is used to protect the hands and fingers from cuts, scratches, bruises, chemicals, and burns.

- Extreme temperature protection protects against burns (hot or cold).
- Metal mesh protects against knives and sharp objects.
- Rubber, neoprene, and vinyl protect against different types of chemicals.
- Leather protects against rough objects.

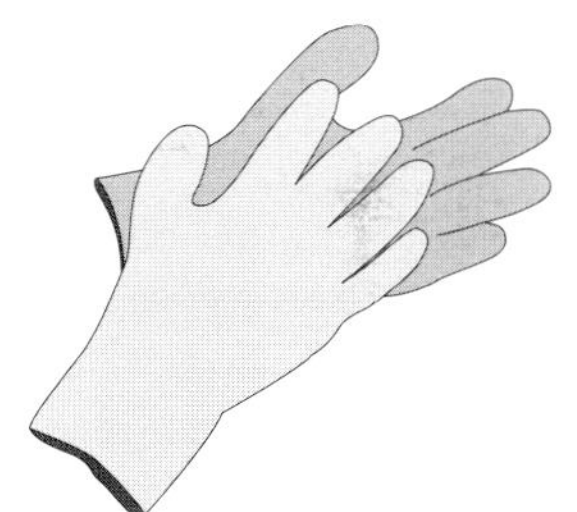

Figure 12-14: Protective gloves

Foot Protection

Foot protection is used to protect the feet and toes against falling or rolling objects.

A variety of footwear can be required for you to use, based on the type of job or task: For example, you might be required to wear safety shoes or boots, rubber boots, flat-soled shoes, high-tops, etc. Some footwear is not permitted (e.g. open-toed shoes or sandals).

Figure 12-15: Safety shoes and boots

THE OSHA VOLUNTARY PROTECTION PROGRAM (VPP)

OSHA established the **Voluntary Protection Program (VPP)** to recognize and promote effective safety and health management. The following outlines the cooperative relationship between management, employees and OSHA to implement a VPP:

1. Management agrees to operate an effective program that meets an established set of criteria.
2. Employees agree to participate in the program and work with management to ensure a safe and healthful workplace.
3. OSHA initially verifies that a site's program meets the VPP criteria.
4. OSHA publicly recognizes the site's exemplary program and removes the site from routine scheduled inspection lists (Note: OSHA may still investigate major accidents, valid formal employee complaints and chemical spills).

There are two OSHA VPP Ratings: Star and Merit.

- **Star**—Participants meet all VPP requirements.
- **Merit**—Participants have demonstrated the potential and willingness to achieve Star program status and are implementing planned steps to fully meet all Star requirements.

Periodically (every three years for the Star program and every year for the Merit program), OSHA reassesses the site to confirm that it continues to meet VPP criteria.

THE ISO 14000 STANDARD

The International Organization of Standardization (ISO), headquartered in Geneva, Switzerland, consists of a network of national standards institutes from over 140 countries.

ISO has published more than 13,700 International Standards. ISO standards are voluntary, since the organization is non-governmental and has no legal authority to enforce the standards. The standards that impact the process industries most are ISO 9000 and ISO 14000.

ISO 9000 provides a framework for quality management by addressing the processes of producing and delivering products and services. The chapter on Quality covers ISO 9000.

ISO 14000 addresses how to incorporate environmental aspects into operations and product standards. It requires a site to implement an

Environmental Management System (EMS) using defined, internationally recognized, standards as described in the ISO 14000 specification.

ISO 14000 specifies requirements for:

- Establishing an environmental policy.
- Determining environmental aspects and impacts of products, activities and services.
- Planning environmental objectives and measurable targets.
- Implementing and operating programs to meet objectives and targets.
- Checking against the standard and making corrective actions.
- Performing management review.

ISO 14001, one of the sub classifications of ISO 14000, addresses the following:

- Sites must document and make available to the public an Environmental Policy.
- Procedures must be established for ongoing review of environmental aspects and impacts of products, activities, and services.
- Environmental goals and objectives must be established that are consistent with the environmental policy and programs must be set in place to implement goals and objectives.
- Internal audits of the EMS must be conducted routinely to ensure that non-conformances to the system are identified and addressed.
- Management review must ensure top management involvement in the assessment of the EMS and, as necessary, address the need for change.

The Environmental Management System (EMS) document is the central document that describes the interaction of the core elements of the system.

The Environmental Policy and Environmental Aspects/Impacts provide the following:

- Analysis, including legal and other requirements.
- Direction for the environmental program by influencing the selection of specific, measurable, environmental goals, objectives, and targets.
- Recommendations for specific programs and/or projects that must be developed to achieve environmental goals, objectives, and targets.
- Ongoing management review of the EMS and its elements to help ensure continuing suitability, adequacy, and effectiveness of the program.

PHYSICAL AND CYBER SECURITY REQUIREMENTS

Physical security is intended to protect specific assets such as pipelines, control centers, tank farms, and other vital areas. **Cyber security** is intended to protect information and information technology (e.g. computing systems, networks) from unauthorized access and use.

Physical and cyber-security threats can come from:

- Terrorist organizations and hostile nation-states
- Insiders
- Criminal elements

Process technicians should recognize threats to physical security and cyber security. These include:

- Terrorist threats and acts
- Workplace violence
- Criminal acts
- Industrial espionage

To reduce the hazards of such threats, companies can create threat response and emergency action plans. In these plans, companies analyze their critical resources and operations, determine what vulnerabilities and threats may exist, and identify what level of risk exists. The plan then describes the processes and procedures on how to lessen or eliminate threats.

Physical security aims to protect a company's critical assets from unauthorized access, thereby preventing them from being damaged or stolen. Physical security involves using measures to protect and monitor assets and resources. Such measures can include physical access barriers (e.g., fences and doors), monitoring devices (e.g., cameras and motion detectors) and security patrols or guard stations.

Along with access and perimeter security, physical security also includes operations planning, communications planning, personnel background checks, and more.

Cyber security involves protecting information assets and capabilities from unauthorized access, modification or destruction. It is a way of securing a company's information and information technology (IT) infrastructure, to ensure that sensitive data and computer networks are protected.

Cyber security involves:

- The prevention of unauthorized computer access (e.g., through passwords protection and computer fire walls).

- Making sure those with access are trustworthy, and monitoring computer access and usage.
- The prevention of physical threats (e.g., direct access to sensitive information and computer networks),
- Communication safeguards (e.g., cell phone and PDA protection).
- Preventing groups or individuals from using sensitive information and computer networks to carry out physical threats (e.g., hacking into a pipeline monitoring network and taking control of the system).

SUMMARY

The process industries can have many safety, health and environmental hazards. These hazards can be chemical, physical, ergonomic or biological.

In order to prevent or minimize many of these hazards, the U.S. government has created several agencies. The Environmental Protection Agency (**EPA**) protects human health and the environment. The Occupational Safety and Health Administration (**OSHA**) establishes workplace safety and health standards, conducts workplace inspections, and proposes penalties for noncompliance. The Department of Transportation (**DOT**) develops and coordinates policies for an efficient and economical national transportation system. The Nuclear Regulatory Commission (**NRC**) protects public health and safety through regulation of nuclear power and nuclear materials.

In an attempt at protecting workers and the environment, OSHA, the DOT and the EPA have created many regulations. These regulations help minimize the consequences of catastrophic releases of toxic, reactive, flammable, or explosive chemicals; prevent worker exposure to potentially hazardous substances; ensure that the hazards of all produced or imported chemicals are evaluated and that information relating to the hazards is provided to employers and employees; establish emergency response operations for the releases of hazardous substances; set requirements for handling hazardous materials.

Process technicians play a vital role when it comes to safety and health. Technicians must always maintain a safety conscious attitude and behave in a safe, responsible, and appropriate manner. By being familiar with government regulations, following plant policies and procedures, attending training, learning to manage hazards appropriately, and more, technicians facilitate compliance and create a safer and healthier work place.

A fire triangle is a list of the three essential elements required for combustion: Fuel, oxygen, and heat. If any one of these elements are removed, combustion cannot occur. A fire tetrahedron adds a fourth element, a chain reaction as part of the combustion process. Chain reactions are what cause fires to build on themselves and spread.

Failure to comply with regulations can have legal, moral, ethical, safety, health, and environmental consequences. In order to prevent these consequences, process industries often employ engineering controls (technology and engineering improvements), administrative controls (programs, procedures and activities) and personal protective equipment (specialized gear used to protect the body) to make a safer workplace.

The intent of OSHA's Voluntary Protection Program (VPP) is to help recognize and promote effective safety and health management. ISO 14000 addresses how to incorporate environmental aspects into operations and product standards.

Physical and cyber security measures are used to protect assets from internal and external threats. Physical security focuses on protecting pipelines, control centers and other vital areas from damage or theft. Cyber security protects information assets and computing systems.

CHECKING YOUR KNOWLEDGE

1. Define the following key terms:
   a. Administrative controls
   b. Engineering controls
   c. Hazardous agent
   d. Biological hazard
   e. Chemical hazard
   f. Physical hazard
   g. Ergonomic hazard
   h. Fire triangle
   i. Chain reaction
   j. Cyber security
   k. Physical security

2. Which type of hazard involves noise and radiation?

3. Which type of pollution occurs when there is an emission of a potentially harmful substance or pollutant into the atmosphere?

4. Which government regulation deals with establishing emergency response operations for releases of, or substantial threats of releases of, hazardous substances?

5. Which of the following is addressed by OSHA regulation 1910.119?
   a. Process Hazard Analysis
   b. Process Waste Analysis
   c. Process Communications Analysis
   d. Process Safety Analysis

6. Personal attitudes and behaviors towards safety can play a significant part in preventing accidents or incidents. What is the definition of an attitude?

7. What are the three parts of a fire triangle?
   a. Air, wood, and a match
   b. Fuel, combustion, and air
   c. Earth, wind, and fire
   d. Fuel, oxygen, and heat

8. What type of eye protection should be worn to guard against flying objects or splashing?

9. The intent of OSHA's Voluntary Protection Program is to:
   a. Regulate all environmental activities of the process industries.
   b. Recognize and promote effective safety and health management.
   c. Gauge the impact that the process industries have on the safety and health of process technicians.
   d. Help companies implement operation programs that are within EPA guidelines.

10. What is the central document that describes the interaction of the core elements of ISO 14001?

11. Define physical security and cyber security.

ACTIVITIES

1. Select one of the following government regulations and write a one-page report. Use any available resources. Describe the regulation, its history, what area it impacts most in the process industries (safety, health, or environmental) and why it is an important regulation. Be sure to include a list of all your reference sources.
   - OSHA Process Safety Management PSM (29 CFR 1910.119)
   - OSHA Personal Protective Equipment PPE (29 CFR 1910.132)
   - OSHA Hazard Communication HAZCOM (29 CFR 1910.1200)
   - OSHA Hazardous Waste Operations and Emergency Response HAZWOPER (29 CFR 1910.120)
   - DOT Hazardous Materials (49 CFR 173.1)
   - EPA Resource and Conservation Recovery Act RCRA (40 CFR 264.16)

2. Keep a journal for a week, describing how you use a safe attitude and behaviors on a daily basis (e.g., checking my blind spot before changing lanes, or being alert to my surroundings).

3. For each of the items below, indicate which category it falls into (i.e., engineering control, administrative control, or personal protective equipment).

Item	Engineering Control	Administrative Control	PPE
a. Adding a machine guard to a piece of equipment.			
b. Writing a procedure for dealing with hazardous materials.			
c. Training workers on hearing protection.			
d. Providing hearing protection.			
e. Documenting how to perform a hazardous task safely.			
f. Wearing a face shield.			
g. Improving ventilation in a work area.			
h. Setting up shifts to limit exposure to hearing hazards.			
i. Writing up fall protection procedures.			
j. Training workers on fire extinguishers.			
k. Making respirators available.			
l. Training on using respirators.			
m. Adding soundproofing around a pump.			
n. Wearing a chemical protective suit.			
o. Creating procedures for handling spills.			

Chapter 13

Quality

OBJECTIVES

Upon completion of this chapter you will be able to:

1. Identify responses in the process industries to quality issues.
2. Describe the role each of the following played in quality implementation:
   - W.E. Deming
   - Joseph Juran
   - Philip Crosby
3. Describe the four components of Total Quality Management (TQM) and how it is applied in today's workplace.
4. Describe the application of the International Organization for Standardization, ISO 9000 series, as it relates to the petrochemical and petroleum industry.
5. Describe the use of Statistical Process Control (SPC) in the workplace.
6. Describe the roles and responsibilities of the process technician in supporting quality improvement within the workplace.

KEY TERMS

- **Assignable Variation**—states that when a product's variation goes beyond the limits of a natural variation, it is the result of a cause that can be identified.
- **Attributes**—also called discrete data, or data that can be counted and plotted as distinct or unconnected events (such as percentage of late shipments or number of mistakes made during a process).
- **ISO**—taken from the Greek word isos, which means equal, ISO is the International Organization for Standardization, which consists of a network of national standards institutes from over 140 countries.
- **Pareto Principle**—a quality principle, also called the 80-20 rule that states 80 percent of problems come from 20 percent of the causes.
- **PPM**—Predictive/Preventive Maintenance, a program to identify potential issues with equipment and use preventive maintenance before the equipment fails.
- **Six Sigma**—a relatively new TQM approach, Six Sigma is considered advanced quality management using a data driven approach and methodology to eliminate defects.
- **SPC**—Statistical Process Control uses mathematical laws dealing with probability. Companies utilize SPC to gather data (numbers) and study the characteristics of processes, then use the data to make the processes behave the way they should.
- **TPM**—Total Productive Maintenance is an equipment maintenance program that emphasizes a company-wide effort to involve all levels of staff in various aspects of equipment maintenance.

- **TQM**—Total Quality Management is a collection of philosophies, concepts, methods and tools used to manage quality; TQM consists of four parts: Customer Focus, Continuous Improvement, Manage by Data and Facts, and Employee Empowerment.
- **Variables**—also called continuous data, or data that can be measured and plotted on a constant scale (such as flow through a pipeline or liquid in a tank).
- **Zero Defects**—a quality practice with the objective of reducing defects, which can increase profits.

INTRODUCTION

Quality is an important part of the process industries. Without quality measures, products and services could be deficient or unsatisfactory. Unsatisfactory products lead to unhappy customers, increased waste, inefficiencies, increased costs, reduced profits, and an inability to maintain a competitive edge.

In order to maintain a competitive edge, many companies have adopted theories and philosophies from famous quality pioneers. By incorporating these philosophies, and acting in ways that support them, companies and process technicians can improve processes, reduce waste, increase efficiency, reduce costs, produce superior products, and maintain a competitive edge in a global marketplace.

What Is Quality?

The term quality has different meanings to different people, but generally speaking in regard to the process industries, quality has two major definitions:

- A product or service free of deficiencies.
- The characteristics of a product or service that bear on its ability to satisfy stated or implied needs.

This chapter will explore these definitions, as well as other meanings that influential people have given to the term quality.

Industry Response to Quality Issues and Trends

Why are organizations concerned with quality? Because producing quality products and services is critical to those organizations for a variety of reasons:

- To satisfy and retain customers.
- To maintain a competitive advantage and respond to rapidly changing markets.
- To capture a leading position in the global marketplace.

- To improve profitability.
- To manage change more effectively.
- To maintain or bolster the organization's reputation.

Companies in the process industries have implemented quality processes to address additional business needs, such as:

- Offering standardized products on a consistent basis.
- Improving efficiency of operations and maintenance.
- Reducing waste of resources such as utilities and feedstocks.
- Decreasing downtime of people and equipment.
- Ensuring certifications necessary to trade internationally.
- Tapping new technologies and methods.

DID YOU KNOW?

Shewhart developed his ideas about quality when he was tasked to develop a radio headset for use during World War I.

After measuring people's heads to determine a standard size for the headset, he used charts to show a pattern in the measurements.

The Quality Movement and its Pioneers

Today, most companies take quality seriously, showing great concern for providing quality products and services to their customers. But not too long ago, companies were not as committed to quality. Competition was almost non-existent and customers did not have a large number of choices. As more competitors emerged (not just locally but internationally) and economic conditions changed, the effort companies made to improve quality gained momentum.

During the start of the 20th century, F.W. Taylor pioneered the concept of "scientific management", which sought to improve manufacturing by using engineers to develop plans that supervisors and workers executed. Taylor's system successfully raised productivity, but human relations were negatively impacted.

Following Taylor, Dr. Walter Shewhart developed the theory of statistical quality control in the 1920s. While working for Bell Telephone Labs, Shewhart recognized that not all products created during work processes were exactly alike. Shewhart stated that every process produces variation in its products due to random, natural causes.

According to Shewhart, when a product's variation goes beyond the limits of a natural variation, it is the result of a worker-related cause. He called this cause an **assignable variation**, meaning it could be attributed to a specific worker's action. Shewhart developed control charts that were used to spot assignable causes of variation. By identifying and correcting the variation, the company improved the quality of its products.

Taylor and Shewhart laid the groundwork for the quality movement during the early part of the century. From the mid-1940s on, three specific people made a significant impact on quality with their ideas, approaches and tools:

- Dr. W. Edwards Deming
- Joseph Juran
- Philip Crosby

Many organizations have adopted some or all of these pioneers' definitions of quality into their corporate culture. Let's look at the role each of these men played in the quality movement.

Dr. W. Edwards Deming

Figure 13-1: Dr. W. Edwards Deming

Courtesy of The W. Edwards Deming Institute ®

Dr. Deming studied statistical process control under Dr. Walter A. Shewhart in 1927 and adopted his theories for quality control. During World War II, he worked with the War Department, applying Shewhart's statistical quality control principles to the production of materiel. He assisted the war effort by recommending that engineers be trained in the basics of applied statistics.

After the war, Deming began working in Japan to help rebuild its war-torn industries. Deming convinced Japanese companies to apply statistical methods to help improve the quality of products and services. During the decade that followed, the Japanese companies implemented his principles and established themselves as leaders in quality manufacturing.

Deming developed a theory on quality control, referred to as the Deming Cycle. The Deming Cycle states that every task or every job is part of a process. Specifically, a system is a group of interrelated components that work toward optimization of the system, even if the result does not benefit the individual component.

His fourteen point Theory of Management emphasized the need to build customer awareness, reduce variation, and foster constant change and improvement.

He defined quality as meeting and exceeding the customer's needs and expectations then continuing to improve. He also urged conformance to specifications.

DID YOU KNOW?

The Union of Japanese Scientists and Engineers awards the Deming Prize for quality to companies that meet or exceed customer's needs based on Deming's fourteen points. It is the highest quality award in Japan.

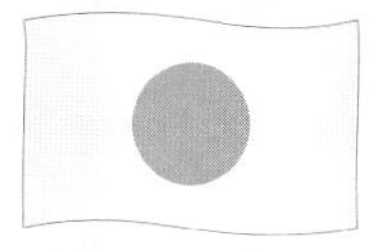

Figure 13-2: Joseph M. Juran
Courtesy of J.M.Juran

Joseph M. Juran

Like Deming, Joseph Juran worked with Japanese companies after the war. He was instrumental in assisting the Japanese as they rebuilt their economy.

In the mid-1950s, the Union of Japanese Scientists and Engineers group invited Juran to conduct quality control courses for middle and top management. These courses extended the philosophy of quality control to every aspect of an organization's activities. He emphasized using quality control as a management tool. In the mid-1960s, he predicted that the Japanese would lead the world in quality.

Juran's best-known quality teachings emphasized the **Pareto Principle**, an idea of separating the vital few from the trivial many. The 80-20 Rule is an example of the Pareto principle; this means that 80 percent of our problems come from 20 percent of the causes.

DID YOU KNOW?

Joseph Juran based his 80-20 rule on the work of Vilfredo Pareto, an Italian economist, who actually used the rule to describe the distribution of wealth in his country: 20% of the people had 80% of the wealth.

His other contributions to the quality movement are involving top management and stressing the need for widespread training in quality. The main components of his Trilogy of Managerial Processes quality system are:

- Quality Planning
- Quality Control
- Quality Improvement.

He defined quality as fitness for use as perceived by customers.

Philip B. Crosby

Philip Crosby is best known for creating the concept of zero defects in the early 1960s. Zero defects is a practice with the objective of reducing defects, which can increase profits.

In the 1970s, Crosby became a leader in the quality movement. He served as vice president for International Telephone and Telegraph (ITT) before forming his own management-consulting firm.

He published a book titled, *Quality Is Free*. The single word Crosby uses to sum up quality is "prevention." He stated that quality involved doing "it right the first time." He advocated a proactive approach to quality; fix the process so errors will not occur.

Crosby's Quality Improvement Program promotes fourteen points covering management commitment, zero defects, training, goals, teams, corrective action and removal of error causes.

He defined quality as conformance to requirements and the elimination of variation.

Figure 13-3: Philip B. Crosby
Courtesy of Phillip Crosby Associates

The Japanese Influence

Kaoru Ishikawa, considered a pioneer of the Japanese quality movement, studied statistical quality control with Deming and quality management with Juran. He played a major role in the growth and concept of the quality circle, which demonstrates causes and effects using diagrams.

In the 1980s, in response to Japan's quality movement, many American companies instituted quality programs or revamped their existing programs. Dr. Genichi Taguchi, a Japanese engineer and scientist, began working with Ford Motor Company to provide seminars to its managers. Dr. Taguchi contributed his expertise in the field of industrial research, focusing on the design of experiments, to the quality movement.

QUALITY INITIATIVES

Some of the different quality initiatives that process industries have adopted are:

- Total Quality Management (TQM)
- ISO 9000 series
- Statistical Process Control (SPC)
- Six Sigma
- Self-directed or Self-managed work teams
- Malcolm Baldridge award standards
- Maintenance programs such as Total Productive Maintenance and Predictive/Preventive Maintenance

The following sections describe these initiatives.

TOTAL QUALITY MANAGEMENT (TQM)

The quality principles from Deming, Crosby, Juran and other experts in the field have evolved into a concept referred to as Total Quality Management (TQM). TQM is not a specific, well-defined program, but a collection of philosophies, concepts, methods and tools used to manage quality. Consequently, not every company practices TQM the same way, or even calls it by the same name.

TQM, whether referred to as Customer Satisfaction, Reengineering, or some other name, encompasses four major components:

- **Customer Focus**—the customer viewpoint determines what quality is. Companies use customer satisfaction with its products and services as a measure of quality.
- **Continuous Improvement**—companies improve the activities used to produce products and services, resulting in higher quality products and services. The philosophy is to create a process that makes

it easy to do things the right way and difficult to do them the wrong way.

- **Management by Data and Facts**—companies gather information to understand how their processes work, what can be produced and where improvements can be made.
- **Employee Empowerment**—top management must demonstrate their buy-in, commitment and involvement to quality improvements. Every employee is free to question, challenge and help change the way that products and services are produced.

ISO

Figure 13-4: ISO Logo
Courtesy of ISO

ISO, the International Organization for Standardization, is headquartered in Geneva, Switzerland and consists of a network of national standards institutes from over 140 countries. The American National Standards Institute (ANSI) is a member of ISO.

Between its founding after World War II and the present day, ISO has published more than 13,700 International Standards. These standards address everything from screw sizes to symbols, computers to shipping containers, and more.

ISO standards are voluntary since the organization is non-governmental and has no legal authority to enforce the standards. However, without the consent of both government and private sectors to use ISO standards, many vital economic segments around the world would be affected, including: manufacturing, trade, science, technology, and many others.

The standards that impact the process industries most are ISO 9000 and ISO 14000. ISO 9000 provides a framework for quality management by addressing the processes of producing and delivering products and services. The ISO 14000 standard addresses environmental management systems, helping organizations improve their environmental performance.

This section addresses ISO 9000 in more detail. The "Safety, Health and Environment" chapter describes ISO 14000 in more detail.

ISO 9000 provides global standards of product and service quality. These standards form a quality management system that aims to fix quality system defects and make products and services conform to stated standards.

More than half a million organizations, in over 60 countries, have or will implement ISO 9000. Most companies in the process industries have ISO

DID YOU KNOW?

The term ISO is derived from the Greek word isos, meaning equal; that way whatever the country or language, the short form of the organization's name is always ISO.

ISOS = Equal

9000 certification. Their ISO 9000 certification efforts usually overlap with TQM programs.

While TQM policies and practices do compare in some ways to the ISO 9000 model, TQM is defined in many different ways, while ISO standards are consistent and have been adopted by hundreds of thousands of companies around the world.

Since ISO certification is recognized internationally, organizations can participate in the global marketplace and be assured of conformity. Both manufacturers and suppliers know that quality standards are being met when dealing with ISO certified organizations.

Companies seek ISO 9000 certification for a variety of reasons:

- **Contractual**—many companies are requiring certification from suppliers.
- **Liability**—certification can result in improved product liability procedures and documentation.
- **Cost savings**—companies with certification report a significant increase in operational efficiency, which results in improved profitability.
- **Competition**—certification allows companies to compete on a global basis and levels the "playing field" between small and large companies.
- **Customer satisfaction**—customer confidence is increased with companies that are certified.

ISO 9000 provides companies with a common approach for documenting and maintaining a quality system. An accredited registrar audits a company's quality system to determine if that system compiles with the ISO 9000 standard. In other words, ISO 9000 certification targets the quality system itself, not products or services.

ISO 9000 certification guarantees that a plan for continuous improvement is in place. Re-certification relies on continuous adherence to the standards and is based on regular audits.

To obtain ISO 9000 certification, a company must have a quality program in place that meets documentation and operational criteria.

As a process technician, your role is to make sure that all processes and documentation are carried out in accordance with your company's quality program.

STATISTICAL PROCESS CONTROL AND OTHER ANALYSIS TOOLS

Walter Shewhart developed the concept of Statistical Process Control during World War I, with Dr. W. E. Deming improving and expanding it. Their work influenced the quality initiatives of many companies, which integrated the approach of SPC into their quality programs.

SPC uses mathematical laws dealing with probability, or how often certain events could occur under certain conditions. Data plays a crucial role in SPC, allowing companies to monitor, control, correct and improve their processes. Companies utilize SPC to gather data (numbers) and study the characteristics of processes, then use the data to make the processes behave the way they should.

Many companies use SPC or statistical thinking to make decisions, basing them on meaningful data and statistics. It basically employs a scientific method to decision making. Decisions are not made unless some type of proof (data or statistics) is available to substantiate the changes.

Specifically, SPC is a way of determining whether a process is producing predictable results. SPC is also a basic tool for identifying both immediate systemic problems and opportunities for improvement.

Data derived through SPC can predict how a process will function in the future, making it possible to avoid off-spec products and unnecessary process changes. Data can be placed into one of two categories:

- **Attributes**—also called discrete data, or data that can be counted and plotted as distinct or unconnected events (such as percentage of late shipments or number of mistakes made during a process).
- **Variables**—also called continuous data, or data that can be measured and plotted on a constant scale (such as flow through a pipeline or liquid in a tank).

Data is critical to the operation and maintenance of a plant. Data helps process technicians understand how the processes in a plant are performing. So, taking accurate samples and readings is vital to quality.

Data observed, collected and recorded can describe processes and products, infer what might be happening to them, and predict what can be done to improve them. Part of a process technician's daily tasks include collecting data to monitor, improve, control and correct processes and products.

For example, technicians must track crucial data called process variables, such as temperature, pressure, flow and level. They must also ensure that your data is accurate, thorough and timely.

SPC utilizes specific tools to analyze data. These include:

- Control charts
- Flow charts
- Cause and effects diagrams
- Fishbone
- Pareto charts
- Histograms
- Scatter plots

Control Charts establish limits for the amount of variation in a process. Figure 13-5 shows an example of a control chart.

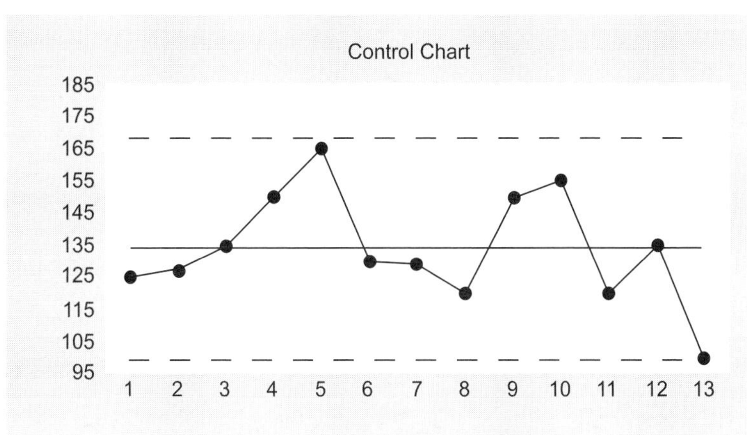

Figure 13-5: Control Chart Example

Flow Charts are RA charts that represent a sequence of operations schematically (or visually).

Figure 13-6 shows an example of a flow chart.

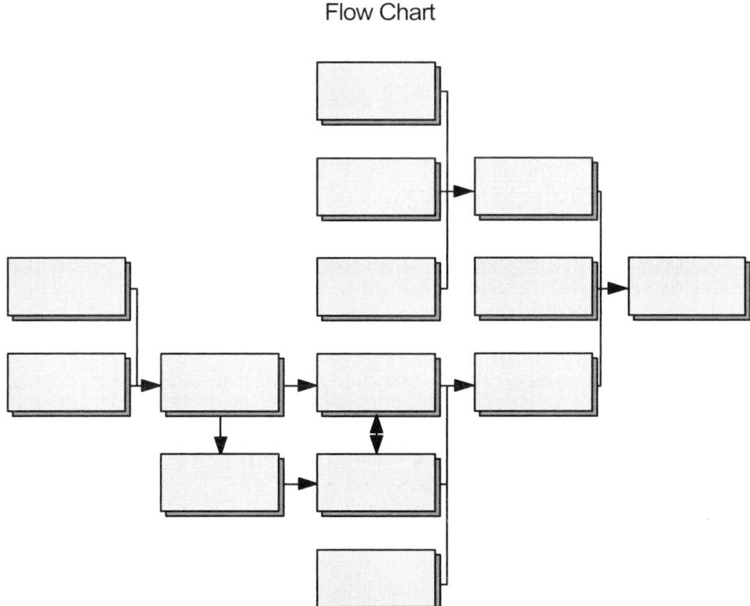

Figure 13-6: Flow Chart Example

Cause and Effects Diagrams are graphics that show the relationship between a cause and effect. On a cause and effect diagram, activities (causes) are connected to attributes (effects) by arrows. Figure 13-7 shows an example of a cause and effect diagram.

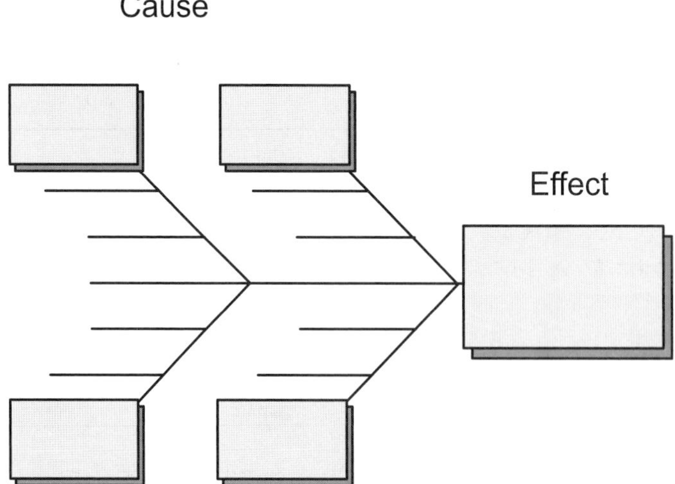

Figure 13-7: Cause and Effect Diagram Example

Fishbone diagrams are cause/effect diagrams (sometimes called an Ishikawa, after Kaoru Ishikawa) and are used to help identify possible causes of a problem. Figure 13-8 shows an example of a Fishbone diagram.

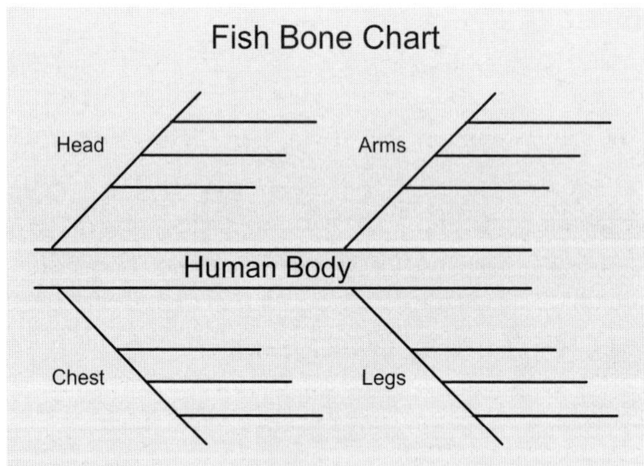

Figure 13-8: Fishbone Diagram Example

Pareto Charts are graphics that rank causes from most significant to least significant; they represent the 80/20 rule described by Juran, stating that most effects comes from relatively few causes. Figure 13-9 shows an example of a Pareto Chart example.

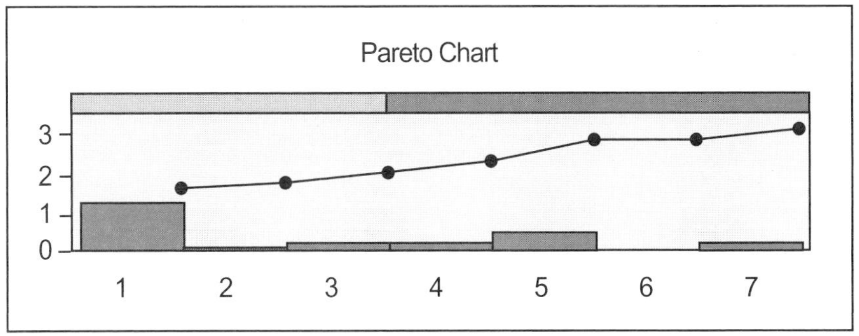

Figure 13-9: Pareto Chart Example

Histograms are bar graphs of a frequency distribution in which the widths of the bars are proportional to the classes into which the variable has been divided and the heights of the bars are proportional to the class frequencies. Figure 13-10 shows an example of a Histogram.

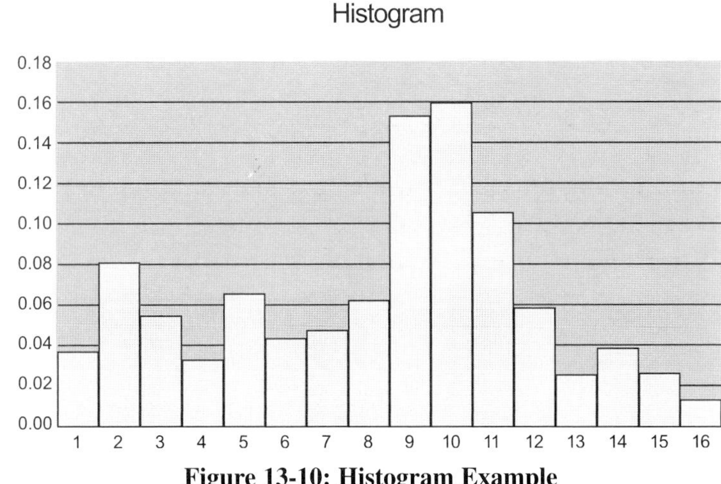

Figure 13-10: Histogram Example

Scatter Plots are graphs drawn using dots or a similar symbol to represent data. Figure 13-11 shows an example of a Scatter Plot.

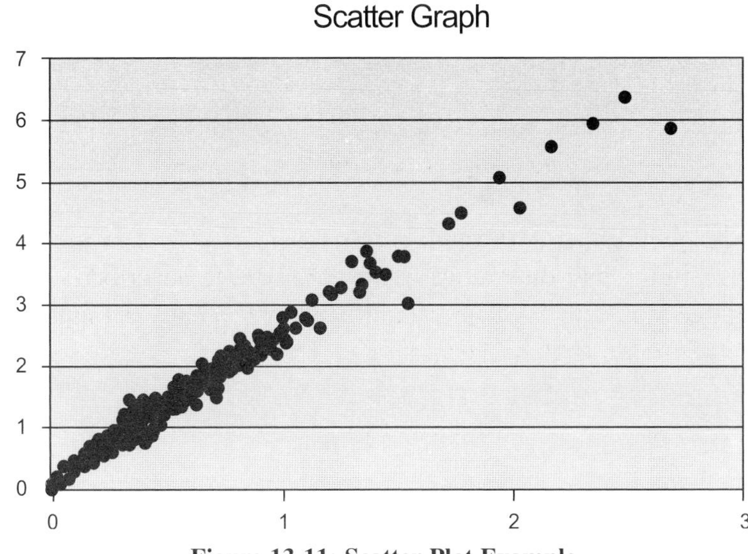

Figure 13-11: Scatter Plot Example

DID YOU KNOW?

Leaders in six sigma efforts are sometimes referred to using karate belt-like designations, such as green belt or black belt.

SIX SIGMA

Compared to other quality initiatives, Six Sigma is a relatively new TQM approach used in the process industries. Based on Juran's work in quality management, Six Sigma is considered advanced quality management using a data driven approach and methodology to eliminate defects. Six Sigma concentrates on measuring product quality and improving process engineering.

Six Sigma was developed in the 1980s using the concept of zero defects promoted by Philip Crosby. Six Sigma aims for a measure of quality that is near perfection. Sigma (Σ), a Greek letter, represents standard deviation (or variation) in a process. Companies utilizing Six Sigma aim to reduce process variation so that no more than 3.4 defects per one million opportunities (DPMO) result (99.99966% with no variation).

The system requires that all processes are inspected, errors (and where they occur) are identified and then corrected, and measures are implemented to control the processes. Along with TQM efforts, Six Sigma focuses on strategy and leadership development.

Six Sigma uses two sub-methodologies for process improvement:

- **DMAIC**—Define, Measure, Analyze, Improve, Control; an improvement system utilized for existing processes that fall below specifications and will be improved incrementally.
- **DMADV**—Define, Measure, Analyze, Design, Verify; an improvement system utilized to develop new processes or products at quality levels that meet Six Sigma. This system can also be used for existing processes that require more than incremental improvement.

SELF-DIRECTED OR SELF-MANAGED WORK TEAMS

Self-directed and self-managed Teams are not quality-specific initiatives, but they can involve such efforts since teams are a key component to various quality systems (like Crosby's Quality Improvement Program).

Companies often refer to the use of such teams as "empowerment." These teams work together to perform a specific function, such as create a product or provide a service, while also managing that work (i.e., performing tasks such as scheduling and recognition). This frees up managers and supervisors from tasks such as directing and controlling, allowing them to take on a facilitation role by coaching, developing and teaching.

Work becomes restructured around the whole process of providing the product or service. Team members are involved in all aspects of the process, from design to development to deployment. This approach integrates the needs of the team members with the work to be done.

There are differences between self-directed teams and self-managed teams. A self-directed team involves a group of people working together to a set of common goals. The team defines these goals, along with determining compensation and discipline for team members. Self-directed teams determine their own future, acting like a profit center within the company. For example, if the team holds down costs, improves quality, and increases profitability, and then the team is rewarded for its efforts.

A self-managed team is different from a self-directed team. It consists of a group of individuals working their own way to achieve a set of common goals defined outside of the team. For example, the team receives goals from management and then handles its own processes, training, scheduling, rewards, and more.

MALCOLM BALDRIDGE NATIONAL QUALITY AWARD

Created by President Ronald Reagan in 1987, the Malcolm Baldridge Quality Improvement Act established an annual U.S. National Quality Award.

Figure 13-12: Malcolm Baldridge
Courtesy of National and Atmospheric Administration/Department of Commerce

Named after the 26th Secretary of Commerce, the award honors Baldridge (who died in 1987) for his managerial excellence, which contributed to long-term improvement in efficiency and effectiveness of government. The award, managed by the National Institute of Standards and Technology (NIST), aims to establish a standard of excellence that can help U.S. organizations achieve world-class quality.

The U.S. President presents this award to manufacturing and service businesses of all sizes (as well as educational and health care organizations), for their achievements in quality and performance.

Baldridge award recipients must be judged outstanding in seven areas:

1. Leadership
2. Strategic planning
3. Customer and market focus
4. Information and analysis
5. Human resources focus,
6. Process management, and
7. Business results

The award focuses on performance excellence for the entire organization, not just for products or services. Organizations applying for the award must undergo an intense self-assessment process, identifying and tracking organization results in customer products and services, financials, human resources and organizational effectiveness.

DID YOU KNOW?

Malcom Baldridge, who died while competing in a rodeo in 1987, believed that quality management was a key to the United State's prosperity and long-term strength.

This award has proven very influential to U.S. businesses, as a model of high standards against which they can evaluate their own organizations. Many companies use the criteria from the seven areas to measure their own efforts, even though they may not apply for the award.

MAINTENANCE PROGRAMS

U.S. plants spend billions in maintenance costs annually. Since the 1980s, maintenance costs have doubled. At least one-third of those costs were wasted due to inefficient maintenance programs (e.g., repairing equipment after it fails instead of preventing the failure).

There are four types of maintenance approaches:

- **Corrective**—waiting for a failure to occur, and then fixing the equipment as quickly as possible to restore production.
- **Preventive**—conducting regular maintenance tasks to keep failures from happening.
- **Predictive**—monitoring the condition of equipment and interpreting the data to identify possible failures and prevent them from occurring.
- **Detective**—checking equipment such as alarms and detectors on a regular basis to make sure they work.

Two practical maintenance systems that can reduce costs and improve quality are Total Productive Maintenance and Predictive/Preventive Maintenance.

TOTAL PRODUCTIVE MAINTENANCE (TPM)

Total Productive Maintenance (TPM) employs a series of methods to ensure every machine in a process is always able to perform so that production is never interrupted. It emphasizes a company-wide equipment management program that involves all levels of staff in various aspects of equipment maintenance.

TPM also utilizes continuous improvement techniques, one of the four components of Total Quality Management.

Small teams perform activities aimed at maximizing equipment effectiveness, such as prioritizing problems, applying problem solving, evaluating processes (in an effort to simplify them), and measuring data. Each team establishes a thorough system of preventive maintenance for the life span of the equipment.

TPM uses a concept of Overall Equipment Effectiveness (OEE), taking into account six big losses:

1. Equipment downtime
2. Engineering adjustment
3. Minor stoppages
4. Unplanned breaks
5. Time spent making non-conforming products
6. Waste

Three elements are evaluated to determine OEE: Availability (Time), Performance (Speed) and Yield (Quality), resulting in the following equation:

OEE = Time x Speed x Quality

The TPM system sets goals for OEE and measures variations. The team then seeks to eliminate problems and improve performance.

PREDICTIVE/PREVENTIVE MAINTENANCE (PPM)

Predictive/Preventive Maintenance (PPM) bases equipment maintenance on conditions, not schedules. Some maintenance programs are time-driven. Maintenance tasks are based on elapsed running hours of equipment and historical/statistical data.

Instead of relying on manufacturer's recommended maintenance schedules or plant-established maintenance schedules, PPM uses actual operating conditions and direct monitoring of the equipment to determine maintenance tasks and frequency.

The philosophy of PPM is to improve productivity and efficiency, optimize plant operations and improve quality, by using operating conditions to predict potential troubles before they occur. Preventive maintenance is then performed on an as-needed basis.

PPM allows plants to:

- Prevent equipment deterioration.
- Reduce potential equipment failures.
- Decrease the number of equipment breakdowns.
- Increase the life span of equipment.

A PPM approach uses direct monitoring of equipment, checking operating conditions (such as heat or vibration), efficiency, and other indicators to predict when failures or loss of efficiency can occur. Data and facts are gathered using direct observation and/or tools such as thermal monitors and vibration sensors.

After monitoring conditions and gathering data, the information is analyzed. Potential problems are detected before they become serious, and appropriate preventive maintenance is planned and scheduled. Preventive maintenance tasks include:

- Inspection
- Lubrication
- Cleaning
- Replacement of worn parts
- Minor adjustments
- Repairs
- Rebuilds

It is the process technician's responsibility to maintain the health of the equipment being used.

THE PROCESS TECHNICIAN AND QUALITY IMPROVEMENT

The company a process technician works for determines the quality terms, methods, processes and tools that will be used as part of the technician's everyday tasks.

A process technician's responsibilities typically include:

- Familiarizing oneself with the quality program.
- Practicing good quality habits, using the company's quality manual as a basis.
- Providing good customer service (even if a technician does not deal with customers directly, there are always "internal" customers to satisfy).
- Gathering data for use with statistical quality control tools (e.g., flowcharts and cause and effect diagrams) and using those tools.
- Following documented procedures, such as operating procedures and work instructions.
- Monitoring and controlling processes and operations.
- Assisting with equipment maintenance tasks such as TPM or PPM.
- Identifying and troubleshooting problems with a goal of continuous improvement.
- Communicating effectively.
- Keeping quality records.
- Working with teams to meet quality goals.
- Participating in quality-oriented training sessions.
- Utilizing skills in time management, organization, planning and prioritization.

6. What do the initials TQM stand for?

7. Which of the following is one of the four major components of TQM?
   a. The Pareto Principle (separating the vital few from the trivial many)
   b. Every task or job is part of a process
   c. ISO Certification
   d. Customer focus

8. Describe the four main components of TQM.

9. What is the purpose of ISO 9000 certification?

10. Define Statistical Process Control.

11. Which of the following tools are used for statistical process control (select all that apply):
    a. Control charts
    b. Cause and effect diagrams
    c. ISO reports
    d. Histograms

12. Which of the following is an aim of Total Productive Maintenance (TPM)?
    a. To maximize equipment effectiveness (overall effectiveness)
    b. To use control charts for quality control
    c. To achieve ISO-9000 certification
    d. To involve only management in TPM

13. PPM involves:
    a. Using SPC for TQM to achieve ISO
    b. Creating a quality tree for all staff members to follow
    c. Promoting it through TQM
    d. Preventing potential equipment failures

ACTIVITIES

1. Write a one-page paper comparing how the following three individuals shaped the quality movement in the process industries. Also, indicate which one you feel had the most significant impact and list why you selected him.
   - Dr. W. Edwards Deming
   - Joseph Juran
   - Philip Crosby

2. Discuss how ISO 9000 and TQM are similar and different.

3. For Statistical Process Control, tell which category (attribute or variable) the following data belongs in:

Data	Attribute	Variable
Pounds per Square Inch (PSI)		
Gallons Per Minute		
Temperature reading		
Bad batches per shift		
Flow through a pipe		
Downtime for equipment		
Days lost to accidents		
Level in a tank		
Amount of electricity used		
Percent of delayed deliveries		

4. Draw examples of the following types of charts and provide descriptions of each:
   a. Control chart
   b. Flow chart
   c. Cause and effect diagram
   d. Fishbone diagram
   e. Pareto chart
   f. Histogram
   g. Scatter plot

5. You are a process technician working for a plant and your boss has asked you to review XYZ process that produces product ABC. Brainstorm and create a list of how a quality initiative can impact the product. Then, outline a brief plan on three tasks that can be done to improve quality for that product.

Chapter 14
Teams

OBJECTIVES

Upon completion of this chapter you will be able to:

1. Describe the differences between work groups and teams.
2. Describe the different types of teams encountered in the process industries.
3. Identify the characteristics of a "High Performance" or an effective team.
4. Define the terms:
   - Synergy
   - Team Dynamics
5. Describe the steps or stages through which a team evolves (forming, storming, norming, performing and adjourning).
6. Identify factors that contribute to the failure of a team including:
   - Failure to achieve the defined outcome
   - Failure as a team to work together and achieve full synergy
7. Define workforce diversity and its impact on workplace relations:
   - In a team environment
   - Work group (coworker)

KEY TERMS

- **Criticism**—a serious examination and judgment of something; criticism can be positive (constructive) or negative (destructive).
- **DESCC Conflict Resolution Model**—a model for resolving conflict, comprised of the following steps: Describe, Express, Specify, Contract and Consequences.
- **Diversity**—the presence of a wide range of variation in qualities or attributes; in the workplace, it can also refer to anti-prejudice training.
- **Dynamics**—describes interpersonal relationships; how workers get along with each other and function together.
- **Ethnocentrism**—belief in the superiority of one's own ethnic group; belief that others should believe and interpret things exactly the way you do.
- **Feedback**—evaluative or corrective information provided to the originating source about a task or a process.
- **Prejudice**—attitude towards a group or its individual member based on stereotyped beliefs.
- **Process**—a method for doing something, which generally involves tasks, steps or operations, which are ordered and/or interdependent.
- **Stereotyping**—Beliefs about individuals or groups based on opinions, habits of thinking, or rumors, which are then generalized to every member of a group.

- **Synergy**—the total effect of a whole is greater than the sum of its individual parts.
- **Task**—a set of actions that accomplish a job.
- **Team**—a small group of people, with complementary skills, committed to a common set of goals and tasks.
- **Work group**—a group of people organized by a logical grouping within a company, with a designated leader that handles routine tasks.

INTRODUCTION

In the process industries, many companies place people into work groups. Work groups are logical organization of workers with similar skills that typically handle routine tasks. A supervisor usually heads up the work group, handling tasks assignments, monitoring worker performance and resolving conflicts.

Teams usually consists of a small number of people. These people are usually picked for the team because they have skills that complement the other team members' skills. In effective teams, each team member is committed to a common purpose and is responsible for specific tasks and projects. In addition, each team member holds the other members mutually accountable for the success of the team.

When working as part of a team, it is important for process technicians to recognize and appreciate others for their contributions to the workplace, while not discounting them because of their differences. It is vital for workers to understand diversity and practice its principles in the workplace.

COMPARING WORK GROUPS AND TEAMS

Many companies organize workers into groups, usually based on a logical grouping within the overall hierarchy (structure) of the company. For example, a company may divide workers into regions, divisions, departments, shifts, units and so on. These are called work groups.

A **work group** is a group of people organized by logical structures within a company, having a designated leader, and performing routine tasks. Work groups may be subdivided into teams. Not all work groups are teams, however.

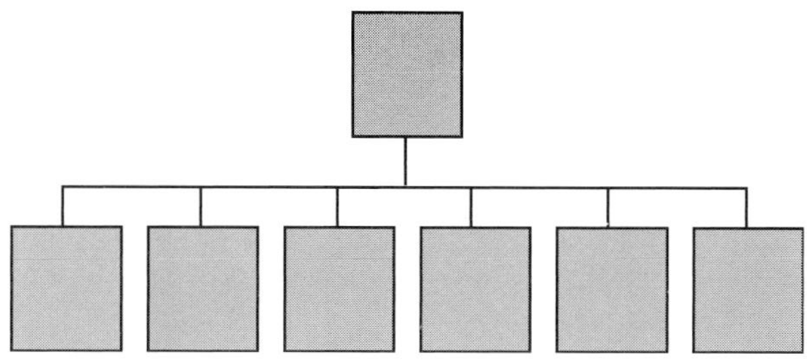

Figure 14-1: Workgroup Concept

A **team** is a small group of people, with complementary skills, committed to a common set of goals and tasks. For example, a process technician, an instrument technician, a mechanical technician, a safety, health and environmental representative, a mechanical services contractor, vendor representatives, and a work scheduler may all be called together to plan a plant turn-around.

Teams are generally formed to handle specific projects or tasks, and operate differently than a work group. Teams can be formed for a limited time to complete a project, or they can be ongoing, without a defined start or stop date, or a specific project. Team leadership is shared. The lead person and team composition may change as a project progresses through its individual phases.

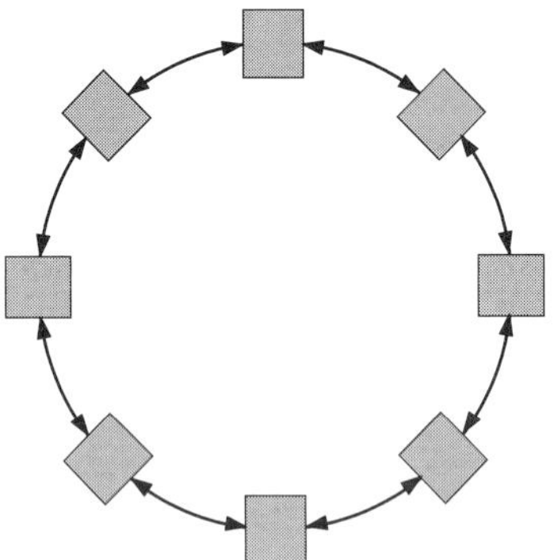

Figure 14-2: Team Concept

The following chart compares the defining characteristics of work groups and teams.

Work Groups...	Teams...
■ One leader	■ Shared leadership
■ Purpose and tasks decided by supervisor	■ Purpose and tasks selected by team leaders
■ Members answer to supervisor	■ Members answer to each other
■ Supervisor resolves conflicts between members	■ Members resolve conflicts

TYPES OF TEAMS IN THE PROCESS INDUSTRIES

Process technicians are often required to work as part of one or more teams. Teams may vary based on factors such as organization, processes, skills, tasks and deadlines. Some common types of teams used in the process industries include:

- Audit
- Commissioning
- Investigation/troubleshooting
- Maintenance
- Operations
- Process or quality improvement
- Safety
- Turn-around
- Cross discipline (e.g. management, operations and maintenance)

CHARACTERISTICS OF HIGH PERFORMANCE TEAMS

In the early 1990s, two researchers showed that high performance teams are effective teams. Jon Katzenbach and Douglas Smith researched teams in high-performing organizations such as Motorola, Hewlett-Packard, the Girl Scouts and the U.S. military (during Operation Desert Storm) to identify characteristics of the highest performing teams.

Their research resulted in valuable findings, and showed that high performance teams were similar in three main areas:

1. **Composition**—how the team chose its members.
2. **Technique**—how the team approached its tasks.
3. **Process**—how the team operated.

Composition

With regard to composition, and how teams chose their members, Katzenbach and Smith learned the following about high performance teams:

- No team started out with all the needed skills. They had to learn them along the way.
- The higher performing teams were made up of less than 10 members.
- Team members possessed a mix of complementary skills in areas such as:
  - Technical or functional expertise.
  - Problem-solving and decision-making skills.
  - Interpersonal skills (relating to others).

Technique

With regard to technique, or how high performing teams approached their tasks, Katzenbach and Smith found that high performance teams:

- Spent time getting consensus (agreement) about team purpose.
- Specified measurable goals, objectives, achievements and deadlines.
- Specified goals that helped the team focus on the task at hand and allowed them to enjoy "small wins" that strengthened commitment and motivation.
- Monitored team progress towards achieving their goals.

DID YOU KNOW?

High performance teams:

- Usually consist of 10 people or less.
- Seldom start out with all of the skills they need.
- Consist of team members with complimentary skills.
- Contain team members that support each other as they work toward a common goal.

Process

Relating to process, or how high performance teams operate, Katzenbach and Smith's research showed that team members:

- Did equivalent amounts of work; there were no "free-riders".
- Were open about individual members' skills, and chose the best fit for the task.
- Responded constructively to views expressed by others.
- Gave others the benefit of the doubt.
- Provided support for, and recognized the interests and achievement of, others.

SYNERGY AND TEAM DYNAMICS

When speaking of teams, it is important to understand the concepts of synergy and dynamics.

Synergy is when the whole is greater than the sum of the individual parts. For example, steel is stronger than the metals that go into it (iron and carbon). In the same way, each member of a team provides talents, perspectives, and skills that make the team stronger than any of its individual members.

Dynamics describes how team members get along with each other and function together (a concept called interpersonal dynamics). As new members are added or old members are removed, team dynamics change. These changes can either be positive or negative.

Both synergy and dynamics are vital to teams. Strong synergy and positive dynamics can result in high performance. Poor synergy and negative dynamics can result in team failure. Process technicians working on teams should be aware of the synergy and dynamics, and do his or her part to improve or maintain positive dynamics.

TEAM TASKS VERSUS PROCESS

When working on tasks as a team, team members must understand the difference between a task and the process. A **task** is a set of actions which accomplish a job by a mutually agreed upon deadline. A **process** is a method for doing something that generally involves tasks, steps, or operations, which are ordered and/or interdependent.

When approaching tasks, teams must:

- Decide how to accomplish the task.
- Describe the process required to accomplish the task, to ensure that the team can work effectively.

In order to accomplish a task, a typical process might be to:

- Decide what exactly is to be accomplished.
- Come up with ideas on how to accomplish the task.
- Select the best idea.
- Achieve mutual buy-in.
- Divide up the responsibilities.
- Set deadlines.
- Track progress toward achieving the goal.
- Support fellow team members and their efforts.
- Share in the rewards and recognize individual and team accomplishments.

Although this process sounds simple, a team can fail if it does not properly address the task and its related process.

Ineffective teams often have problems with their members and do not address these problems through their processes. Highly effective teams rely on their processes to work through people problems while still achieving tasks and meeting objectives.

STAGES OF TEAM DEVELOPMENT

Teams go through different stages of development, encountering various issues along the way. Five general stages (forming, storming, norming, performing, and adjourning) have been identified, with each stage typically representing characteristic issues. For example, a period of conflict is not unusual after the initial formation of the team, or at the midpoint of the project.

Recognizing what stage your team is in, and understanding what issues might surface in that stage, can help you contribute to the team and improve its processes.

DID YOU KNOW?

Teams typically go through five stages of development:

1. **Forming:** Initial team development
2. **Storming:** Team conflict resolution
3. **Norming:** Development of work standards
4. **Performing:** Getting the job done
5. **Adjourning:** Preparing to disband the team

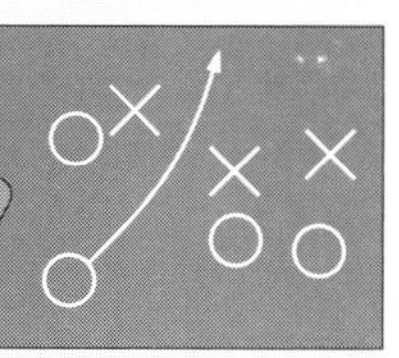

Stage 1: Forming

The fundamental issue during the forming stage is the development of team trust. Characteristics of this stage include:

- Tentative interactions or guarded discussion.
- Careful behavior; trying not to offend anyone.
- Mild tension; uncomfortable feelings or polite discourse.
- Concern over ambiguity and what roles the members will play.
- Concern over how a particular member will be accepted by the group.

Stage 2: Storming

The fundamental issue during the storming stage is the resolution of team conflicts. Characteristics of this stage include:

- Individual actions are resisted by, or incompatible with, other members.
- Disagreements are more frequent.
- Hostility.
- Conflicts with roles and procedures.

Stage 3: Norming

The fundamental issue during the norming stage is the development of teamwork standards. Characteristics of this stage include:

- Emerging sense of group unity and positive relationships among members.
- Development of procedures, group norms and roles.
- Lower levels of anxiety.

DID YOU KNOW?

There are many factors that can cause a team to fail. Some of the most common contributors include:

- Inappropriate tasks
- Lack of clear purpose
- Having the wrong members
- Poor team dynamics

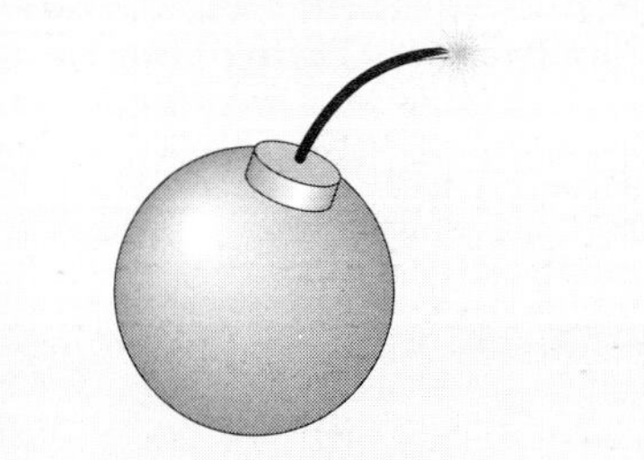

Stage 4: Performing

The fundamental issue during the performing stage is the team getting the job done properly. Characteristics of this stage include:

- Good, strong decision-making.
- Creative problem solving.
- Mutual cooperation and buy-in.
- Strong feelings of commitment to the team's success.

Stage 5: Adjourning

Adjourning occurs when the project is either successfully completed, deferred, or cancelled. The fundamental issue during the adjourning stage is how the team is dealing with its impending breakup. Characteristics of this stage include:

- Growing feelings of independence.
- Increased anxiety.
- Regret.
- Blame.
- Sorrow.
- Withdrawing emotionally from other members.

FACTORS THAT CAUSE TEAMS TO FAIL

For a variety of reasons, teams can fail to achieve the desired outcome. For example, members may fail to work together and achieve synergy, or it could be that the team was not properly constructed to begin with (i.e., the team member were not the "right fit" for the job).

When a team fails, it is often a highly visible failure, especially if the team was tasked with a special project. It is critical that process technicians understand what causes teams to fail and watch for warning signs. If identified early and dealt with properly, potential problems can be resolved and the team can stay on track.

The following are some factors that can contribute to the failure of a team:

- **Tackling tasks that are not appropriate for teams**

  Example:
  - Situations that require fast decisions.
  - Tasks that require a higher degree of skill than what is present on the team.
  - Painful or difficult decisions, such as deciding who gets laid off when times are bad.
- **Lacking a clear purpose**

  Example:
  - Faltering without a clear direction and mission.
  - Conflicts over purpose and roles.
  - Lack of individual buy-in.
  - Taking too long to figure out a purpose.
- **Having the wrong members**

  Example:
  - Team members are assigned who are not the right fit for the team.
  - Having members who appear "right" for the team but turn out to not be.
  - Not having processes in place to rectify (or correct) having the wrong members
- **Poor team dynamics**

  Example:
  - Paying attention to the wrong issues.
  - Focusing too closely on the tasks and not the team dynamics.
  - Lacking processes to handle poor team dynamics.

RESOLVING CONFLICT

Conflict can be a disruptive force, or a driving force, depending on how a team handles it. If the team handles conflict poorly, team dynamics and performance can suffer. If the team can resolve the conflict and focus it into a driving force, it can lead to improved team performance. The spirit of teamwork should be "How can the issue be resolved, the situation made better, the end goal brought back into focus?"

Some conflicts occur at fairly predictable times when working as a team. When conflict arises, it must be dealt with swiftly and effectively. One tool that is ideal for conflict resolution is the **DESCC Conflict Resolution Model**. DESCC stands for:

- Describe
- Express
- Specify
- Contract
- Consequences

The following sections describe each aspect of DESCC.

Describe Phase

In the describe phase, team members should:

- Ask "What is not happening that should be?"
- Ask "What is happening that shouldn't be?"
- Describe the situation in specific, behavioral terms (i.e., what someone can see or hear).
- Avoid vague descriptions such as "bad communication," "lack of commitment," and "bad attitude."
- Pay attention to the response; LISTEN.

Express Phase

In the express phase, team members should:

- Express how the situation affects you.
- Ask "How does it make you feel?"
- Pay attention to the response; LISTEN.

Specify Phase

In the specify phase, team members should:

- Identify what must happen for them to be satisfied?
- Identify what the improved situation should look like?
- Be very specific and use behavioral terms.
- Pay attention to the response; LISTEN.

Contract Phase

In the contract phase, team members should:

- Negotiate an agreement as to what will change.
- Be specific in the agreement terms.
- Pay attention to the response; LISTEN.
- Achieve individual member buy-in and sign-off.

Consequences Phase

In the consequences phase, team members should:

- Explain the anticipated outcome if the changes are made.
- Explain the anticipated outcome if the changes are not made.
- Initiate consequences only when people have not lived up to the contract.
- Pay attention to the response; LISTEN.

GIVING FEEDBACK

Teams not only have to work on tasks and projects, but they must also deal with work **processes** (how team members work together to get the job done). One key issue related to work processes is feedback. When working with team members, vendors, clients, and other individuals, process technicians must learn how to give and receive feedback.

Often, people equate feedback with criticism. Feedback and criticism are actually quite different. **Feedback** is evaluative or corrective information provided to the originating source about a task or a process. Feedback is intended to be helpful and redirecting. **Criticism** is a serious examination and judgment of something. While criticism can be positive (constructive) or negative (destructive), it is often meant to be negative, hurtful, or punishing.

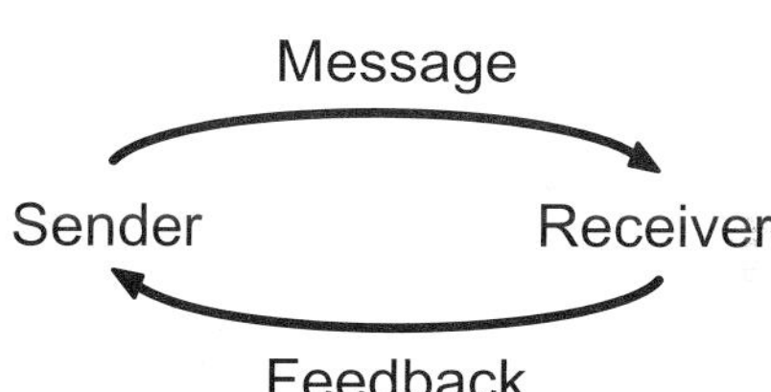

Figure 14-3: Communication Model with Feedback

To better illustrate the concept of constructive feedback, think about driving a car. When driving, the car is seldom pointed directly at the destination. Instead, we start in a general direction and then make corrections along the way until our destination is reached. Teamwork operates under very similar principles.

In teamwork, the team starts out pointed in the general direction of the goal. As the project progresses, each team member must make small corrections in their individual processes to keep the team headed in the right direction. These corrections come from feedback.

Feedback can become emotionally loaded when conflict arises, and there is always some conflict in any team project. The following tips are intended to help process technicians and other team members give feedback that is objective, fact-based, and less likely to create or increase conflict.

When giving feedback:	Example:
Speak only for yourself; only use "I" messages.	"I don't like it when you send large print documents to the shared printer."
Critique the problem, not the person.	"When large documents are sent to the shared printer, it ties the unit up so no one else can use it."
Let the other person respond; do not interrupt.	Remember, you would not want to be interrupted when you are talking.
Speak from your perspective.	"I think it would be better if you sent large print jobs to the print shop downstairs."
Offer alternatives.	"Hey, that's a great idea. Have you also thought about..."

When receiving feedback:	Example:
Listen carefully. Do not just wait for your turn to speak.	Concentrate on what the other person is saying.
Be patient.	Relax and stay focused.
Consider the other person's viewpoint.	Apply the old adage "Walk a mile in another person's shoes".
Restate what the other person said, to make sure you understand it.	"So what you're saying is you don't like it when I send large print jobs to the shared printer."
Discuss ways to provide a positive resolution.	"The print shop downstairs costs too much. How about if I send it to Susan's printer instead?"
Never take feedback personally; focus on the issue to be resolved.	Check your ego. Don't let emotions get in the way.
Remember, feedback is good for the team and for you.	If someone cares enough to provide feedback, you should care enough to listen.

WORKFORCE DIVERSITY

Diversity is the presence of a wide range of variation in qualities or attributes. In today's work environment, the term diversity is generically used to refer to the ways that we are different and unique from one another, and as a shorthand reference for anti-prejudice training.

Diversity goes beyond this though. It also involves considering individual backgrounds, personality differences, learning styles, approaches to a task and group dynamics.

When dealing with teams, work groups, or other people on the job, a firm understanding and acceptance of diversity is crucial.

From the Melting Pot to the Salad Bowl

At one time, the ideal vision of the U.S. was a melting pot, where immigrants were expected to drop their language, customs, beliefs, traditional clothes, and other aspects of their culture in order to blend in with other Americans and become virtually "indistinguishable."

However, organizations and businesses realize that people are most productive and creative when they feel valued and believe that their individual and group differences are accepted and taken into account. This new way of thinking is often referred to as the "Salad Bowl" concept.

The "Salad Bowl" concept uses a vision of a mixed-up bowl of colorful ingredients and flavors that complement each other well, yet each retains its own colors, textures and flavors. In other words, the individuality of each component is respected and accepted for its own unique characteristics.

In the workplace, the "Salad Bowl" approach means that individual workers must be accepting of different cultures, beliefs, races and religions. In other words, they must be accepting of diversity. However, striving for a common language is important to clear communications. It is important that everyone speaks and understands a "common language". Clearly spelled out terms and definitions helps. Another tool is to use paraphrase what the other person said: "So, what you're saying is . . ."

DIDYOU KNOW?

Diversity is an important part of the workplace. In order for employees to feel satisfied and productive, diversity must be accepted and respected.

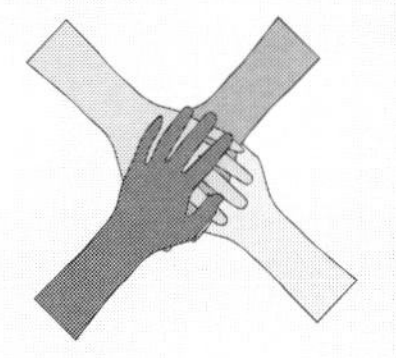

Negative beliefs and behaviors (e.g., stereotyping, ethnocentrism, and prejudice) cause a person to feel devalued and limit a person's opportunities for success.

Terms Associated with Diversity

Diversity is about allowing people to do their jobs without having to leave parts of themselves out of the workplace. This means being comfortable around people who are different from you and may not necessarily want to be like you. However, some individuals are not comfortable and they exhibit negative behaviors such as stereotyping, ethnocentrism, and prejudice.

The following are descriptions of each of these key terms:

- **Stereotyping**—beliefs about individuals or groups based on opinions, habits of thinking, or rumors, that are then generalized to every member of a group.
  - Many of our beliefs about other groups come from stereotypes.
  - This way of thinking assumes that all members of one group have similar qualities.
  - Some stereotypes are errors that are repeated until they seem true.
  - Stereotyping greatly limits a person's opportunities.

- **Ethnocentrism**—when a person believes that other groups and individuals should think and interpret things exactly the way they do.
    - Ethnocentrism generally decreases as individuals spend time outside their own culture.
    - People who are ethnocentric rarely recognize it in themselves.
    - A major cause of ethnocentric misunderstandings is lack of communication.
- **Prejudice**—an attitude towards a group or its individual members based on stereotyped beliefs.
    - Prejudice can result in selective perception (i.e. noticing only the bad qualities of a group, while not noticing the same qualities in one's own group).
    - Another result of prejudice is a heightened sensitivity to "trigger" behaviors or words uttered by one group about another (e.g., a male manager calling professional women "girls"). Even if the communicator intends no harm, trigger words generate hostility. Uninformed or insensitive people often mean no harm and feel hurt or puzzled by the negative reaction they receive.
    - Prejudice can cause breakdowns in communication.

RESPECTING DIVERSITY

In any workplace, there will be others who are different from you. Be open-minded and sensitive to other views, opinions, and approaches. Look for common ground on projects and tasks. Seek out "win-win" solutions and approaches for you and your co-workers no matter how different they are.

Your company's management will not only expect you to get along with co-workers who are different from you, but to be successful when working with them as part of a work group or a team.

If you are part of a work group whose people have difficulties dealing with diversity, the problem can compound itself since you are forced to work together almost every day. Although the situation probably has less pressure than a team environment, it can still result in conflicts that impact performance, job satisfaction, and stress.

If you are part of a team whose members have difficulties dealing with diversity, the problem can increase to a "boiling point" since teams are often faced with intense pressure and tight deadlines.

In either situation, it is critical that everyone in the work environment understands and appreciates diversity. Diversity goes beyond obvious issues such as gender, race and religion; it involves all of the ways that we are unique as individuals.

SUMMARY

A **work group** is a group of people organized by a logical grouping within a company, with a designated leader that handles routine tasks. A **team** is a small group of people, with complementary skills, committed to a common set of goals and tasks. Not all work groups are teams.

The process industries contain many different types of teams (e.g., audit, maintenance, safety, and operations). Each of these teams is created with a specific task or function in mind.

When building teams, it is important to know the characteristics of high performance teams. Namely, that high performance teams usually consist of 10 people or less, seldom start out with all of the skills they need, consist of team members with complimentary skills, and contain team members that support each other as they work toward a common goal.

Two other key concepts that pertain to teams are synergy and dynamics. **Synergy** is when the whole is greater than the sum of the individual parts. Synergy occurs when team members combine their individual expertise, talents and contributions for a greater outcome than is individually possible. **Dynamics** describes how team members get along with each other and function together. Both synergy and dynamics are vital to teams.

As teams grow and develop, they go through several stages (forming, storming, norming, performing, and adjourning). Each of these stages has its own unique characteristics.

There are many factors that can cause a team to fail. These include tackling tasks that are not appropriate for teams, lacking a clear purpose, having the wrong members, and poor team dynamics.

Regardless of the team and its purpose, all employees should understand and respect team member diversity and differences, and avoid negative and limiting behaviors such as stereotyping, ethnocentrism, and prejudice.

CHECKING YOUR KNOWLEDGE

1. Define the following key terms:
   a. Diversity
   b. Ethnocentrism
   c. Feedback
   d. Prejudice
   e. Stereotyping
   f. Team
   g. Work group

2. Which of the following is a characteristic of a team?
   a. Members answer to each other
   b. Has one leader
   c. Supervisor resolves conflict between members
   d. Handles only safety audits

3. During studies of highly effective teams, researchers found:
   a. There was no synergy
   b. Team members argued often
   c. Team members responded constructively to the views of others
   d. There were "free riders"

4. ___________ is when the total effect of a whole is greater than the sum of its individual parts.

5. Which of the following is the best definition of team dynamics:
   a. How energetic the team members are
   b. How the teams function interpersonally
   c. How the teams cheer each other on
   d. How the team responds to criticism

6. Which of the following is NOT a stage of team development?
   a. Forming
   b. Storming
   c. Quoruming
   d. Norming

7. List three reasons why teams fail.

8. List and describe the components of the DESCC conflict resolution model?
   a. D ______________
   b. E ______________
   c. S ______________
   d. C ______________
   e. C ______________

9. Diversity is best described as:
   a. Ethnocentrism
   b. Selective perception
   c. Stereotyping
   d. Ways that we are different and unique from each other

10. The ________ ________ approach states that the U.S. should be like a mixed-up container of colorful ingredients and flavors that complement each other well, while each retains its own colors, textures and flavors.

Match the term to its description

11. Stereotyping	a. A belief that other groups think and interpret things exactly the way you do.
12. Ethnocentrism	b. An attitude towards a group or its individual members.
13. Prejudice	c. America has changed from "melting pot" to "salad bowl."
14. Diversity	d. All members of a particular group are perceived as having the same qualities.

ACTIVITIES

1. Think about a team you have been a part of. How and why was the team formed? What were its good qualities? What could have been improved? Did the team meet its goals? Why (or why not)?

2. List the types of teams that you might be a member of in a process industry company. List the skill sets each team needs, and discuss possible goals the team might have.

3. Think of a team that you consider high performance. List 10 characteristics that make that team high performance. Explain your choices.

4. Discuss some real world examples of synergy and dynamics (you can draw on your experience, sports teams, business, or other organizations).

5. Get together with a classmate and present a demonstration of how to perform a particular task or skill (approximately 5 minutes long). Then, have your classmate provide feedback on your presentation. Next, switch roles. Finally, try your presentation again, using the feedback from your classmate.

6. With your classmates, discuss all the ways you are similar. Then, describe ways that you are different. How does it make you feel when you have things in common with your classmates? How does it make you feel when you are different?

Chapter 15
Process Drawings

OBJECTIVES

Upon completion of this chapter you will be able to:

1. Describe the purpose or function of process systems drawings.
2. Identify the common components and information within process systems drawings.
3. Identify the different drawing types and their uses:
   - Block Flow Diagrams
   - Process Flow Diagrams (PFD)
   - Piping and Instrument Diagrams (P&ID)
   - Utility Flow Diagrams (UFD)
   - Electrical Diagrams
   - Wiring Diagrams
   - Schematics
   - Isometrics
4. Identify the different components and their symbols in each type of drawing:
   - Block Flow Diagrams
   - Process Flow Diagrams (PFD)
   - Piping and Instrument Diagrams (P&ID)
   - Utility Flow Diagrams (UFD)
   - Electrical Diagrams
   - Isometrics

KEY TERMS

- **Block Flow Diagram (BFD)**—a very simple drawing that shows a general overview of a process, indicating the parts of a process and their relationships.
- **Electrical Diagram**—a drawing that show electrical components and their relationships.
- **Elevation Diagram**—a drawing that represents the relationship of equipment to ground level and other structures.
- **Equipment Location Diagram**—a drawing that shows the relationship of units and equipment to a facility's boundaries.
- **Isometric**—a drawing that shows objects as they would be seen by the viewer (like a 3-D drawing, the object has depth).
- **Loop Diagram**—a drawing that shows all components and connections between instrumentation and a control room.

- **Piping & Instrument Diagram (P&ID)**—also called a Process & Instrument Drawing. A drawing that shows the equipment, piping and instrumentation of a process in the facility, along with more complex details than a Process Flow Diagram (see next definition).
- **Process Flow Diagram (PFD)**—a basic drawing that shows the primary flow of product through a process, using equipment, piping and flow direction arrows.
- **Schematic**—a drawing that shows the direction of current flow in a circuit, typically beginning at the power source.
- **Symbol**—figures used to designate types of equipment.
- **Utilities Flow Diagram (UFD)**—a drawing that shows the piping and instrumentation for the utilities in a process.
- **Wiring Diagram**—a drawing that shows electrical components in their relative position in the circuit and all connections in between.

INTRODUCTION

Process drawings are as critical to a process technician as a topographical map (a map that shows features such as hills, streams and trails) to a hiker in the deep woods. Just as the hiker must be able to read a map, a process technician must be able to read process systems drawings to understand what is happening at a process facility.

Process facilities use process drawings to assist with operations, modifications and maintenance. These drawings visually explain the components of the facility and how they interrelate.

TYPES OF PROCESS DRAWINGS

Process technicians are exposed to different types of industrial drawings on the job. The two most common types of drawings are Process Flow Diagrams and Piping & Instrument Diagrams.

Process Flow Diagrams (PFDs) are basic drawings that use symbols and direction arrows to show the primary flow of a product through a process. **Piping & Instrument Diagrams (P&IDs)** are more detailed drawings that graphically represent the equipment, piping and instrumentation contained within a process in the facility.

Figures 15-1 and 15-2 show examples of PFDs and P&IDs.

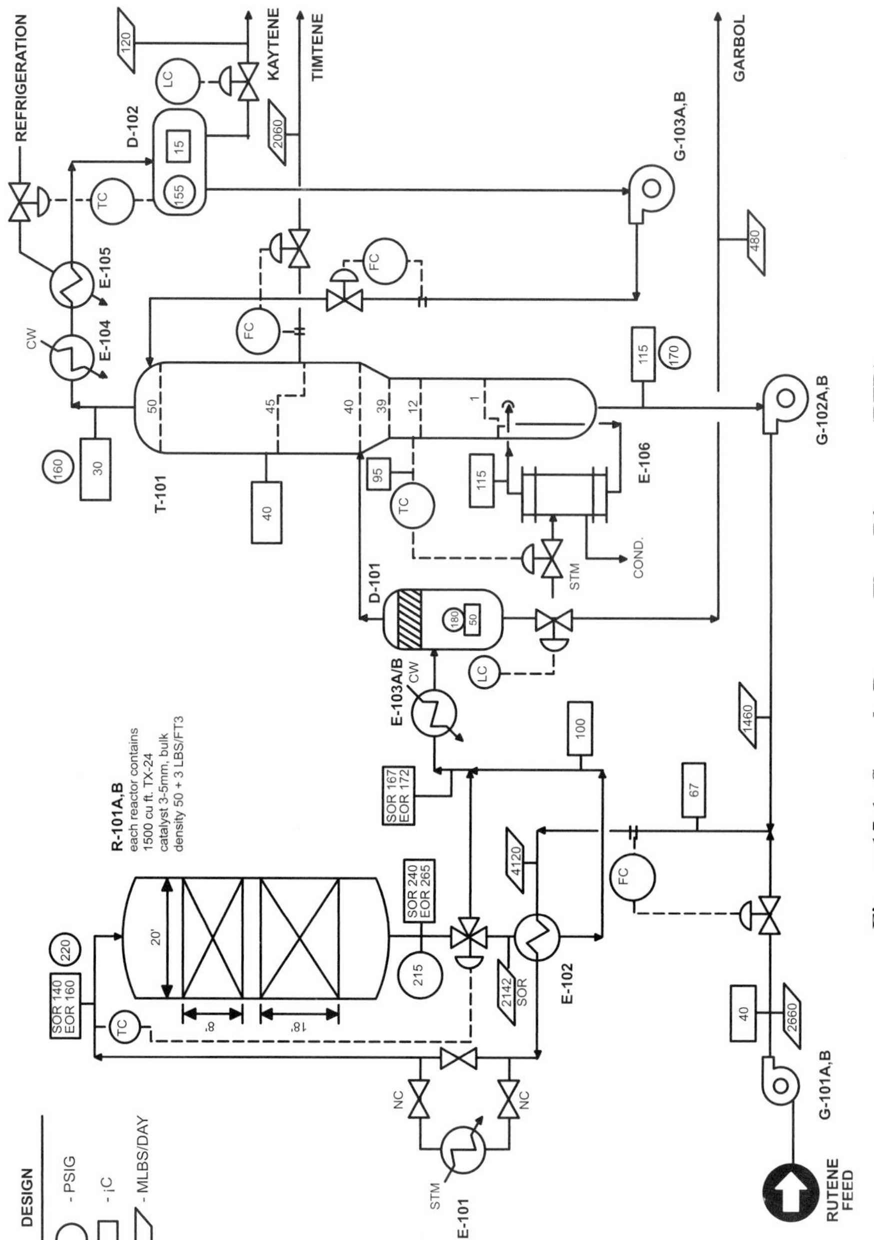

Figure 15-1: Sample Process Flow Diagram (PFD)

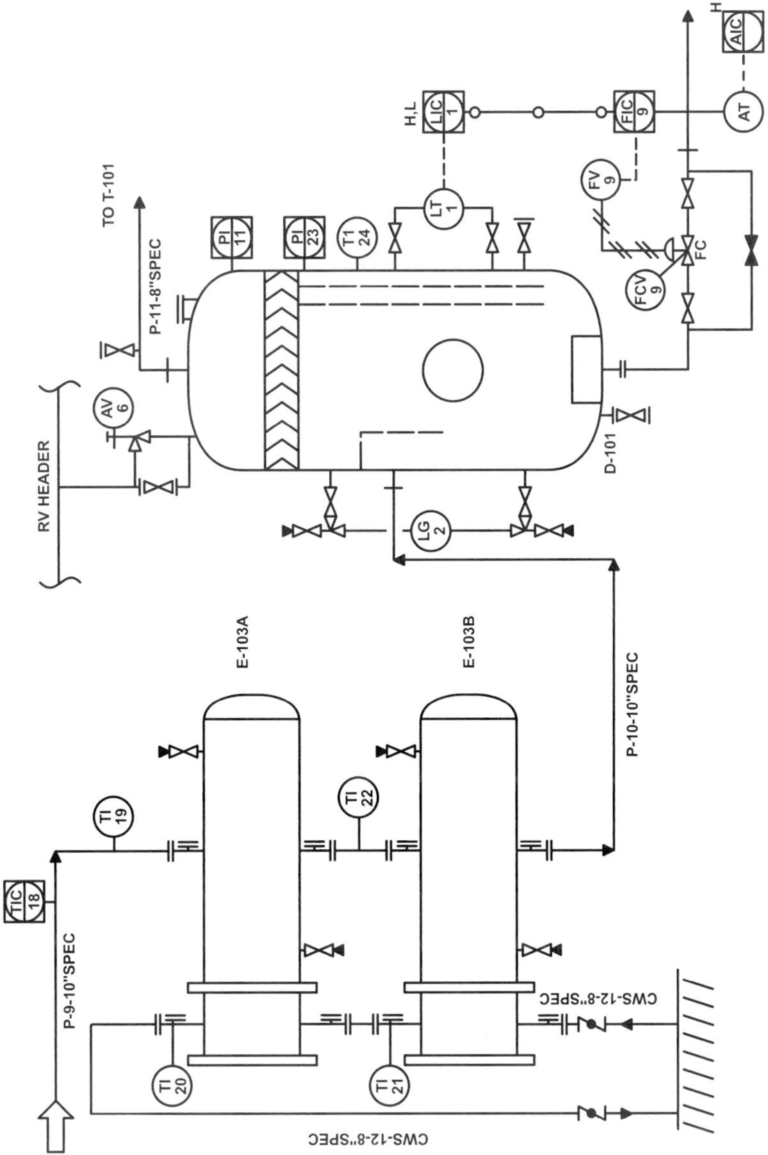

Figure 15-2: Sample Piping & Instrument Diagram (P&ID)

Other types of drawings technicians use include:

- **Block Flow Diagrams (BFDs)**—very simple drawings that show a general overview of a process, the components within the process, and their relationships.
- **Utility Flow Diagrams (UFD)**—drawings that show the piping and instrumentation for the utilities in a process.
- **Electrical Diagrams**—drawings that represent how electrical components and devices are connected.
- **Isometrics**—drawings that show objects as they would be seen by the viewer (i.e., the object has depth and dimension like a 3-D drawing). Isometrics are drawn on graph paper and show compass points and line relationships to the equipment.
- **Other Types**—other drawing types include 3-D, CAD or simulations (realistic three dimensional representations), cutaways (showing internals) and so on.

Figures 15-3 to 15-6 show examples of BFDs, UFDs, electrical diagrams, and isometric drawings.

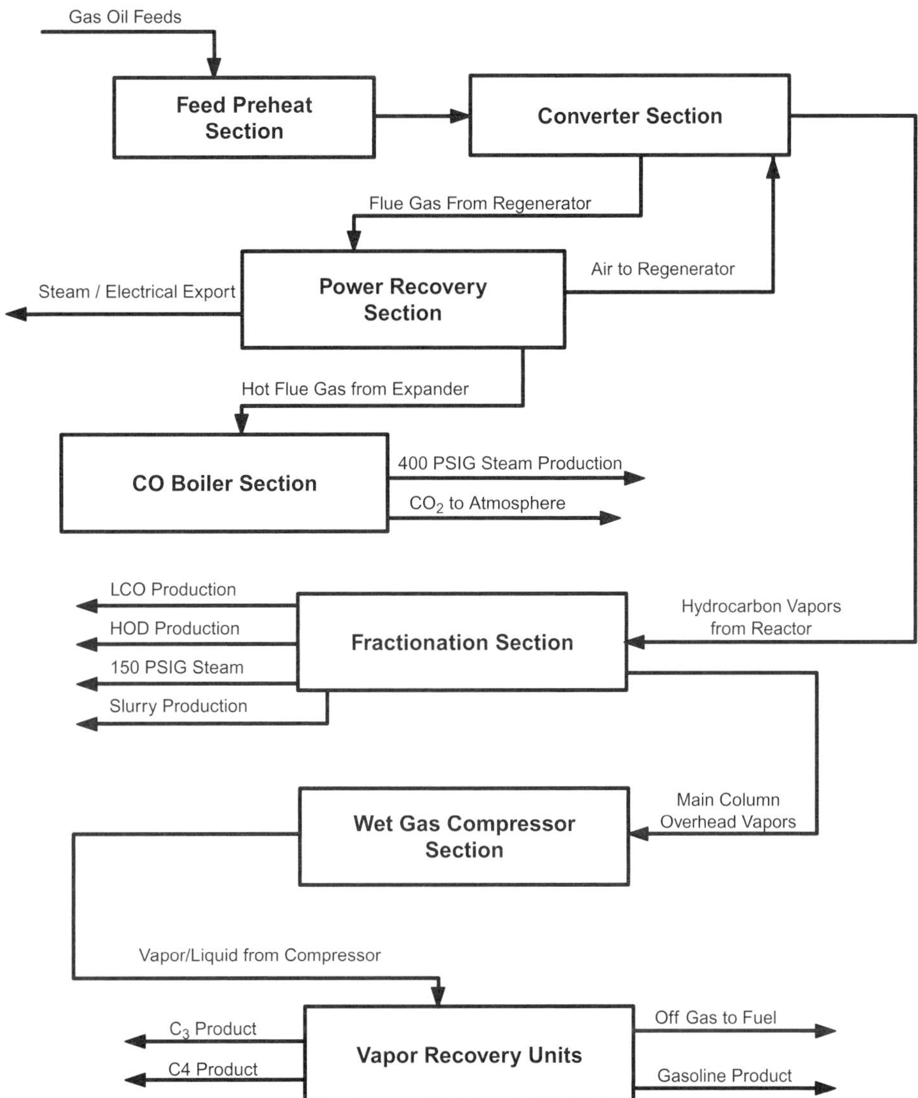

Figure 15-3: Block Flow Diagram (BFD)

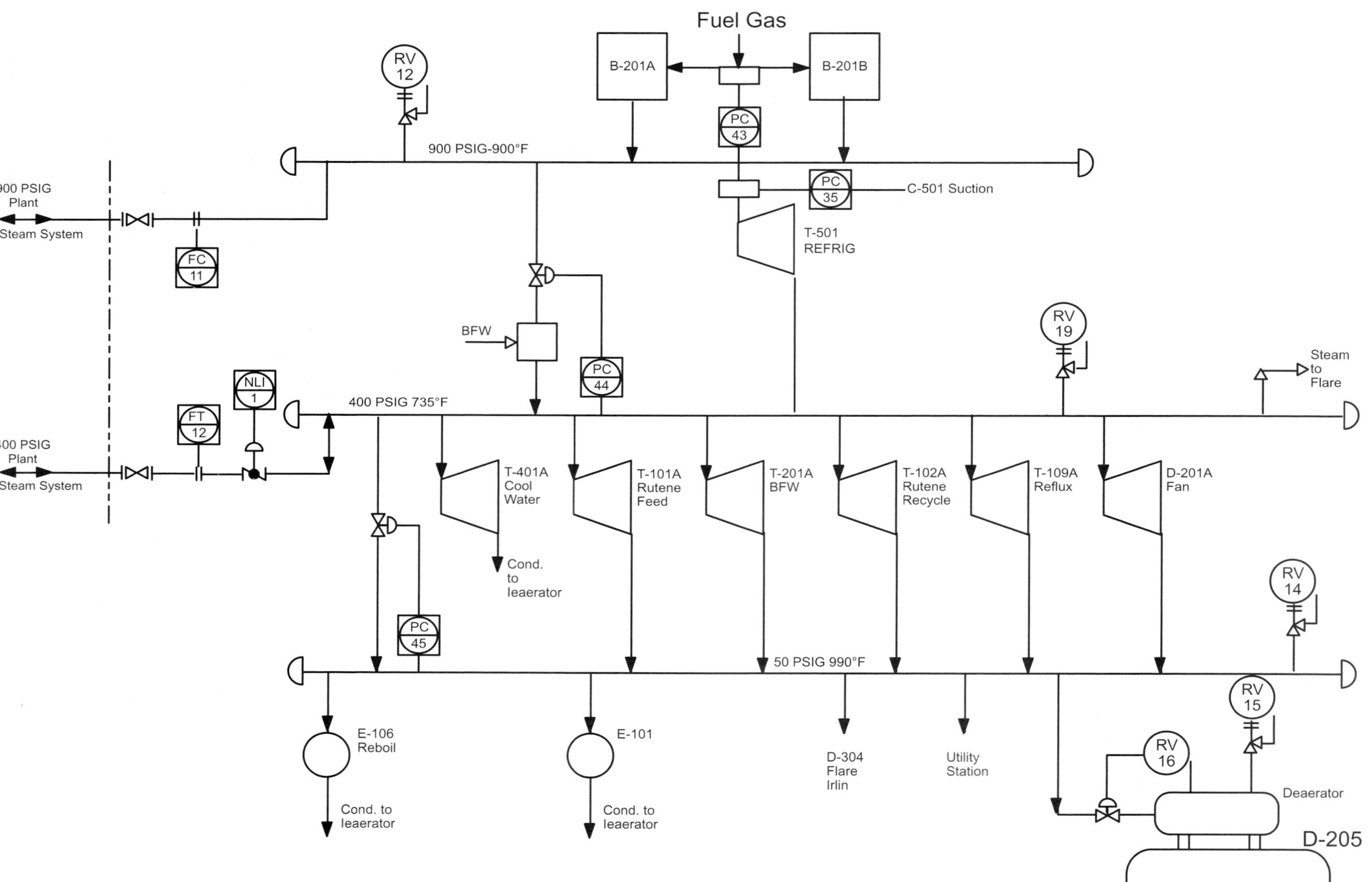

Figure 15-4: Utility Flow Diagram (UFD)

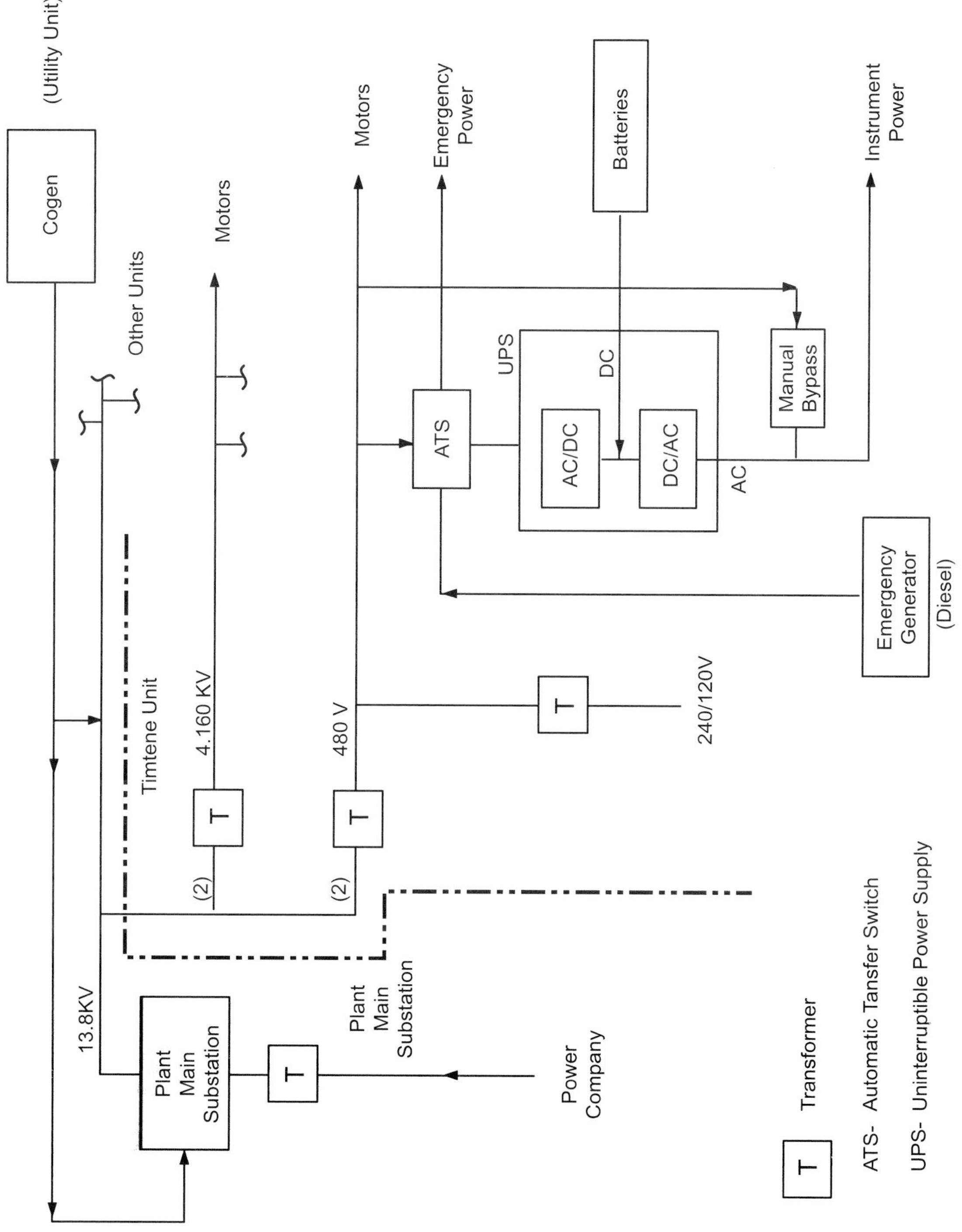

Cogen
(Utility Unit)
Other Units
Motors
Motors
Emergency Power
Batteries
Instrument Power
UPS
DC
AC/DC
DC/AC
AC
Manual Bypass
ATS
Emergency Generator (Diesel)
Timtene Unit
4.160 KV
480 V
240/120V
T
(2)
13.8KV
Plant Main Substation
Plant Main Substation
Power Company
T Transformer
ATS- Automatic Tansfer Switch
UPS- Uninterruptible Power Supply

Figure 15-5: Electrical Diagram

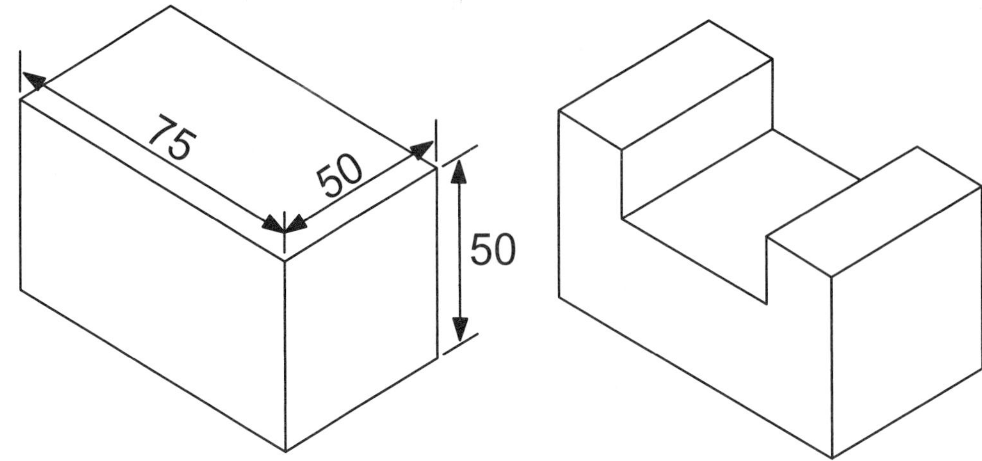

Figure 15-6: Isometric Drawing

PURPOSE OF PROCESS DRAWINGS

Process drawings provide a process technician with a visual description and explanation of the processes, equipment, and other important items in a facility.

As you learned in the last section, there are many different types of drawings. Each drawing type represents different aspects of the process and levels of detail. Looking at combinations of these drawings provides a more complete picture of the processes at a facility.

Without process drawings, it can be difficult for process technicians to understand and operate the processes, and even more difficult to make repairs.

So, the main purposes of process drawings are to allow a process technician to:

- Become familiar with the process in a safe environment.
- Understand or explain a process.
- Understand relationships and repair equipment.

All drawings have three things in common. Namely, they are intended to:

- **Simplify**—using common symbols to make complicated processes easy to understand.
- **Explain**—describes how all of the parts or components of a system work together; drawings can quickly and clearly show the details of a system that might otherwise take many written pages to explain.
- **Standardize**—uses a common set of lines and symbols to represent components; if a process technician knows these symbols, he or she can interpret drawings at any facility.

COMMON COMPONENTS AND INFORMATION WITHIN PROCESS DRAWINGS

Process drawings must meet several requirements to be considered a proper industrial drawing. These requirements include specific, universal rules on:

- How lines are drawn.
- How proportions are used.
- What measurements are used.
- What components are included.
- What is the drawing's industrial application.

All process drawings have common components that contain a great deal of useful information. These components include symbols, a legend, a title block, and an application block.

Figure 15-7 shows each of these drawing components and where they are located.

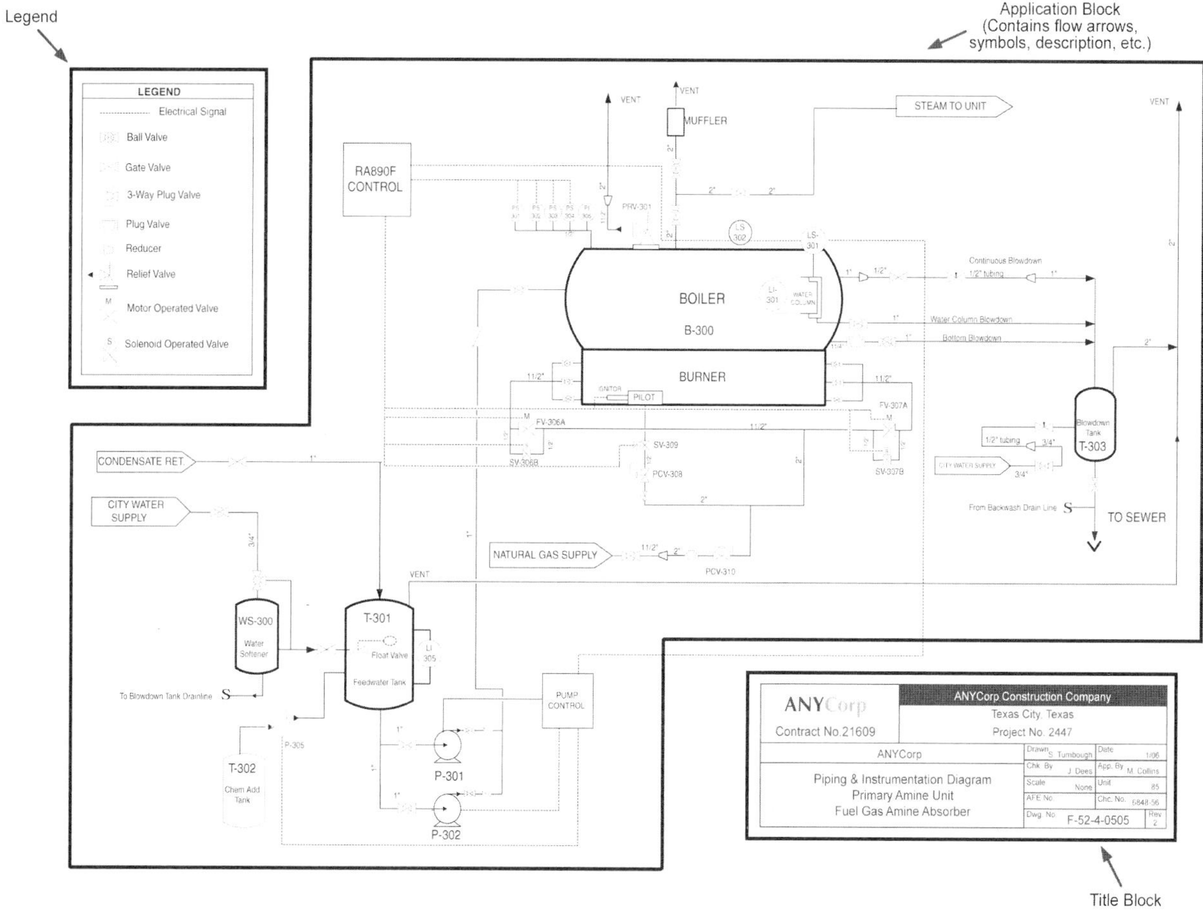

Figure 15-7: Process drawing with components labeled

SYMBOLS

Symbols are figures used to designate types of equipment. A set of common symbols has been developed to represent actual equipment, piping, instrumentation and other components. While some symbols may differ from plant to plant, many are universal with only subtle differences.

Figure 15-8 shows a few common equipment symbols.

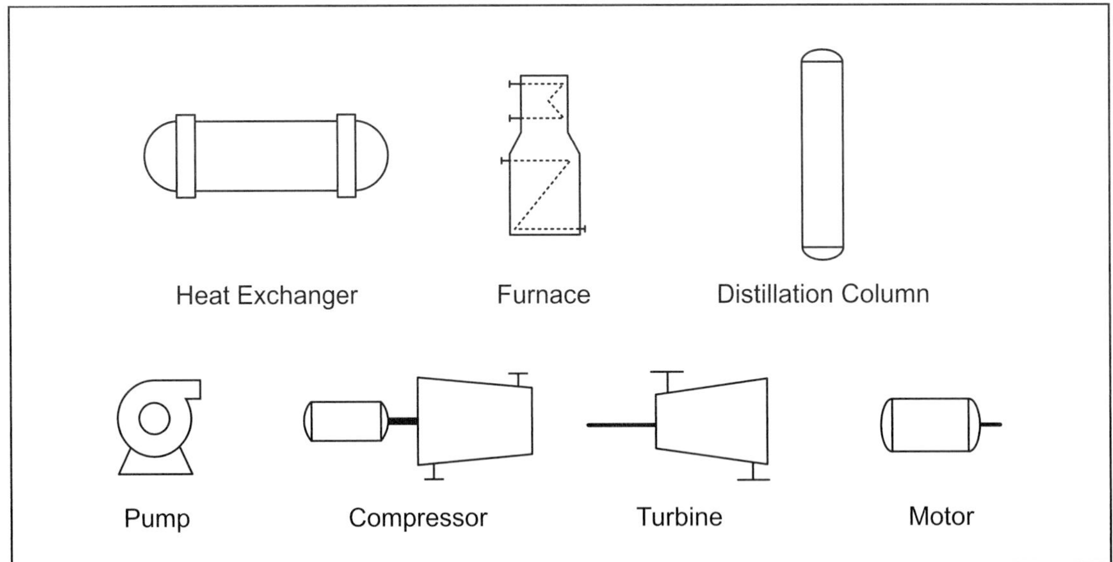

Figure 15-8: Examples of Common Symbols

It is critical that process technicians recognize and understand these symbols. Each equipment chapter in this textbook includes the symbols that represent the equipment on a drawing.

Legend

A **legend** is a section of a drawing that explains or defines the information or symbols contained within the drawing (like a legend on a map). Legends include information such as abbreviations, numbers, symbols, and tolerances.

Figure 15-9 shows an example of a legend.

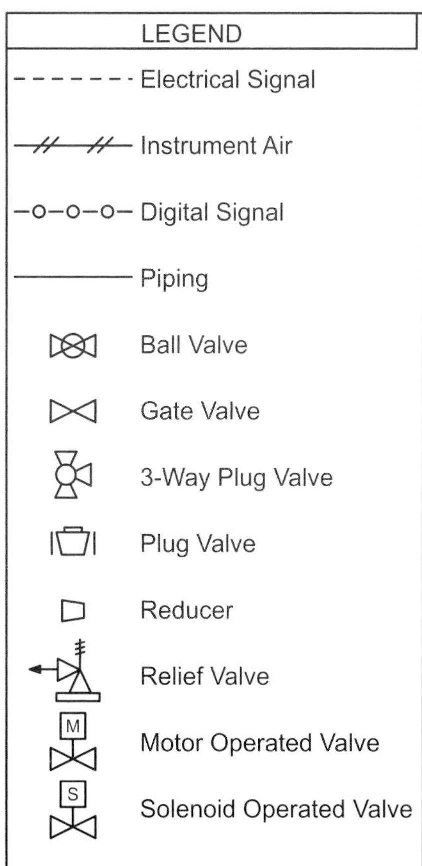

Figure 15-9: Legend example

Title Block

The **title block** is a section of a drawing (typically located in the bottom right corner) that contains information such as drawing title, drawing number, revision number, sheet number, and approval signatures.

Figure 15-10 shows an example of a title block.

<table>
<tr><td colspan="2">ANYCorp
Contract No.21609</td><td colspan="3">ANYCorp Construction Company
Texas City, Texas
Project No. 2447</td></tr>
<tr><td colspan="2">ANYCorp</td><td>Drawn S. Turnbough</td><td colspan="2">Date 1/06</td></tr>
<tr><td colspan="2" rowspan="4">Piping & Instrumentation Diagram
Primary Amine Unit
Fuel Gas Amine Absorber</td><td>Chk. By J. Dees</td><td colspan="2">App. By M. Collins</td></tr>
<tr><td>Scale None</td><td colspan="2">Unit 85</td></tr>
<tr><td>AFE No.</td><td colspan="2">Chc. No. 6848-56</td></tr>
<tr><td colspan="2">Dwg. No. F-52-4-0505</td><td>Rev 2</td></tr>
</table>

Figure 15-10: Title example

Application Block

An **application block** is the main part of a drawing that contains symbols and defines elements such as relative position, types of materials, descriptions, and functions.

Figure 15-10 shows an example of an application block.

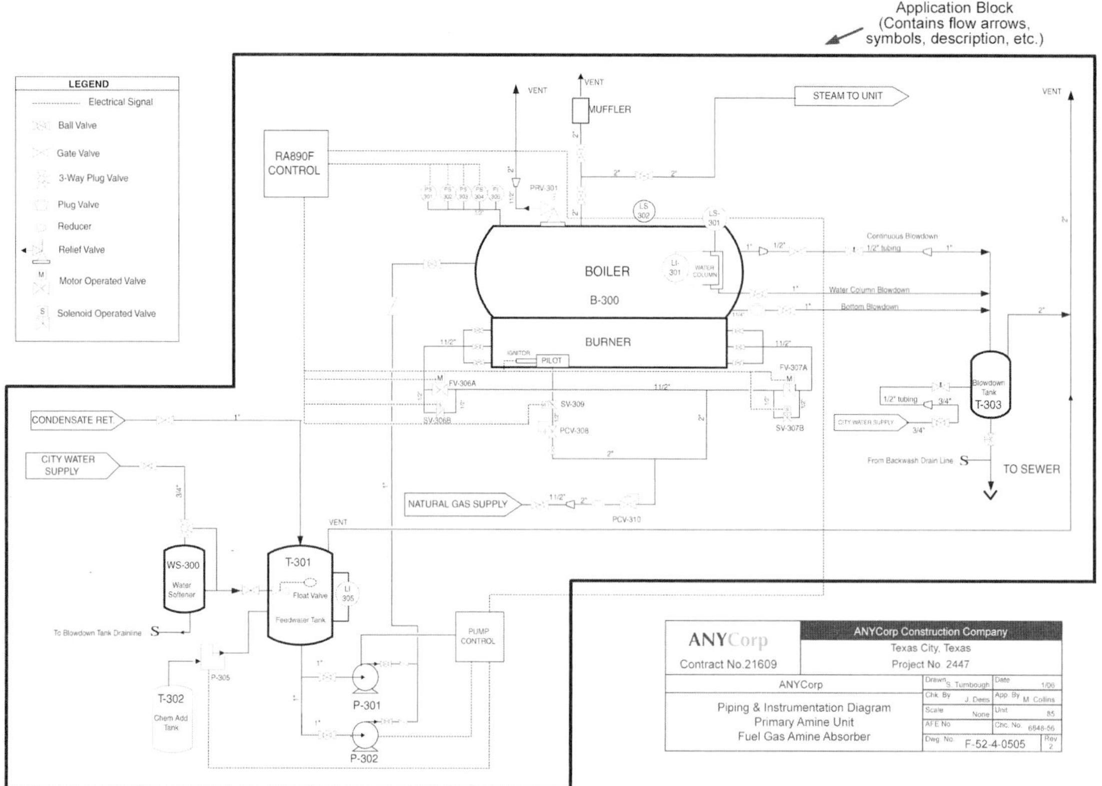

Figure 15-11: Application block example

COMMON PROCESS DRAWINGS AND THEIR USES

Process technicians must recognize a wide variety of drawings and understand how to use them.

These most commonly encountered drawings include:

- Block Flow Diagrams (BFD)
- Process Flow Diagrams (PFD)
- Piping and Instrument Diagrams (P&ID)
- Utility Flow Diagrams (UFD)
- Electrical Diagrams
- Isometrics

The following sections describe each of these drawing and their uses.

Block Flow Diagrams (BFDs)

Block Flow Diagrams (BFDs) are the simplest drawings used in the process industries. They provide a general overview of the process and contain few specifics. BFDs represent sections of a process (drawn as blocks) and use flow arrows to show the order and relationship of each component. Process technicians find BFDs useful for getting a high-level ("big picture") understanding of a process.

Figure 15-12 shows an example of a block flow diagram.

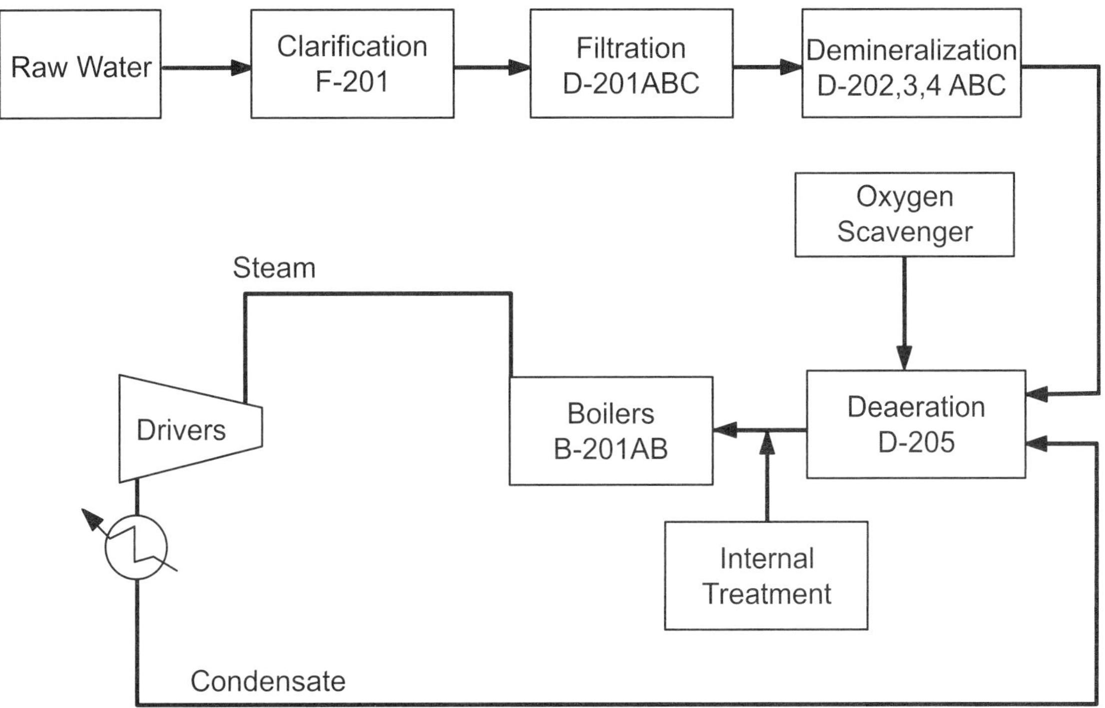

Figure 15-12: Sample Block Flow Diagram

Process Flow Diagrams (PFDs)

Process Flow Diagrams (PFDs) allow process technicians to trace the step-by-step flow of a process. PFDs contain symbols which represent the major pieces of equipment and piping used in the process. Directional arrows show the path of the process.

The process flow is typically drawn from left to right, starting with feed products or raw materials on the left and ending with finished products on the right. Variables such as temperature and pressure may also be shown at critical points.

Figure 15-13 shows an example of a process flow diagram.

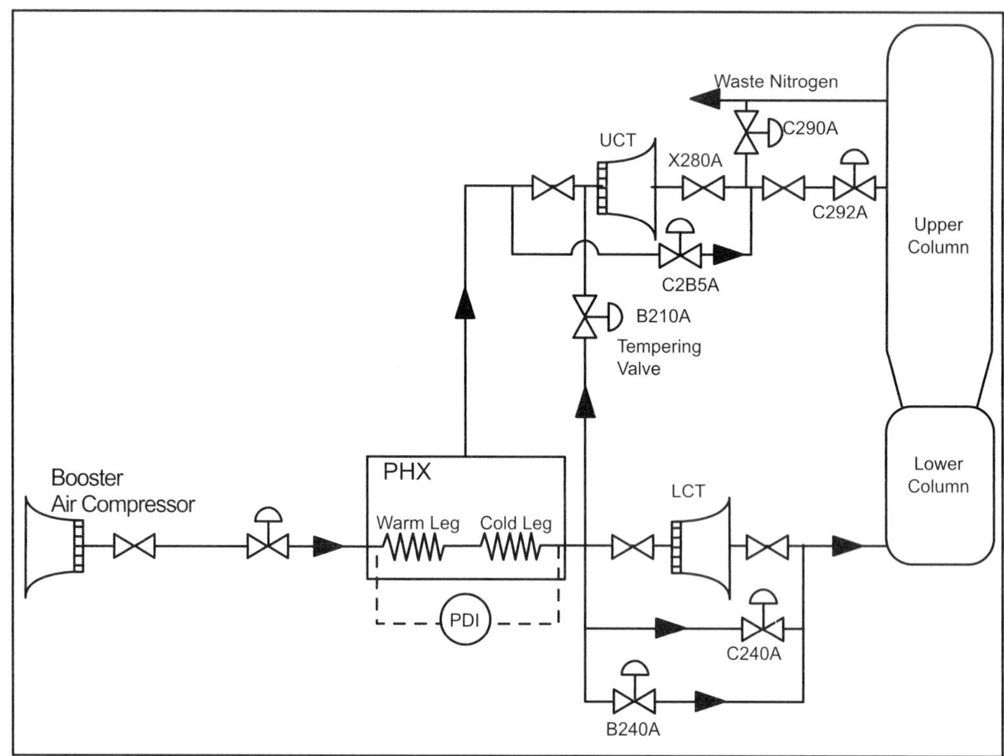

Figure15-13: Sample Process Flow Diagram (PFD)

Piping & Instrument Diagram (P&ID)

Piping & Instrument Diagrams (P&IDs), which are sometimes referred to as Process & Instrument Drawings, are similar to a Process Flow Diagrams, but show more complex process information.

Typical P&IDs contain equipment, piping and flow arrows, and additional details such as equipment numbers and operating conditions, piping specifications, and instrumentation.

For a process technician, a vital part of a P&ID is the instrumentation information. This information gives the technician a firm understanding of how the process is controlled. Using a P&ID, a process technician can understand how a product flows through the process and how it can be monitored and controlled. P&IDs are also critical during maintenance tasks, modifications, and upgrades.

Figure 15-14 shows an example of a P&ID.

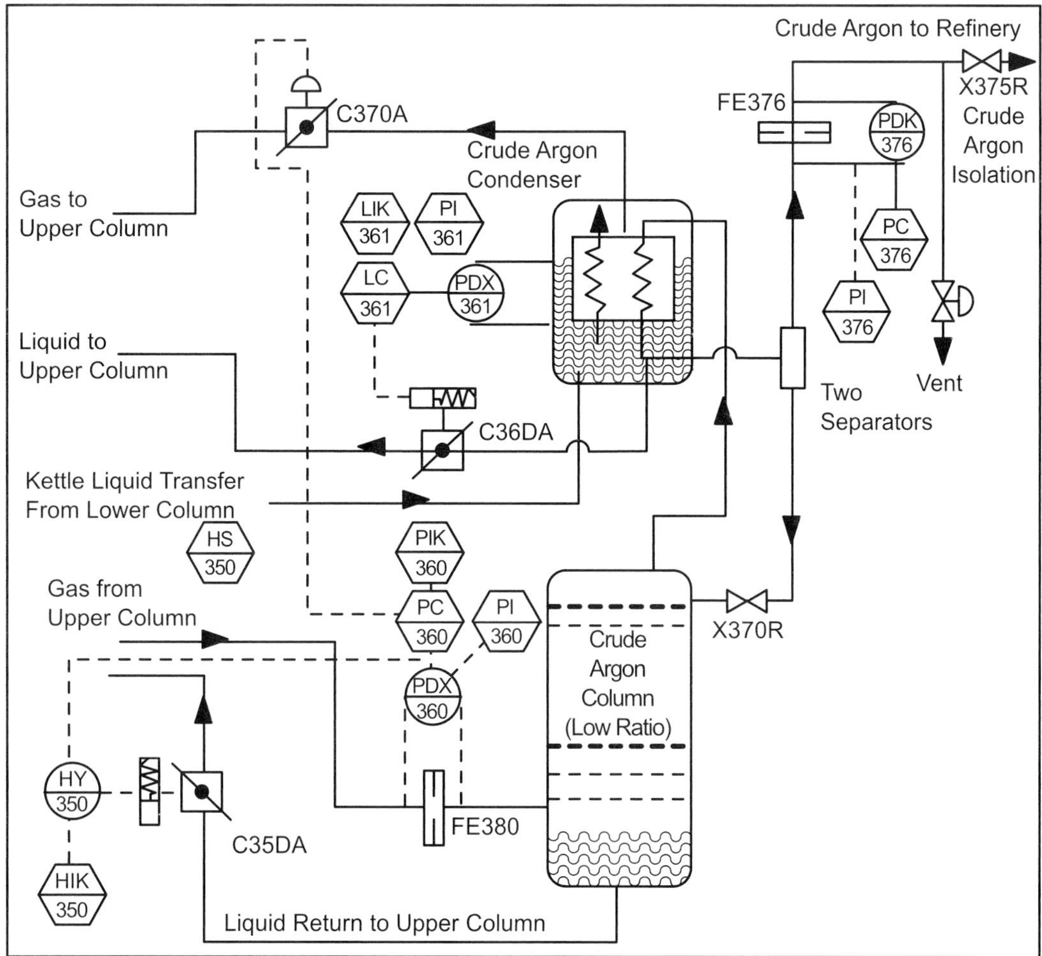

Figure 15-14: Sample Piping & Instrument Diagram (P&ID)

Utility Flow Diagrams (UFDs)

Utility Flow Diagrams (UFDs) provide process technicians a P&ID-type view of the utilities used for a process. UFDs represent the way utilities connect to the process equipment, along with the piping and main instrumentation used to operate those utilities.

Typical utilities shown on a UFD include:

- Steam
- Condensate
- Cooling water
- Instrument air
- Plant air
- Nitrogen
- Fuel gas

Figure 15-15 shows an example of a Utility Flow Diagram (UFD).

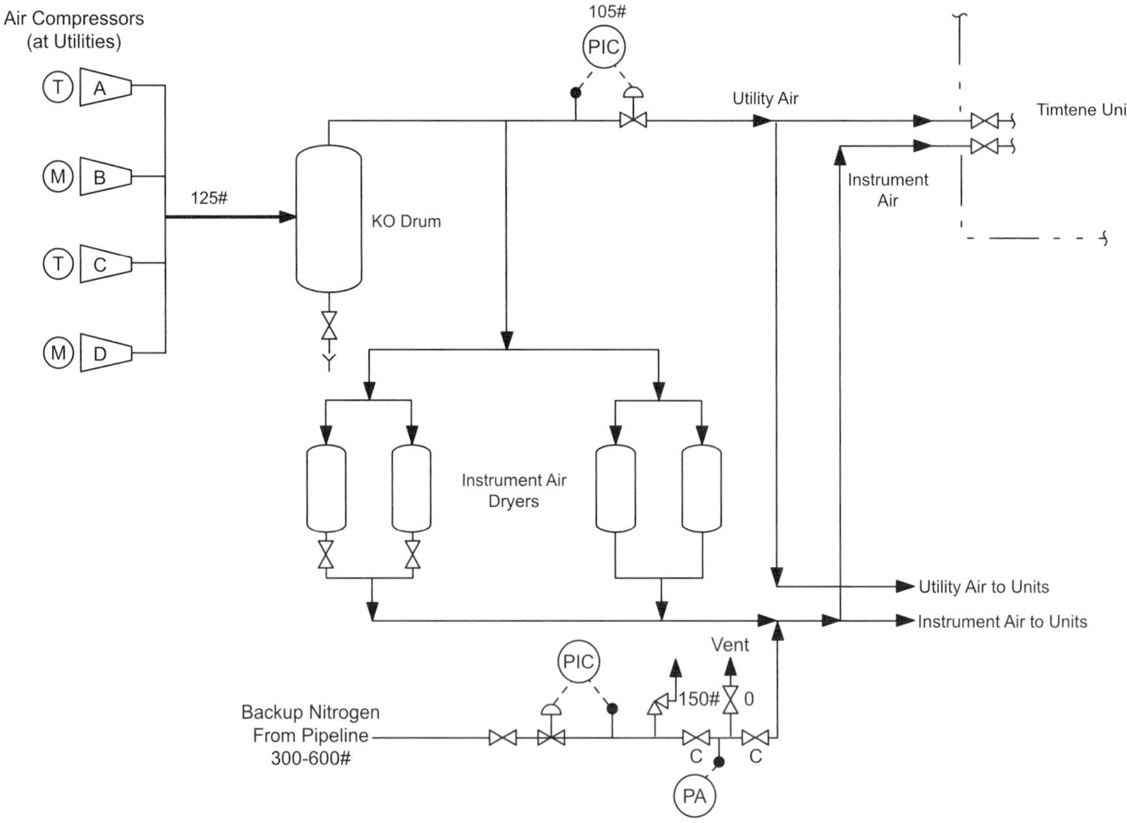

Figure 15-15: Sample Utility Flow Diagram (UFD)

Electrical Diagrams

Many processes rely on electricity, so it is important for process technicians to understand electrical systems and how they work. Electrical diagrams help process technicians understand power transmission and how it relates to the process. A firm understanding of these relationships is critical when performing lockout/tagout procedures (i.e., control of hazardous energy) and monitoring various electrical measurements.

Electrical diagrams show components and their relationships, including:

- Switches used to stop, start, or change the flow of electricity in a circuit.
- Power sources provided by transmission lines, generators or batteries.
- Loads (the components that actually use the power).
- Coils or wire used to increase the voltage of a current.

- Inductors (coils of wire that generates a magnetic field and are used to create a brief current in the opposite direction of the original current) that can be used for surge protection.
- Transformers (used to make changes in electrical power by means of electromagnetism).
- Resistors (coils of wire used to provide resistance in a circuit).
- Contacts used to join two or more electrical components.

Two specific types of electrical diagrams are wiring diagrams and schematics.

WIRING DIAGRAMS

Wiring diagrams show electrical components in their relative position in the circuit, and all connections between components. Process technicians use wiring diagrams to determine specific information about electrical components, how they are connected, and where these components are physically located.

Figure 15-16 shows an example of a wiring diagram.

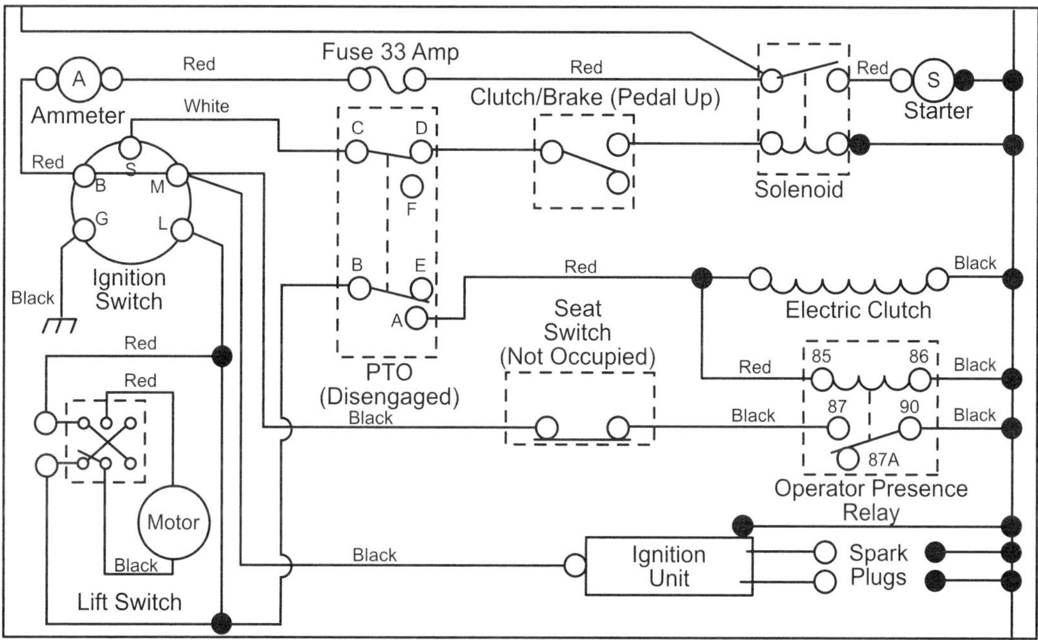

Figure 15-16: Sample Wiring Diagram

SCHEMATICS

Schematics show the direction of current flow in a circuit, typically beginning at the power source. Process technicians use schematics to visualize how current flows between two or more circuits. Schematics also help electricians detect potential trouble spots in a circuit.

ISOMETRICS

Isometric drawings show objects, such as equipment, as they would appear to the viewer. In other words, they are like a 3-D drawing that appears to come off the page. Isometric drawings may also contain cutaway views to show the inner workings of an object.

Isometric drawings show the three sides of the object that can be seen, with the object appearing at a 30-degree angle with respect to the viewer. All vertical lines appear vertical and are parallel to one another. All horizontal lines appear at a 30-degree angle and are parallel to one another.

Isometrics are typically only used during new unit construction. Process technicians are rarely exposed to isometric drawings, unless a new unit is being built. However, such drawings can prove useful to new process technicians, as they learn to identify equipment and understand its inner workings.

OTHER DRAWINGS

Along with the drawings mentioned in the previous sections, process technicians might encounter other types of drawings, such as:

- **Elevation diagrams**—represent the relationship of equipment to ground level and other structures.
- **Equipment location diagrams**—show the relationship of units and equipment to a facility's boundaries.
- **Loop diagrams**—show all components and connections between instrumentation and a control room.

SUMMARY

Process drawings provide process technicians with visual descriptions and explanations of processes, equipment, and other important items in a facility. Without process drawings, it can be difficult for process technicians to understand and operate the processes, and even more difficult to make repairs.

Within the process industries, there are many different types of drawings. Each drawing type represents different aspects of the process and levels of detail.

Block Flow Diagrams (BFDs) are simple drawings that show a general overview of a process, the components within the process, and their relationships.

Piping & Instrument Diagrams (P&IDs) show the equipment, piping and instrumentation of a process in the facility, along with more complex details than a Process Flow Diagram (PFD).

Process Flow Diagrams (PFDs) are basic drawings that show the primary flow of a product through a process, using equipment symbols, piping and flow direction arrows.

Utility Flow Diagrams (UFD) show the piping and instrumentation for the utilities in a process.

Electrical diagrams represent how electrical components and devices are connected.

Wiring diagrams show electrical components in their relative position in the circuit and all connections in between.

Schematics show the direction of current flow in a circuit, typically beginning at the power source.

Isometrics show objects as they would be seen by the viewer (i.e., the object has depth and dimension like a 3-D drawing).

Looking at combinations of these drawings provides a more complete picture of the processes at a facility.

CHECKING YOUR KNOWLEDGE

1. Define the following key terms:
   a. Block
   b. Flow Diagram (BFD)
   c. Electrical Diagram
   d. Elevation Diagram
   e. Equipment Location Diagram
   f. Isometric
   g. Loop diagram
   h. Piping & Instrument Diagram (P&ID)
   i. Process Flow Diagram (PFD)
   j. Schematic
   k. Symbol
   l. Utilities Flow Diagram (UFD)
   m. Wiring diagram

2. What are the two most common types of drawings in the process industries?

3. Which of the following drawing types is the most simple?
   a. PFD
   b. BFD
   c. P&ID
   d. UFD

4. Name at least three components common to all process drawings.

5. What is a PFD?
   a. Piping Flow Diagram
   b. Process Fixtures Diagram
   c. Pipe Fixture Drawing
   d. Process Flow Diagram

6. Where are continuation arrows found on a drawing? What do these arrows represent?

7. What is a vital part of a PFD?
   a. Blocks indicating a part of the process
   b. Instruments
   c. Pressure and temperature variable listings
   d. Utility costs

8. UFDs are similar to what type of drawing, except they show utilities used in a process:
   a. BFD
   b. Schematic
   c. P&ID
   d. PFD

9. A ________ electrical diagram shows the components in their relative position in the circuit while a ________ electrical diagram shows the direction of current flow in a circuit.

10. An Isometric drawing shows an object with respect to the viewer using:
   a. 30 degree angle perspective
   b. 360 degree angle perspective
   c. 90 degree angle perspective
   d. No angle perspective, it is a 2-D style drawing

ACTIVITIES

1. Complete the following chart, writing a three to five sentence description of each drawing and how it is used:

Drawing type	Description and use
Block Flow Diagram	
Process Flow Diagram	
Piping & Instrument Diagram	
Utility Flow Diagram	
Electrical Diagram	
Isometric	

2. Match the symbol to the appropriate equipment name.

Symbol		Equipment Name
1.		a. Compressor
		b. Furnace
		c. Heat Exchanger
2.		d. Motor
		e. Pump
		f. Turbine
3.	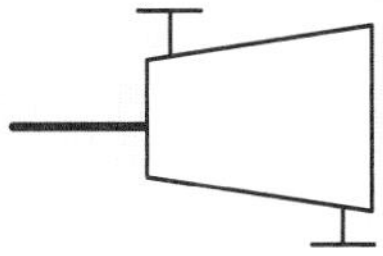	
4.	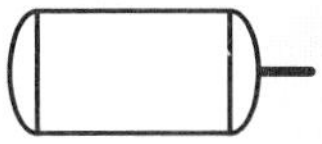	
5.		
6.	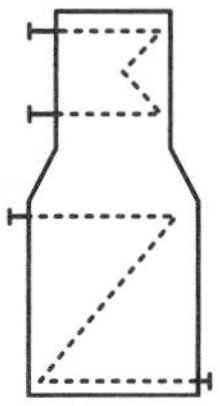	

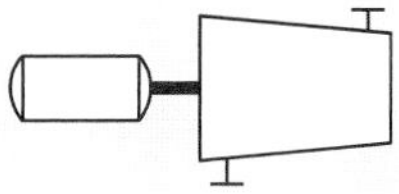

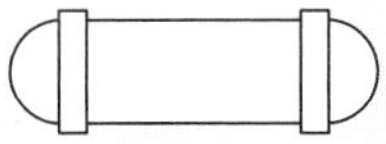

Chapter 16

Piping and Valves

OBJECTIVES

Upon completion of this chapter you will be able to:

1. Describe the purpose and function of piping and valves in the Process industries.
2. Identify the different materials used to manufacture piping and valve components.
3. Identify the different types of piping and valve connecting methods.
4. Identify the different types of pipe fittings used in the industry and their application.
5. Identify the different types of valves used in the industry and their application.
6. Discuss the hazards associated with the improper operation of a valve.
7. Describe the monitoring and maintenance activities associated with piping and valves.
8. Identify the symbols used to represent the different types of piping and valve components presented in this session.

KEY TERMS

- **Alloy**—a material composed of two or more metals that are mixed together when molten to form a solution, not a chemical compound (e.g., bronze is an alloy of copper and tin).
- **Ball Valve**—a flow control element shaped like a hollowed out ball used to start and stop flow; a ball valve only requires a quarter turn to get from fully open to fully closed.
- **Braze**—to solder together using a hard solder with a high melting point.
- **Butt Weld**—a type of weld used to connect two pipes of the same diameter that are butted against each other.
- **Butterfly Valve**—a flow regulating device that uses a disc-shaped flow control element to increase or decrease flow; requires a quarter-turn to go from fully open to fully closed.
- **Check Valve**—a type of valve that only allows flow in one direction and is used to prevent reversal of flow in a pipe.
- **Control Valve**—a valve normally equipped with an actuator to control valve stem movement; these valves are used as the final control element for controlling flow, level, temperature or pressure in a process.
- **Diaphragm Valve**—a flow regulating device that use a chemical resistant, rubber-type diaphragm to control flow instead of a typical flow control element.
- **Fitting**—a piping system component used to connect two or more pieces of pipe together.

- **Flange**—a type of pipe connection that is bolted together.
- **Gasket**—a flexible material placed between flanges to seal against leaks.
- **Gate Valve**—a positive shutoff valve utilizing a gate or guillotine which when moved between two seats causes tight shutoff.
- **Globe Valve**—a type of valve that uses a plug and seat to regulate the flow of fluid through the valve body which is shaped like a sphere or globe.
- **HDPE**—High Density Polyethylene; a plastic material used to create water pipes and drains.
- **Plug Valve**—a type of valve that uses a flow control element shaped like a hollowed out plug to start or stop flow; requires a quarter-turn to go from fully open to fully closed.
- **PVC**—Polyvinyl Chloride; a plastic type material that can be used to create cold water pipes and drains and other low-pressure applications.
- **Relief Valve**—a safety device designed to open if the pressure of a liquid in a closed vessel exceeds a preset level.
- **Safety Valve**—a safety device designed to open if the pressure of a gas in a closed vessel exceeds a preset level.
- **Screwed (Threaded) Pipe**—piping that is connected using male and female threads.
- **Socket Weld**—a type of weld used to connect pipes and fittings when one pipe is small enough to fit snugly inside the other.
- **Solder**—a metallic compound that is melted and applied in order to join and seal the joints and fittings together in tubing systems.
- **Throttling**—partially opening or closing a valve in order to restrict or regulate flow.
- **Tubing**—small copper or stainless steel pipe used extensively in instrument work. Plastic tubing is also used. A seamless type of steel pipe is referred to as tubing.
- **Valve**—a piping system component used to control the flow of fluids through a pipe.
- **Valve Seat**—the internal component of a valve against which the sealing elements presses to stop flow.

INTRODUCTION

Piping and valves are the most prevalent pieces of equipment in the process industries. Some estimates say that piping makes up 30-40% of the initial investment when creating a new plant.

In any plant you will see large segments of pipe going from one location to another. These pipes carry chemicals and other materials into and out of various processes and equipment.

When building a plant it is important to select proper construction materials and connectors, since some materials and connectors are not adequate for certain processes, pressures, or temperatures. Improper operation or improper material selection can lead to leaks, wasted product, or hazardous conditions.

CONSTRUCTION MATERIALS

Industrial pipes and valves can be made of many different materials such as carbon steel, stainless steel, alloy steel, iron, exotic metals, and plastics. The most common type of piping, however, is carbon steel because it is appropriate for a wide range of temperatures and is relatively economical.

When piping and valve systems are designed for a particular process they must be familiar with the process and the substances that will pass through the pipes. Specifically, they need to know the temperature of the substance, its viscosity, how much pressure it exerts, and how flammable, corrosive, or reactive it is.

Some metals become brittle at extremely low temperatures, while others are weakened by high temperatures, high pressures, or the corrosive effects of process substances such as strong acids or bases, or the erosive effects of high velocity fluids.

While some construction materials are pure metals, others may be compounds called alloys. **Alloys** are compounds composed of two or more metals that are mixed together in a molten solution (e.g., bronze is an alloy of copper and tin). Alloys improve the properties of single component metals, and provide special characteristics needed in specific applications.

Table 16-1 contains a list of common construction materials and their applications.

TABLE 16-1: Common piping and valve construction materials and their applications

Construction Material	Used in the following temperature ranges	Description
Carbon Steel	-20° F to 800° F	The most commonly used construction material because of its flexibility, weldability, strength, and relatively low cost.
Stainless Steel	-150° F to 1400+° F	Less brittle than carbon steel at extremely low temperatures; appropriate for use in high temperature applications; more corrosion resistant than carbon steel, especially in acid service.
Brass, Bronze and Copper	-50° F to 450° F	Used for corrosion resistance at low or moderate temperatures; excellent heat transfer characteristics.
Alloys	Varies with alloy	May be used in high-temperature (above 800° F) applications like furnace tubes, and highly corrosive service; high cost.
Plastics	Varies with plastic	Used for low pressure applications and corrosive services; easy to install, low cost and light weight.

CONNECTING METHODS

Pipes and valves can be connected together in a variety of ways. They may be screwed (threaded), flanged, or bonded (e.g., welded, glued, soldered, or brazed).

Figure 16-1 shows examples of each of these connection methods.

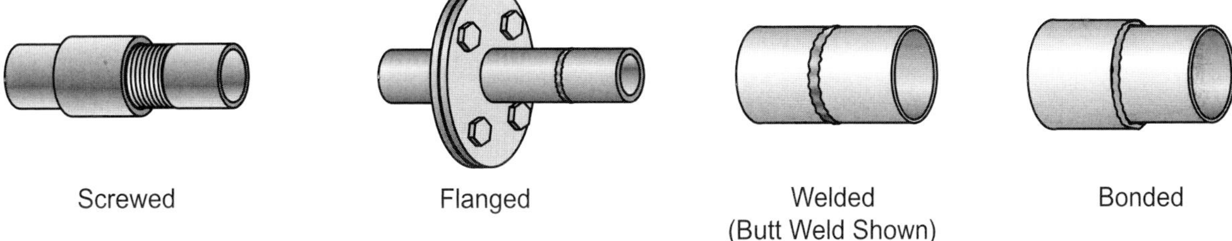

Figure 16-1: Examples of Screwed, Flanged, Welded and Bonded Connections

The factor that determines which connection type is the most appropriate is the purpose of the pipe. For example, if the pipe is used in low pressure water service, a threaded joint might be appropriate because they are cheaper and easier to install than welded or flanged joints.

If the pipe is used in a high pressure, flammable or corrosive service, a welded joint would be a better option because screwed joints, and flanged joints with gaskets, are more likely to leak. This is why welded joints are usually the connection method of choice for critical service piping.

Screwed (Threaded)

Figure 16-2: Screw Connection

Screw-type connections involve the joining together of two pipes through a series of tapered threads like the ones shown in Figure 16-2. In a screw-type connection the pipe is cut with "male" threads and the connector is cut with "female" threads so the two join together. When these threads are cut, they are generally cut with precision to ensure the two pieces fit tightly together to avoid leaks. However, this tightness can make it difficult to connect the two pieces together. That is why threading compound or Teflon® tape are often employed to lubricate the joints, facilitate the connection, and provide a flexible connection to seal against leaks.

Threaded connections are more prone to leak because the connections are weaker (i.e., more easily broken due to vibration or external force). Because of this, threaded connections are typically used for low pressure, non-flammable, non-toxic service.

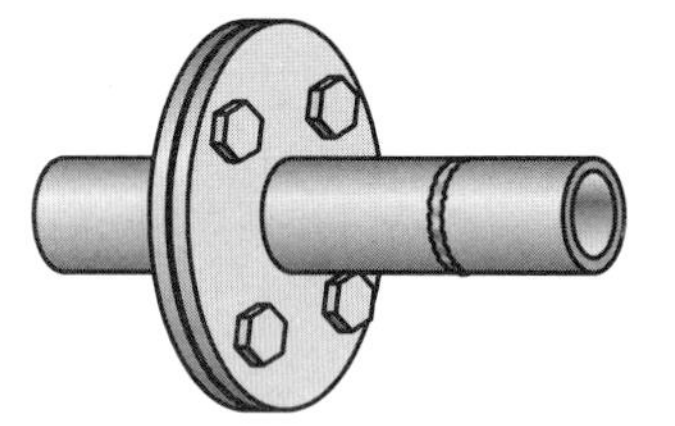
Figure 16-3: Flanged Connection

Flanged

Flanged connections, like the one shown in Figure 16-3, are typically used in instances where the piping may need to be disconnected from another pipe or a piece of equipment.

In a flanged connection, two mating plates are joined together with bolts. Between the two mating plates is a gasket. As the bolts are tightened, the gasket is compressed between the two plates. This compression increases the tightness of the seal and prevents leakage.

DID YOU KNOW?

The thickness of a flange and the number of bolts it contains can be a good "rule of thumb" indicator for process technicians.

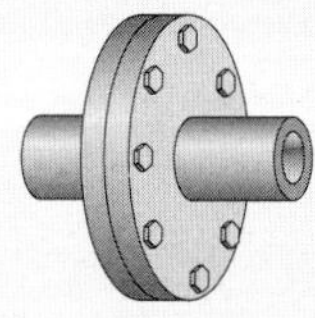

The thicker the flange, and the more bolts used to connect it, the higher the pressure.

Welded

Welding materials are made of similar metallic compounds. In order to create welded joints, welding material must be melted and applied to the pipes being connected.

If the pipes are the same diameter, a **butt weld** is used. If one pipe is small enough to fit snugly inside the other, a **socket weld** is used. Figure 16-4 shows examples of a butt and a socket welds.

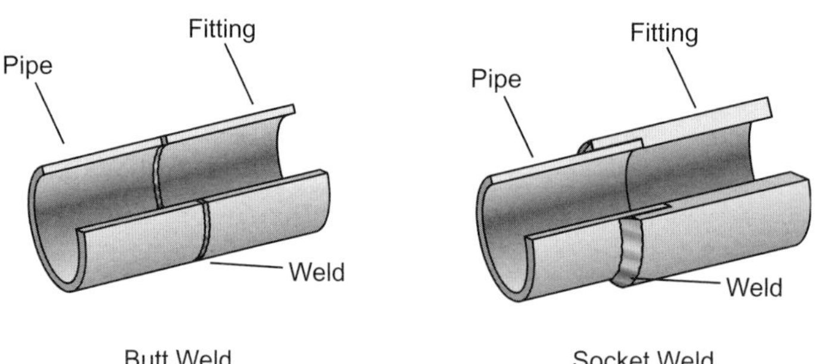

Figure 16-4: Butt and Socket Welds

Bonded

Bonded pipe joints, like the one show in Figure 16-5, can be glued, brazed or soldered.

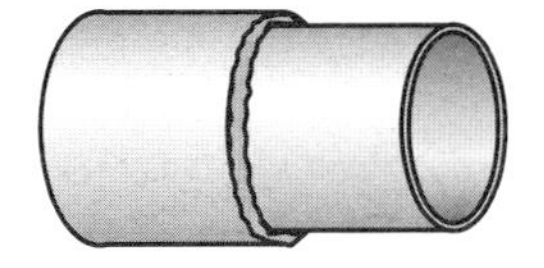

Figure 16-5: Bonded Joint

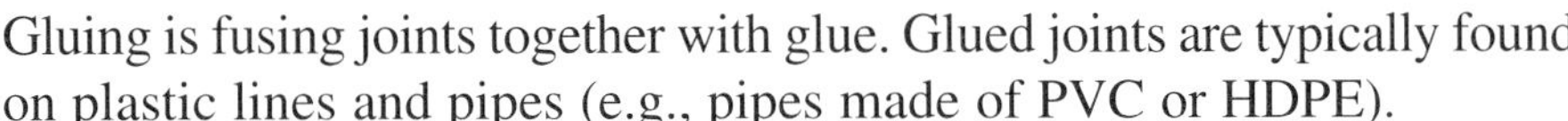

Gluing is fusing joints together with glue. Glued joints are typically found on plastic lines and pipes (e.g., pipes made of PVC or HDPE).

Soldering or **brazing** is fusing joints together with molten metal.

The method used for joining pipes and fittings is determined by the materials being used, and their applications.

FITTING TYPES

Fittings are piping system components used to connect two or more pieces of pipe together. There are many different types of fittings used in the process industries.

Figure 16-6 shows examples of some of the most common fittings. The table below Figure 16-6 lists their applications.

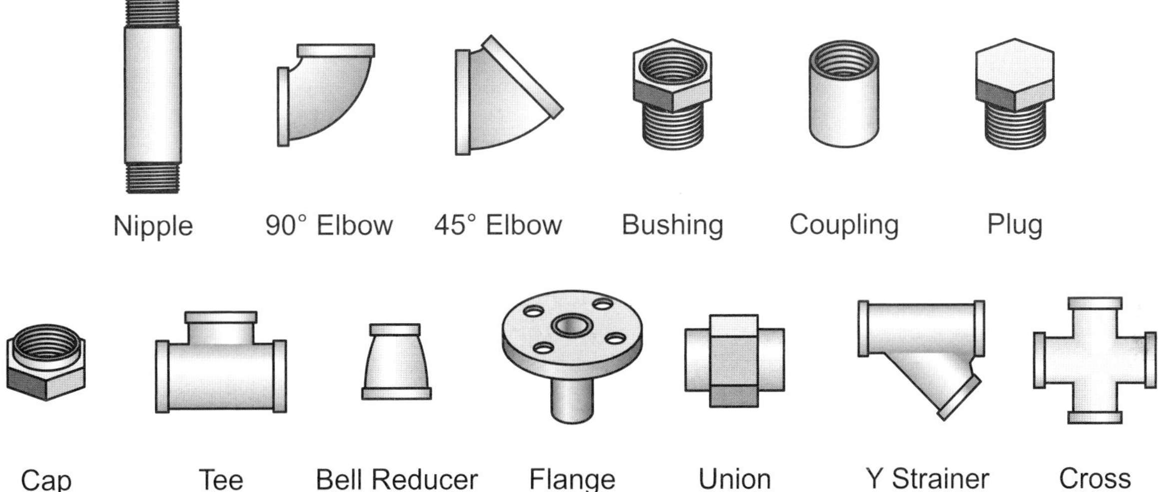

Figure 16-6: Different Types of Fittings and Their Applications

Fitting Name	Description and Application
Nipple	A short length of pipe (usually less than 6 inches) with threads on both ends.
Elbow	A pipe fitting with a 45 or 90 degree angle that is used to change the direction of flow.
Bushing	A pipe fitting, that is threaded on both the internal and external surfaces, which is used to join two pipes of differing sizes.
Coupling	A short piece of pipe or "collar" used to join two lengths of pipe.
Plug	A pipe fitting that fits inside the open end of a pipe to seal it.
Cap	A pipe fitting that fits over the open end of a pipe to seal it.
Tee	A T-shaped pipe fitting that is used to allow flow to two different pipes that are 90 degree apart.
Bell Reducer	A pipe fitting that is used to connect two pipes of different diameters.
Flange	A pipe fitting that consists of two mating plates and a gasket joined together with bolts.
Union	A pipe fitting that joins two sections of threaded pipe but allows them to be disconnected without cutting or disturbing the position of the pipe.
Strainer	A pipe fitting, located before process equipment, which contains fine-mesh which allows fluid flow and holds back solid particles.
Cross	A pipe fitting that allows four pipes to be connected together at 90 degree angles.

Figure 16-6: Different Types of Fittings and Their Applications

VALVE TYPES

Valves are piping system components used to control the flow of fluids through a pipe. Valves work to control, throttle, or stop the flow.

There are many different types of valves used in the process industries. Some of the most common valves include:

- Ball
- Butterfly
- Check
- Diaphragm
- Gate
- Globe
- Plug
- Relief

When determining which type of valve should be used, engineers must take into consideration how the valve will be used and the substances that will pass through it.

For example, if the fluid passing through the valve is very thick (viscous) or corrosive, then a diaphragm valve would be a good choice, since other valve types do not perform well with these types of substances.

Furthermore, if the valve will be used for **throttling** (a condition in which a valve is partially opened or partially closed in order to restrict or regulate the amount of flow), then a globe valve would be a good choice since many other types of valves (e.g., gate valves) can be damaged by throttling.

Process technicians need to be familiar with each of the different valve types and the maintenance and operating characteristics of each.

Ball Valve

A **ball valve** is a flow regulating device that uses a flow control element shaped like a hollowed out ball, attached to an external handle, to increase or decrease flow. Ball valves are quarter-turn valves. In other words, turning the valve's stem a quarter of a turn brings it to a fully open or fully closed position (in comparison to other valves, such as gate valves, which require multiple turns to fully open or fully close).

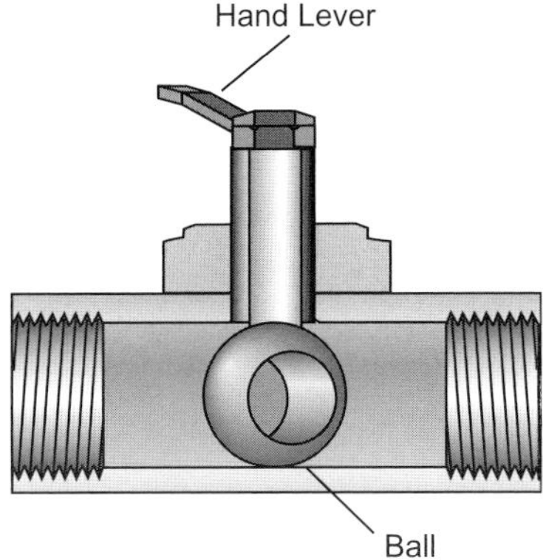

Figure 16-7: Ball Valve (partially open)

Figure 16-7 shows a cutaway of a ball valve.

Ball valves are typically used for on/off service. When a ball valve is open the hollowed out portion of the ball (sometimes referred to as the port) lines up perfectly with the inner diameter of the pipe. When a ball valve is closed, the port aligns with the wall of the pipe.

Plug Valve

A **plug valve** is a flow regulating device that uses a flow control element shaped like a hollowed out plug, attached to an external handle, to increase or decrease flow.

Plug valves are almost identical to ball valves. Both are quarter-turn valves that use a hollowed out object to control flow.

Figure 16-8 shows a cutaway of a plug valve.

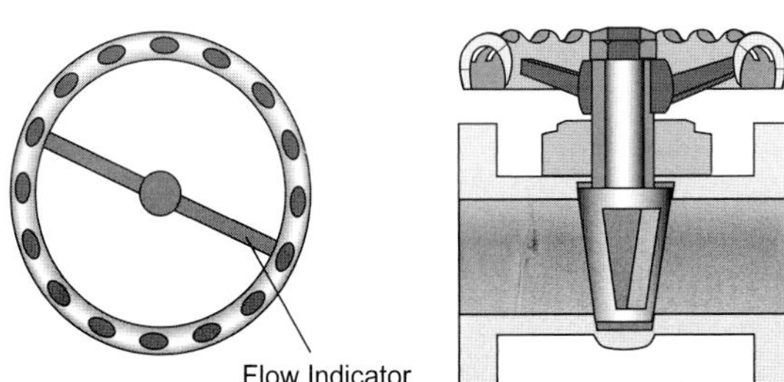

Figure 16-8: Plug Valve (partially open)

Plug valves are designed for on/off service, and are well suited for certain types of applications such as low pressure, slurry, lubrication, and fuel gas.

DID YOU KNOW?

The hot water heater in your home contains a relief valve.

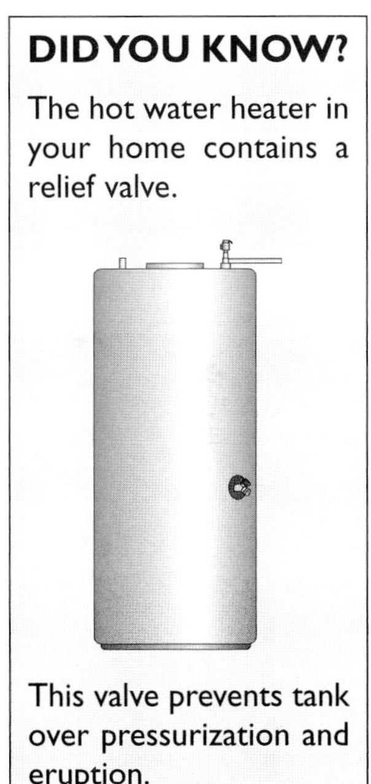

This valve prevents tank over pressurization and eruption.

Butterfly Valve

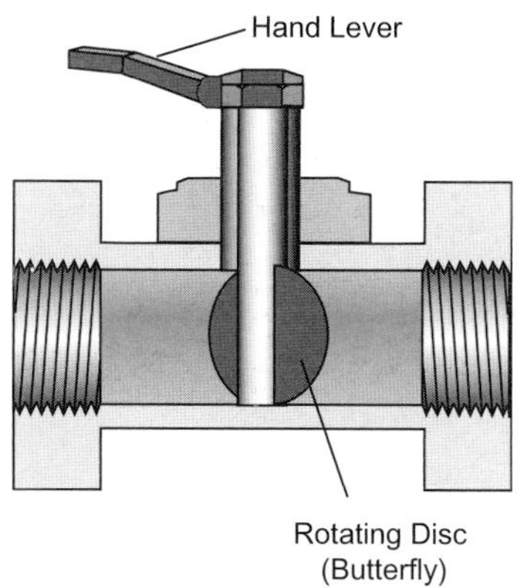

Figure 16-9: Butterfly Valve

A **butterfly valve** is a flow regulating device that uses a disc-shaped flow control element to increase or decrease flow. Like ball valves, butterfly valves can be fully opened or closed by turning the valve handle one-quarter of a turn.

Because of the way they are designed, these valves are most suitable for low temperature, low pressure applications such as cooling water systems or other non-critical service. Figure 16-9 shows a cutaway of a butterfly valve.

Unlike many other valves, butterfly valves can be used for throttling. However, the throttling capabilities of a butterfly valve are not uniform or exact (e.g., opening the valve half way may provide a flow that is near maximum).

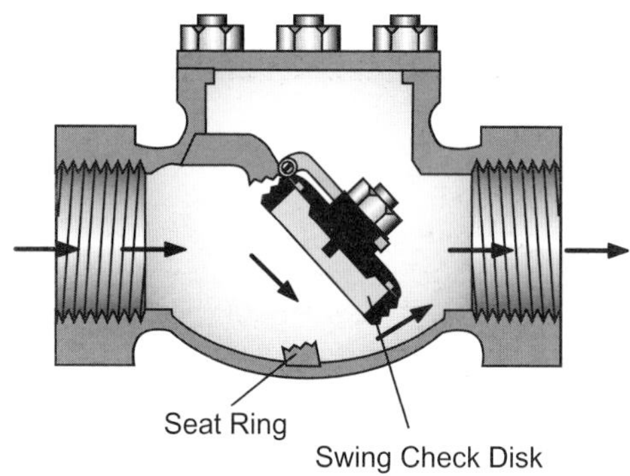

Figure 16-10: Swing Check Valve

Check Valve

Check valves are valves that only allow flow in one direction. These valves eliminate backflow, thereby preventing equipment damage and contamination of the process.

Check valves can be composed of many different materials and can be used in a wide variety of applications. The most common type of check valves include: swing check, lift check, and ball check.

Figure 16-10 shows and example of a swing check valve.

DID YOU KNOW?

The human heart has valves in it that function like check valves.

Without these valves, blood would not circulate properly.

In a swing check valve, the valve disc is lifted as the fluid moves through the valve, and is forced closed if the fluid changes directions (backward flow).

In a lift check valve (Figure 16-11) a disc-shaped flow control element controls the flow of the fluid.

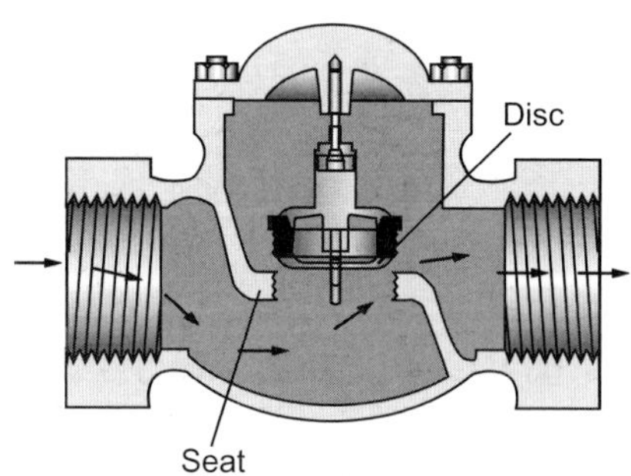

Figure 16-11: Lift Check Valves

In a ball check valve (Figure 16-12) a ball or sphere-shaped flow control element is used.

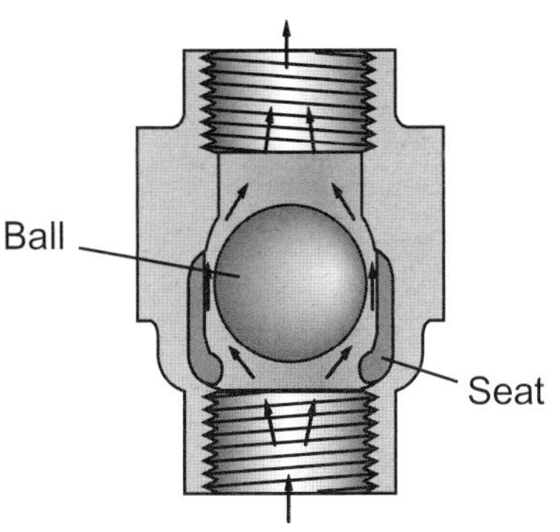

Figure 16-12: Ball Check Valves

Diaphragm Valve

Diaphragm valves are flow regulating devices that use a flexible, chemical resistant, rubber-type diaphragm to control flow instead of a typical flow control element. In this type of valve the diaphragm seals the parts above it (e.g., plunger) from the process fluid.

Figure 16-13 shows an example of a diaphragm valve.

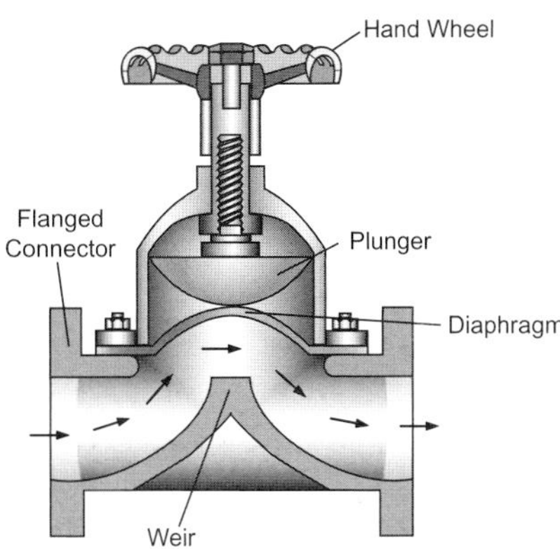

Figure 16-13: Diaphragm Valve

Because of their unique design, diaphragm valves work well with process substances that are exceptionally sticky, viscous, or corrosive. However, they are not adequate for applications with high pressures or excessive temperatures.

Gate Valve

Gate valves are positive shutoff valves that utilizing a gate or guillotine which, when moved between two seats, causes tight shutoff.

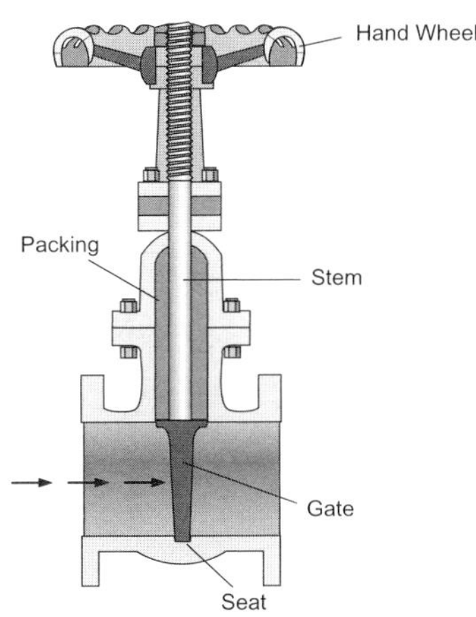

Figure 16-14: Gate Valve

Gate valves are the most common type of valve in the process industries. They are designed for on/off service and are not intended for throttling.

Figure 16-14 shows a cutaway of a gate valve.

In a gate valve, a wedge-shaped disc or "gate" is lowered into the body of the valve with a hand wheel. The body of the valve contains two seat rings which seals around the gate when fully closed and completely blocks flow.

Because of their design, gate valves should not be used for throttling, since throttling can cause metal erosion and seat damage which prevents the valve from sealing properly.

DID YOU KNOW?

Throttling a gate valve can cause the flow control element to vibrate back-and-forth.

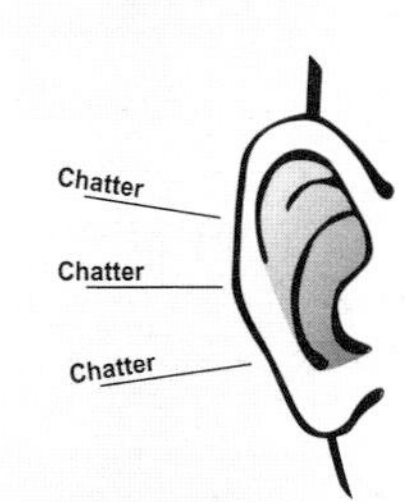

This back-and-forth movement causes an audible sound referred to as "valve chatter."

Globe Valve

Globe valves are valves with a moving seat (usually circular) which moves into an opening to shutoff. These types of valves are designed to regulate flow in one direction usually used in throttling service. They are very common in the process industries.

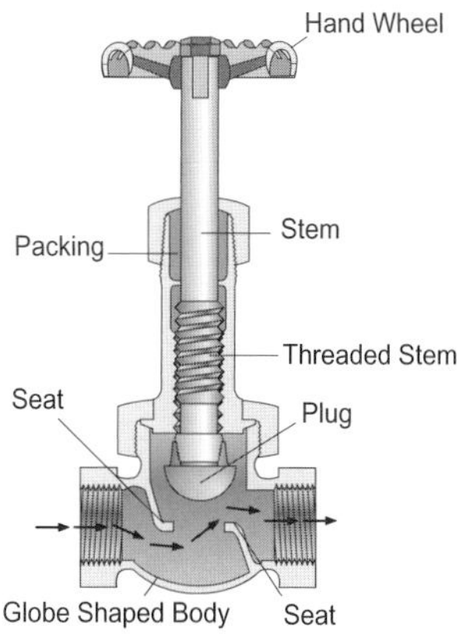

Figure 16-15: Globe Valve

Figure 16-15 shows a cutaway of a globe valve.

In a globe valve, fluid flow is increased or decreased by raising or lowering the plug or flow control element. These flow control elements can come in a variety of shapes including: cylindrical, needle and ball. The seat in these valves is designed to accommodate the shape of the plug.

Figure 16-16 shows an example of these different plug shapes.

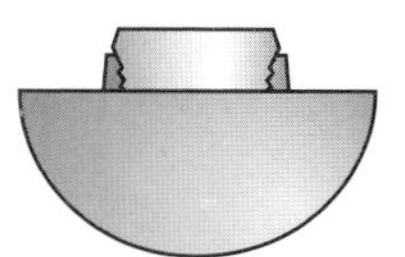

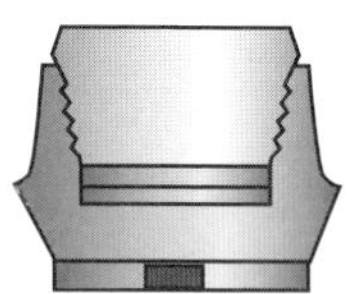

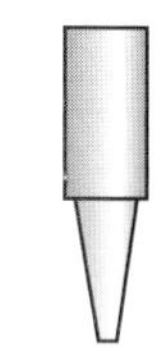

Figure 16-16: Globe Valve Plug Designs

Relief and Safety Valves

Safety and Relief Valves are used to protect equipment and personnel from over pressurization situations. Both of these valves are designed to open and discharge to a collection system when the pressure in a line, vessel, or other equipment exceeds a preset threshold. The difference, however, is the type of service they are intended for (liquid vs. gas) and the speed at which they open.

RELIEF VALVES

Relief valves are safety devices designed to open slowly if the pressure of a liquid exceeds a preset level. These valves open slower and with less volume than safety valves because liquids are virtually non-compressible. With non-compressible substances, a small release is all that is required to correct over pressurization.

Because of the speed at which they open (slow), relief valves are not good for gas service. They are good for pressurized liquid service, however.

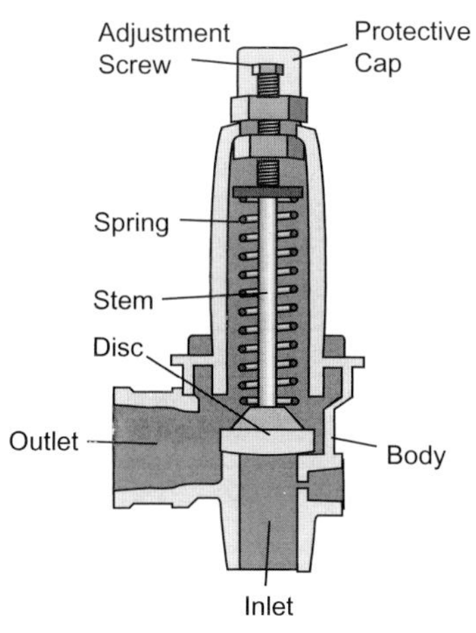

Figure 16-17: Relief Valve

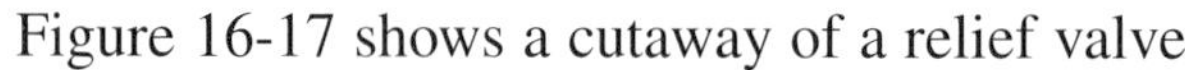

Figure 16-17 shows a cutaway of a relief valve.

In a relief valve, a flow control disc is held in place by a spring. Once the pressure in the system exceeds the threshold of the spring, the valve is forced open (proportional to the increase in pressure) and liquid is allowed to escape into a containment receptacle, flare or other safety system. As the pressure drops below the threshold, the spring gradually forces the flow control element back into the seat, thereby resetting the valve.

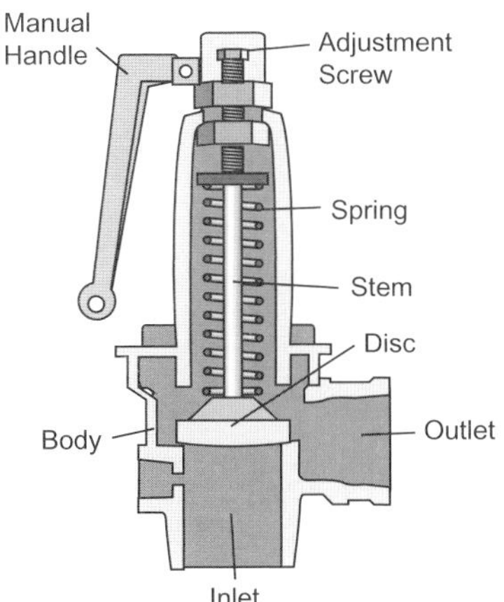

Figure 16-18: Safety Valve with manual handle

SAFETY VALVES

Safety valves are safety devices designed to open quickly if the pressure of a gas exceeds a preset threshold. These valves open quicker, release more volume, and generally have larger outlets than relief valves because gasses are highly compressible. Compressible substances require a much larger release to correct over pressurization.

Figure 16-18 shows an example of a safety valve.

Safety valves are designed to operate quickly and prevent overpressurization that can cause equipment damage or injury.

DID YOU KNOW?

Process technicians often use the slang term "pop valve" or "pop off valve" to refer to a safety valve.

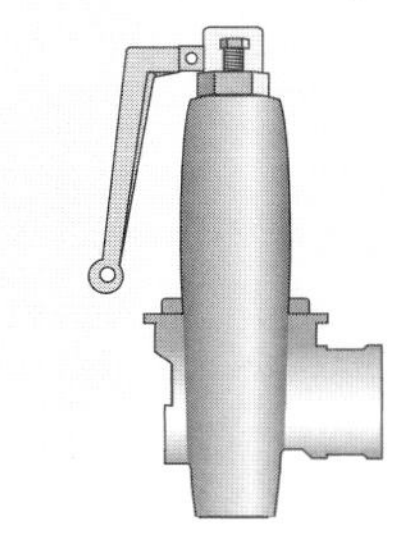

This is because they open quickly or "pop off" once the pressure threshold has been exceeded (as opposed to relief valves which open slowly).

When a safety valve opens, the excess pressure is vented to the flare header or through a large exhaust port into the atmosphere (depending on the substance being vented).

Some safety valves will re-seat themselves after being activated. Others must be taken to a shop to be manually reset.

The maintenance of safety and relief valves must be performed by certified personnel.

HAZARDS ASSOCIATED WITH IMPROPER VALVE OPERATION

Technicians should also avoid using excessive force when opening or closing valves, as this can warp the valve or damage the seat, preventing a good seal. However, properly sized valve wrenches are designed to supply additional force when needed to open or close a valve.

There are many hazards associated with piping and valves. Table 16-2 lists some of these hazards and their impacts.

TABLE 16-2:
Hazards associated with boiler operations

Improper Operation	Possible Impacts			
	Individual	**Equipment**	**Production**	**Environment**
Throttling a valve that is not designed for throttling.		Valve damage to the point that it will not seat and stop flow, even when closed.	Off spec product due to improper flows.	
Use of excessive force when opening or closing a valve.		Damage to the valve seat, the packing, or the valve stem. This causes leakage and makes the valve difficult to open or close.	Off spec product due to improper flows.	
Failure to clean and lubricate valve stems.	Possible injuries as a result of a valve wrench (required because the valve is difficult to open) slipping off of the valve handle.	Valve stem seizure or thread damage. This makes the valve difficult to open and close.		
Improperly closing a valve on a high-pressure line.	Possible injury due to equipment over pressurization.	Equipment damage (e.g., over pressurizing a pump)		Possible leak to the environment.
Failure to wear proper protective equipment when operating valves in high temperature, high pressure, hazardous, or corrosive service.	Burns or other serious injuries.			

MONITORING AND MAINTENANCE ACTIVITIES

When monitoring and maintaining piping and valves, technicians must always remember to look, listen and feel for the following:

Look	Listen	Feel
■ Check valves to make sure there are no leaks. ■ Check valve for excessive wear. ■ Check to make sure valve stems are properly lubricated.	■ Listen for abnormal noises (e.g., valve "chatter")	■ To make sure the valve is not being overly tightened.

Failure to perform proper maintenance and monitoring could impact the process and result in equipment damage.

PIPING AND VALVE SYMBOLS

There are many symbols associated with piping and valves. While there are standards, it is possible that symbols could vary slightly from plant to plant. Figure 16-19 shows some of the more commonly used symbols. Other symbols will be discussed later on in the piping and instrumentation diagram reading portion of this text.

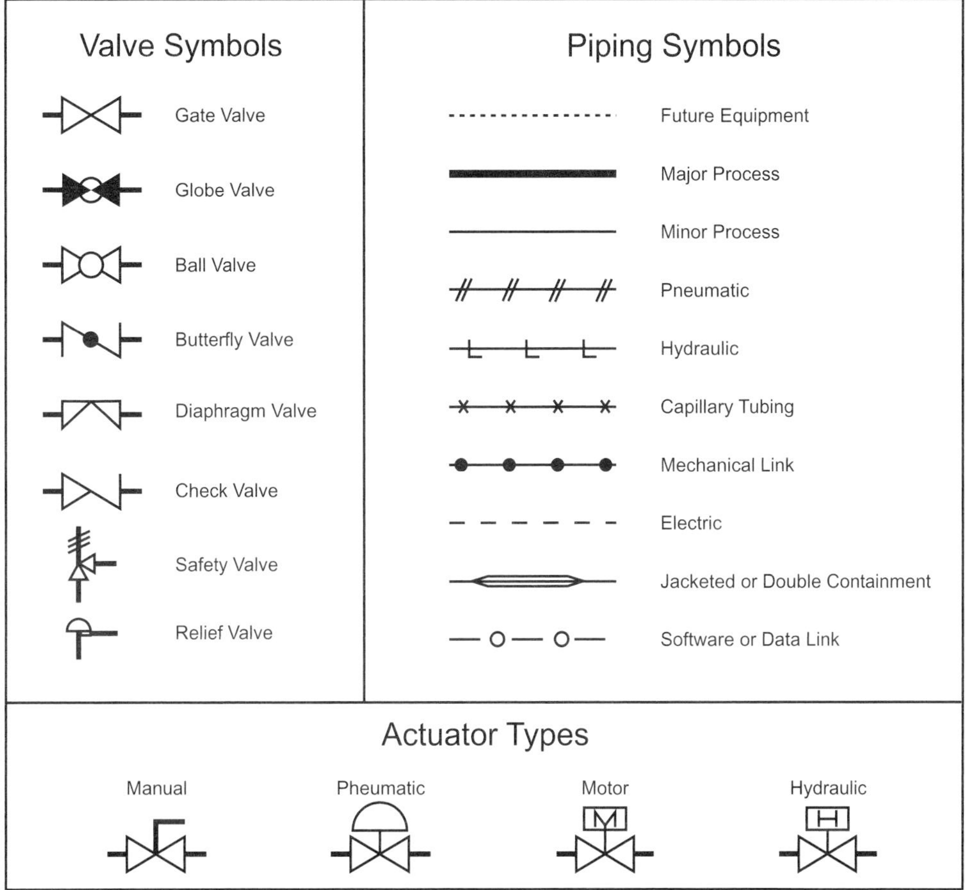

Figure 16-19: Common Piping and Valve Symbols

SUMMARY

Piping and valves are the most prevalent pieces of equipment in the process industries. Their main purpose is to carry chemicals and other materials into and out of other process-equipment.

Piping and valves can be made of many different materials ranging from carbon steel to plastic. Material selection is based on process characteristics (such as temperature, pressure, corrosiveness and erosiveness).

Pipes can be connected using a variety of methods. They can be screwed, flanged, or bonded. Screw joints are less expensive and easier than welded joints. However, they are not as leak proof as a welded joint in high pressure service. Flanged connections are used on pipes that may need to be disconnected from another pipe or piece of equipment. Bonded joints (e.g., glued, soldered, or brazed) are found on low pressure pipes and drain lines.

There are also many different types of valves in the process industries. The most common are: ball, butterfly, check, diaphragm, gate, globe, plug and relief.

Ball valves and plug valves are quarter-turn valves that use a hollowed out ball or plug to increase or decrease flow. Both of these valves are primarily used for on/off service.

Butterfly valves are also quarter-turn valves, but they use a disc-shaped element to control flow. Butterfly valves can be used for throttling. However, the throttling capabilities of butterfly valves are not very linear (e.g., opening the valve half way may provide a flow that is near maximum).

Check valves are regulating devices designed to prevent flow reversal. The most common types of check valves are swing, lift, and ball.

Diaphragm valves use a flexible, chemical resistant, rubber-type diaphragm to control flow instead of a typical flow control element. Because of their design, diaphragm valves are good for low pressure service of viscous or corrosive materials.

Gate valves are the most common type of valve in the process industries. These valves use a metal gate to block the flow of fluids through a valve. They are intended for on/off service and should not be used for throttling.

Globe valves, which are also common in the process industries, use a plug to block the flow of fluid through a valve. These valves can come in a variety of shapes and are designed for throttling.

Relief and safety valves are safety devices designed to open if the pressure of a fluid exceeds a preset threshold.

It is important to operate valves properly. Throttling valves that should not be throttled, using excessive force to open or close a valve, or failing to clean and lubricate valve stems can cause excessive wear and damage the valve. Over time, this damage can cause the valve to leak, seize up, or fail to open or close completely.

Improperly closing a valve on a high pressure line (e.g., blocking or "dead heading" a pump) or operating a steam filled valve without proper protective gear can cause equipment damage or personal injury.

When making rounds, process technicians should always inspect valves for leaks and perform proper maintenance and lubrication procedures to prevent damage or excessive wear. Technicians should also listen for abnormal noises or "valve chatter" that could be an indication of improper throttling.

There are many symbols associated with piping and valves. Process technicians should be familiar with these various symbols and be able to identify them on plant diagrams.

CHECKING YOUR KNOWLEDGE

1. Define the terms alloy, valve and throttling.

2. Explain why design engineers must be familiar with the process when selecting construction materials.

3. If you wanted to connect a piece of pipe to a piece of equipment and you knew you were going to have to disconnect it several times over the next few months, which connection method (screwed, flanged, or bonded) would you select? Why?

4. If you were going to connect two high-pressure, critical service pipes, would it be better to use a screw-type connection, or a welded connection? Why?

5. If you wanted to connect two pipes of the same diameter, would you use a socket weld or a butt weld?

6. Give the proper name for each of the following fittings.

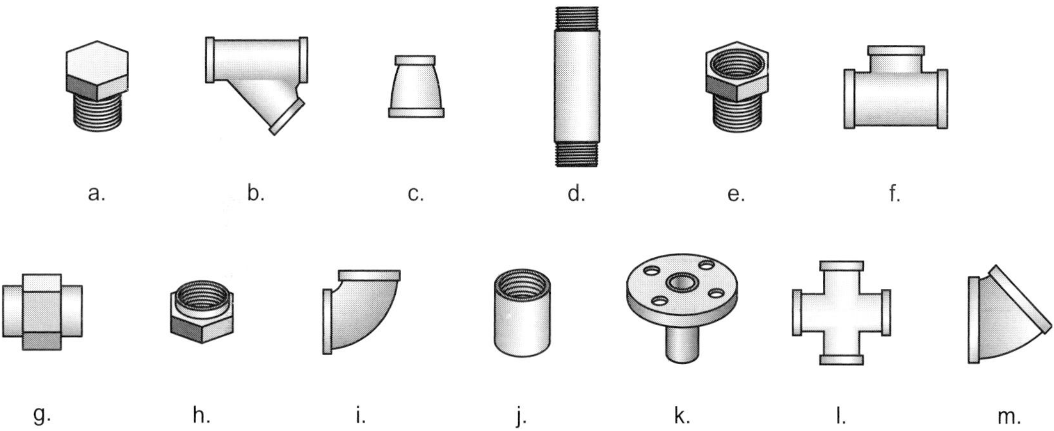

7. *(True or False)* Relief valves are designed to open quickly.

8. *(True or False)* Safety valves are designed for gas service.

9. What can happen to a valve over time if a process technician throttles that valve and it was not designed for throttling?

10. Why should a process technician refrain from using excessive force when opening or closing a valve?

11. Why should process technicians clean and lubricate valve stems regularly?

Match the valve type with its description.

Valve Type	**Description**
12. Ball	a. A safety device designed to open if the pressure of a gas exceeds a preset threshold.
13. Butterfly	b. Uses a disc-shaped flow control element to increase or decrease flow.
14. Check	c. Uses a hollowed out plug to increase or decrease flow.
15. Diaphragm	d. Uses a rubber-type diaphragm to control flow.
16. Gate	e. Uses a metal gate to block the flow of fluids.
17. Globe	f. Uses a hollowed out ball to increase or decrease flow.
18. Plug	g. A safety device designed to open if the pressure of a liquid exceeds a preset level.
19. Relief	h. Uses a spherical or globe-shaped plug to block fluid flow.
20. Safety	i. Used to prevent accidental backflow.

ACTIVITIES

1. Look around your house and identify at least 5 valves (e.g., the valve under your kitchen sink). Tell where each valve is located and try to identify what type of valve you think it might be.

2. Given a drawing of a valve, identify the components (e.g., seat, stem, flow control element).

3. Given a piping and instrumentation diagram, tell how many of the following valves are present:

   a. Gate
   b. Globe
   c. Ball
   d. Butterfly
   e. Diaphragm
   f. Check
   g. Safety
   h. Relief

4. Given a piping and instrumentation diagram, tell how many of the valves are:

   a. Manual
   b. Pneumatic
   c. Motor driven
   d. Hydraulic

Chapter 17
Vessels

OBJECTIVES

Upon completion of this chapter you will be able to:

1. Describe the purpose or function of vessels (tanks, drums and bins) and reactors in the process industries.
2. Explain the relationship of pressure to the vessel shape and wall thickness.
3. Define and provide examples of the following components as they relate to vessels:
   - Floating roof
   - Articulated drain
   - Blanketing
   - Spherical tank
   - Foam Chamber
   - Sump
   - Mixer
   - Gauge Hatch
   - Manway
   - Vapor Recovery System
   - Vortex breaker
   - Baffle
   - Weir
   - Boot
   - Mist eliminator
   - Vane separators
4. Describe the purpose of dikes, firewalls, and containment walls around vessels.
5. Describe the monitoring and maintenance activities associated with vessel operations.
6. Identify and describe the various types of reactors and their purpose.
7. Identify the symbols used to represent the different types of vessels and reactors.

KEY TERMS

- **Articulated Drain**—hinged drains, attached to the roof of a floating roof tank, that raise and lower as the roof and the fluid levels raise and lower.
- **Atmospheric Tank**—enclosed vessels in which atmospheric pressure is maintained; usually cylindrical in shape and equipped with either a fixed or floating roof.
- **Baffle**—a metal plate, placed inside tanks or vessel, which is used to alter the flow of chemicals or facilitate mixing.

- **Bin/Hopper**—vessels which typically hold dry solids.
- **Blanketing**—the process of putting nitrogen into the vapor space above the liquid in a tank to prevent air leakage into the tank (often referred to as a "nitrogen blanket").
- **Boot**—a section at the bottom of a process drum where water is collected and drained to waste. This space is the lowest portion of the drum.
- **Catalyst**—a substance used to change the rate of a chemical reaction without being consumed into the reaction.
- **Containment Wall**—a wall used to protect the environment and people against tank failures, fires, runoff and spills.
- **Continuous Reaction**—a reaction in which raw materials (reactants) are continuously being fed into the reactor and products are continuously being formed and removed.
- **Cylinder**—vessels that can hold extremely volatile or high pressure materials.
- **Dike**—a wall (earthen, shell or concrete) built around a piece of equipment to contain any liquids should the equipment rupture or leak.
- **Drum**—specialized types of storage tank.
- **Firewall**—earthen banks or concrete walls built around oil storage tanks to contain the oil in case of a spill or rupture. Also called bund.
- **Fixed Bed Reactor**—a reactor in which the catalyst bed is stationary as the reactants are passed over it; in this type of reactor, the catalyst occupies a fixed position and is not designed to leave the reactor with the process.
- **Floating Roof**—a type of roof (steel or plastic), used on storage tanks, which floats upon the surface of the stored liquid and is used to decrease the vapor space and reduce the potential for evaporation.
- **Fluidized Bed Reactor**—a reactor in which finely divided solids are suspended by an upward flow of gas.
- **Foam Chamber**—a reservoir and piping that contain chemical foam used to extinguish fires within a tank.
- **Gauge Hatch**—an opening on the roof of a tank that is used to check tank levels and obtain samples of the product or chemical.
- **Inhibitors**—substances which slow or stop a chemical reaction.
- **Manway**—an opening in a vessel that permits entry for inspection and repair.
- **Mist Eliminator**—a device in a tank, composed of mesh, vanes or fibers that collect droplets of mist from gas.
- **Mixer**—a device used to mix chemicals or other substances.
- **Pressurized Tank**—enclosed vessels in which a pressure greater than atmospheric is maintained.
- **Reaction Furnace**—a reactor which combines a firebox with tubing to provide heat for a reaction that occurs inside the tubes.
- **Reactor**—A vessel in which chemical reactions are initiated and sustained.

- **Spherical Tank**—a type of pressurized storage tank that is used to store volatile or highly pressurized material; also referred to as "round" tanks.
- **Stirred Tank Reactor**—a reactor that contains a mixer or agitator mounted to the tank.
- **Sump**—a pit or tank which receives and temporarily stores drainage at a low point.
- **Tank**—a large container or vessel for holding liquids and/or gases.
- **Tubular Reactor**—a tubular heat exchanger used to contain a reaction.
- **Vane Separator**—a device, composed of metal vanes, used to separate liquids from gases or solids from liquids.
- **Vapor Recovery System**—the process of capturing and recovering vapors. Vapors are captured by methods such as chilling or scrubbing. They are then purified and the vapors or products are either sent back to the process, sent to storage, or recovered.
- **Vessel**—a container in which materials are processed, treated, or stored.
- **Vortex**—the cone formed by a swirling liquid or gas.
- **Vortex Breaker**—a metal plate, or similar device, placed inside a cylindrical or cone-shaped vessel, that prevents a vortex from being created as liquid is drawn out of the tank.
- **Weir**—a flat or notched dam or barrier to liquid flow that is normally used for either the measurement of fluid flows or to maintain a given depth of fluid as on a tray of a distillation column.

INTRODUCTION

Vessels are a vital part of the process industries. Without these types of equipment, the process industries would be unable to create and store products in large amounts. By using vessels to create products in large quantities, companies can improve both cost effectiveness and efficiency. Without reactors, companies could not generate the chemical reactions that some processes require. Vessels are also used to provide intermediate storage between processing steps and to provide residence time for reactions to complete or to provide time to settle.

Vessels include tanks, drums, cylinders, hoppers, bins and other similar containers that are used to store materials. Vessels vary greatly in design (e.g., size and shape) based on the requirements of the process. Factors that affect vessel design may include pressure requirements (e.g. high and low), product storage (liquid, gas or solid), temperature requirements (e.g. insulation), corrosion factors and volume.

Reactors are specialized vessels used to contain controlled chemical reactions such as a change of raw materials into finished products. Reaction variables include temperature, pressure, time, concentration, surface area and other factors. Like vessels, reactor designs also vary widely based on the chemical reaction that must occur in the process.

PURPOSE OF VESSELS

Vessels are used to store raw materials and additives (process inputs), intermediate products (products that are not yet finished), final products (process outputs) and wastes (recoverable or non-recoverable off-spec products and byproducts).

Storage **tanks** are a common type of vessel, since every process requires containers to hold input and materials (typically fluids). **Drums** are tanks that are used for storage. **Cylinders** are vessels that can hold extremely volatile or high pressure materials (like propane). **Bins or hoppers** typically hold dry solids.

TYPES OF TANKS

Although many factors can affect the design and manufacture of tanks, pressure, temperature, and chemical properties are key factors which affect wall thickness, materials of construction, and the shape of the tank.

There are two main types of storage tanks: atmospheric and pressurized.

Atmospheric tanks are enclosed vessels in which atmospheric pressure is maintained (i.e., they are neither pressurized, nor placed under a vacuum; they are at the same pressure as the air around them). These tanks are usually cylindrical in shape and are equipped with fixed or floating roofs, or both.

Figure 17-1 shows an example of an atmospheric tank.

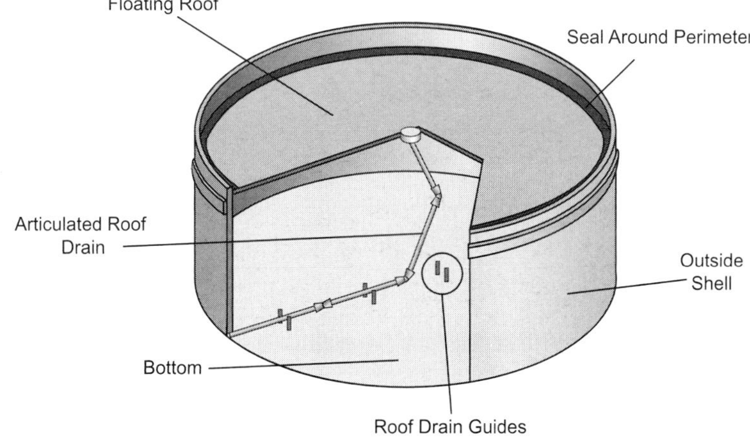

Figure 17-1: Example of a floating roof atmospheric tank

Atmospheric tanks are usually made of steel plates that are welded together in large sections. Because they do not seal as tightly as pressurized tanks, atmospheric tanks are only appropriate for substances that do not contain toxic vapors or high vapor pressure liquids.

Pressurized tanks are enclosed vessels in which a pressure greater than atmospheric is maintained.

DID YOU KNOW?

Spherical tanks are much stronger than non-spherical tanks.

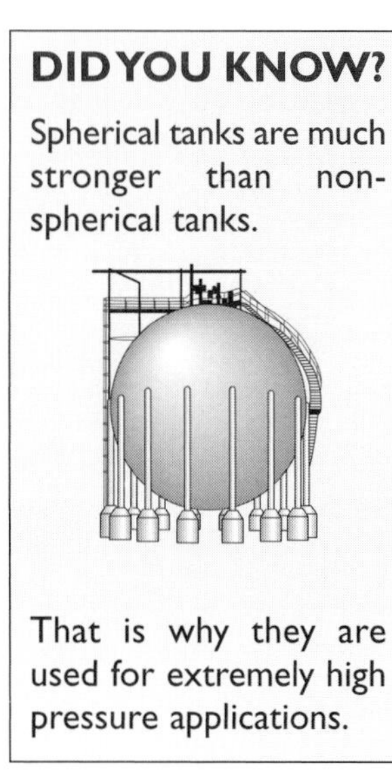

That is why they are used for extremely high pressure applications.

The most common types of pressurized tanks are:

- **Sphere**—used for highly volatile, pressurized substances; spherical shape distributes pressure evenly over every square inch of the vessel.
- **Cylindrical (Bullet)**—used for moderately pressurized contents; rounded ends help distribute pressure more evenly than in a non-rounded tank.
- **Hemispheroid**—used for low-pressure substances.

Figure 17-2 shows examples of spherical, cylindrical (bullet), and hemispheroid tanks.

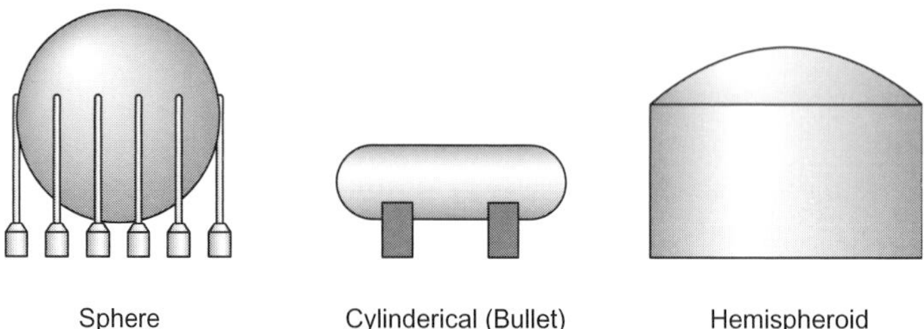

Figure 17-2: Sphere, Cylindrical (bullet), and hemispheroid tanks

COMMON COMPONENTS OF VESSELS

Vessels, including tanks and drums, have many components. Process technicians should be familiar with these components, which include:

- **Floating Roof**—a type of roof (steel or plastic), used on storage tanks, which floats upon the surface of the stored liquid and is used to decrease the vapor space and reduce the potential for evaporation.

  Floating roofs, which can be either internal or external, use a flexible seal to prevent leakage. Because of their design, floating roofs are not appropriate for volatile fluids or pressurized equipment.

  Figure 17-3 shows an example of a floating roof tank with an articulated drain.

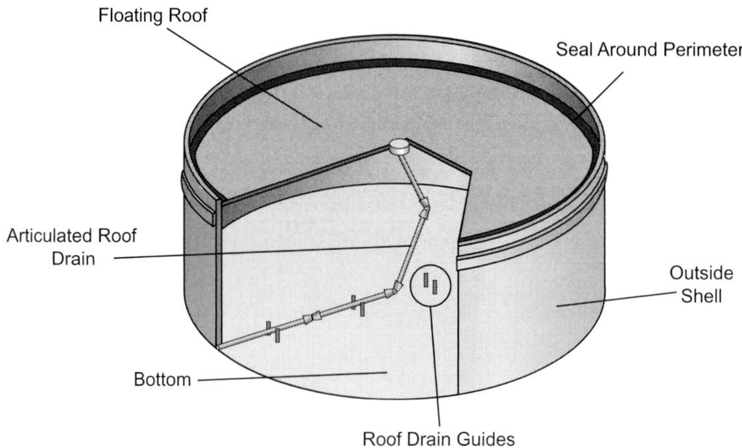

Figure 17-3: Internal floating roof tank

- **Articulated drains**—hinged drains, attached to the roof of a floating roof tank, that raise and lower as the roof and the fluid levels raise and lower.
- **Blanketing**—the process of putting nitrogen into the vapor space above the liquid in a tank to prevent air leakage into the tank (often referred to as a "nitrogen blanket").

  Blanketing a tank reduces the amount of oxygen present and decreases the risk of fire and explosion. Blankets also reduce risk of tank implosion (collapse) by decreasing the amount of vacuum being created as the tank is being emptied.

  Figure 17-4 shows an example of a tank blanketed with an inert gas.

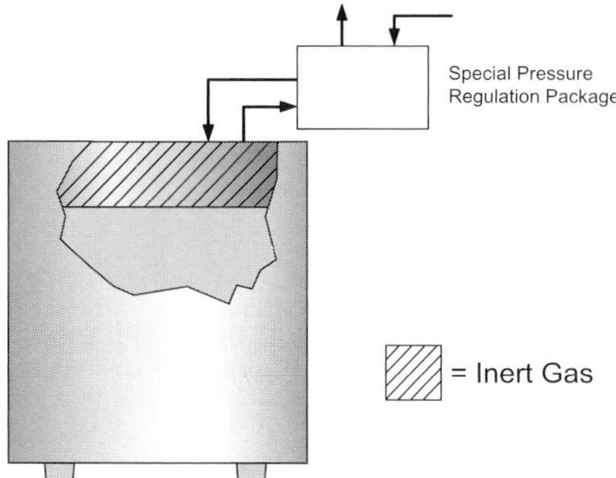

Figure 17-4: Tank blanketed with inert gas

- **Spherical Tank**—a type of pressurized storage tank that is used to store volatile or pressurized material. Spherical tanks are also referred to as "round" tanks.

  Figure 17-5 shows an example of a spherical tank.

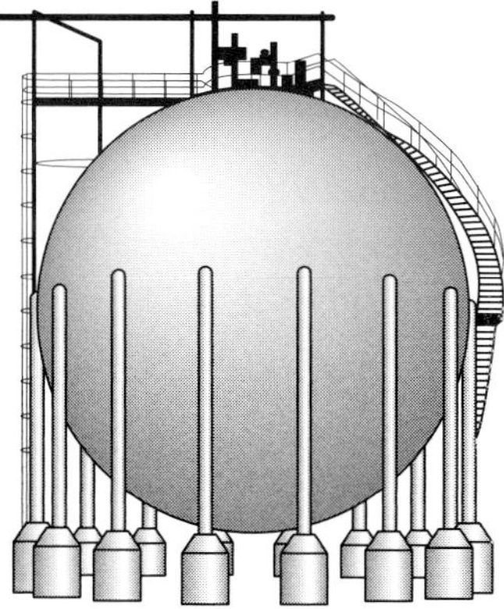

Figure 17-5: Spherical tank

- **Foam Chamber**—A reservoir and piping that contain chemical foam used to extinguish fires within a tank.

  Figure 17-6 shows an example of a foam chamber.

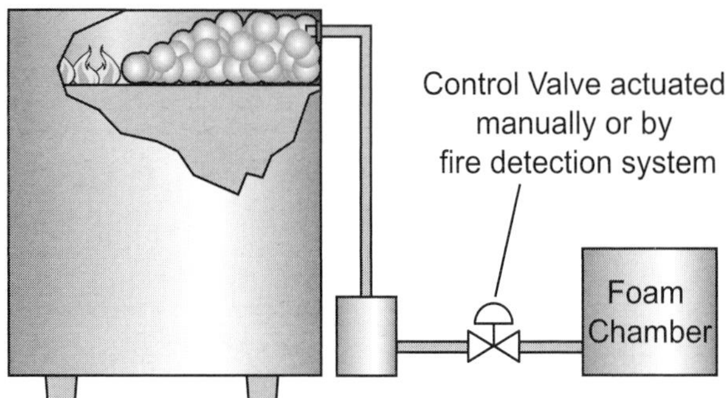

Figure 17-6: Foam Chamber

- **Sump**—a pit or tank which receives and temporarily stores drainage from the bottom of a tank. Sumps, which are sometimes referred to as "Possum Bellies," are usually the lowest point of the drainage system in tanks.

  Figure 17-7 shows an example of a sump.

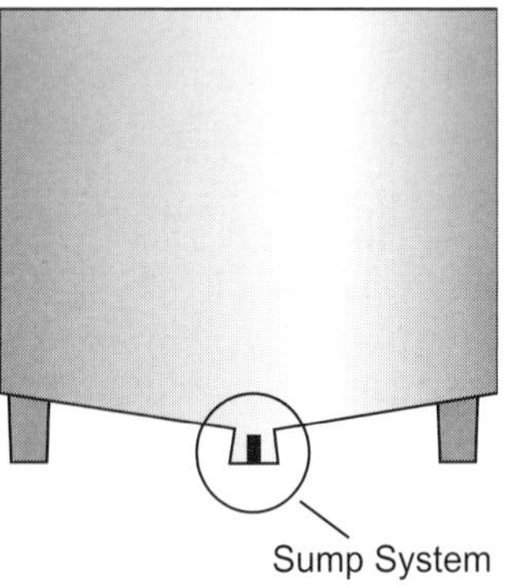

Figure 17-7: Sump

- **Mixer**—a device used to mix chemicals or other substances. Process technicians operate mixers using motors located outside the vessels. Chemical mixers work similarly to a kitchen mixer or a drink blender.

  Figure 17-8 shows an example of a mixer.

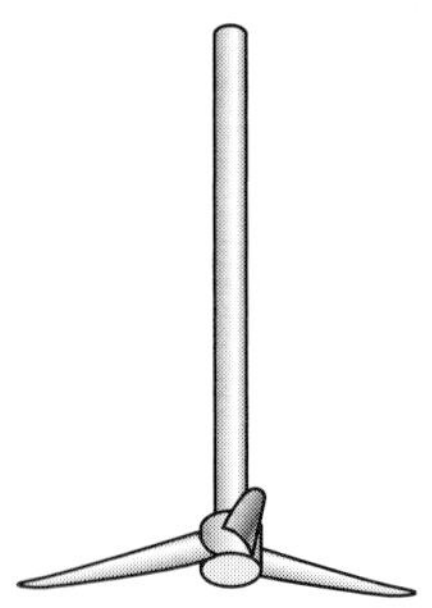

Figure 17-8: Mixer

- **Gauge Hatch**—an opening on the roof of a tank that is used to check tank levels and obtain samples of the product or chemical.

  Figure 17-9 shows an example of a gauge hatch.

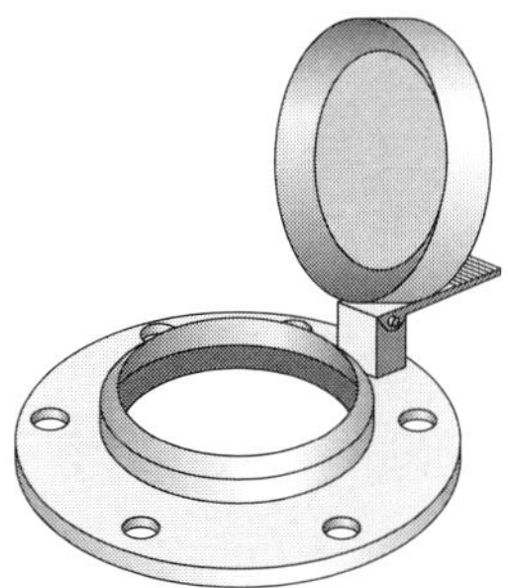

Figure 17-9: Gauge Hatch

- **Manway**—an opening in a vessel that permits entry for inspection and repair. Most manways are flanged openings located on the roof or side of a tank.

  Figure 17-10 shows an example of a manway with bolts and nuts removed in preparation for entry.

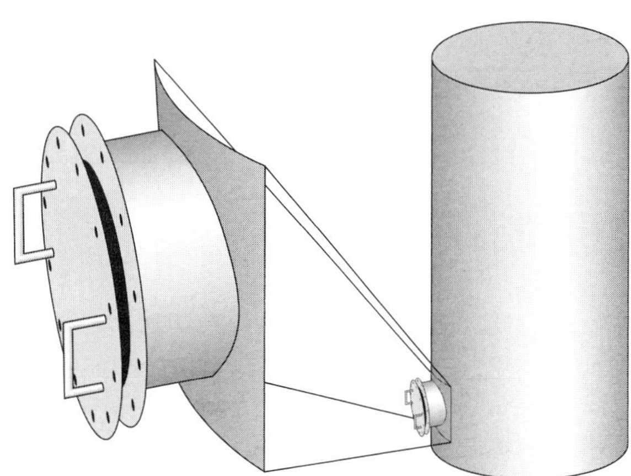

Figure 17-10: Manway

- **Vapor Recovery System**—the system used to capture and recover vapors. Vapors are captured by methods such as chilling or scrubbing. They are then purified and the vapors or products are then sent back to the system.

  Figure 17-11 shows an example of a vapor recovery system.

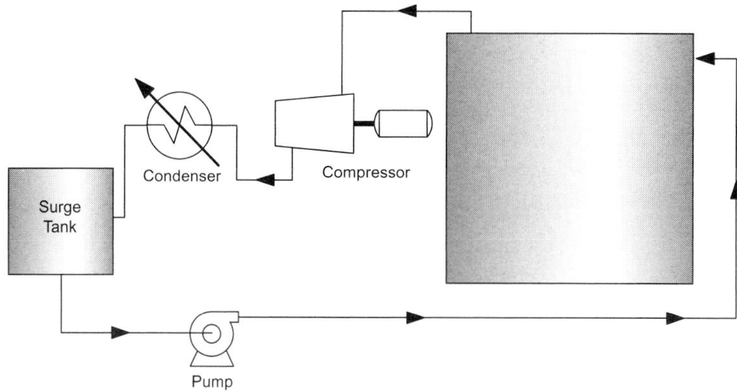

Figure 17-11: Vapor Recovery System

- **Vortex Breaker**—a metal plate, or similar device, placed inside a cylindrical or cone-shaped vessel, that prevents a **vortex** (the cone formed by a swirling liquid or gas) from being created as liquid is drawn out of the tank. Vortex breakers are used to prevent pump cavitations.

  Figure 17-12 shows an example of a vortex breaker.

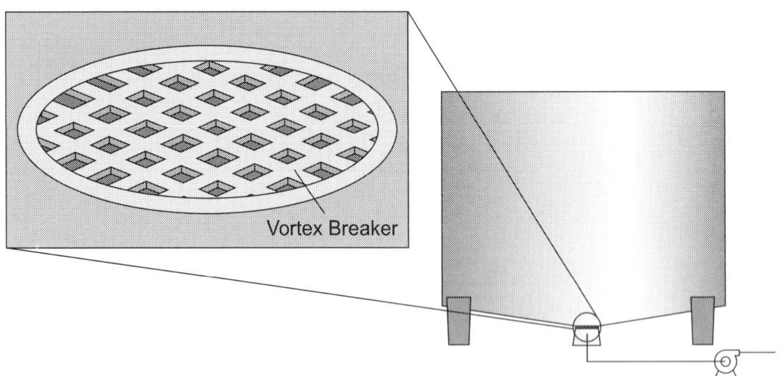

Figure 17-12: Vortex Breaker

- **Baffle**—a metal plate, placed inside tanks or vessel, which is used to alter the flow of chemicals or facilitate mixing.

  Figure 17-13 shows an example of a Baffle.

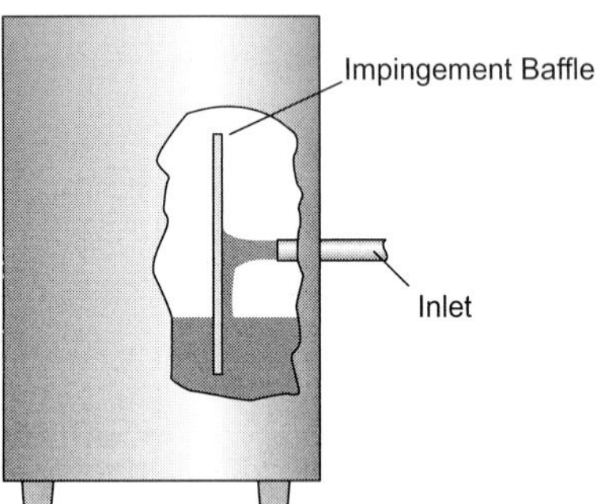

Figure 17-13: Baffle

- **Weir**—a dam or barrier normally used for either the measurement of fluid flows or to maintain a given depth of fluid (setting the level).

  Figure 17-14 shows an example of vessel (decanter) with a weir. In this example, anything that is above the weir flows over. Since gasoline is lighter than water, gasoline is the substance that is separated out.

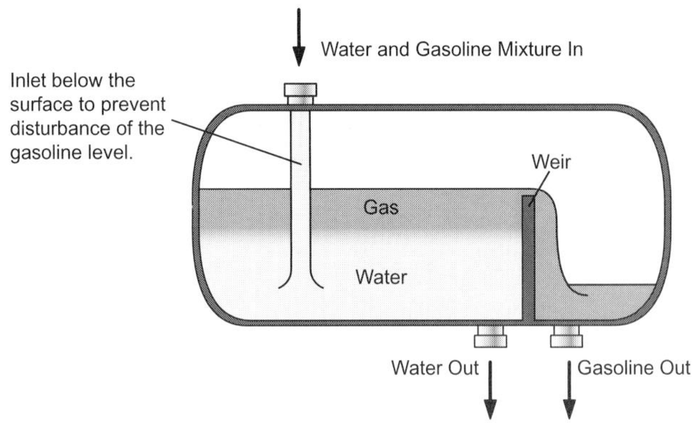

Figure 17-14: Weir in a decanter

A weir is used to separate two substances (gasoline and water) from a mixture. Gasoline, which floats on top of the water, goes over the weir and out.

- **Boot**—a section at the bottom of a process drum where water is collected and drained to waste. This space is the lowest portion of the drum.

  Figure 17-15 shows an example of a boot.

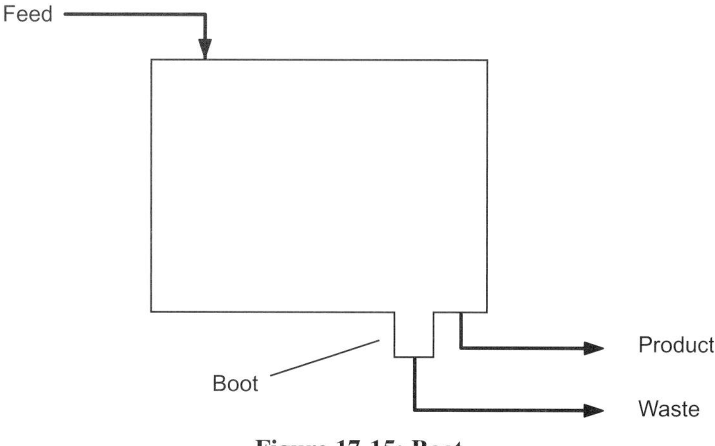

Figure 17-15: Boot

- **Mist Eliminator**—a device in the top of a tank, composed of mesh, vanes or fibers that collect droplets of mist from gas.
- **Vane Separator**—a device, composed of metal vanes, used to separate liquids from gases or solids from liquids.

PURPOSE OF DIKES, FIREWALLS, AND CONTAINMENT WALLS

Some vessels have a containment system built around them, for personal and environmental safety in case of leaks, spills or other accidents. A containment system, usually made from earth or concrete, can take the form of a dike, firewall, or containment wall, and is used to:

- Contain chemical spills in a small area in order to minimize safety risks and trap contaminants before they can spread to other areas.
- Protect the soil and the environment from contaminants.
- Protect humans from potential hazards (e.g., fire or chemical release).
- Contain wastewater and contaminated rainwater until it can be drained into a proper sewage line.
- Contain hazardous materials after a release.
- Protect the environment and people against tank failures.
- Protect against fires.

Figure 17-16 shows an example of a containment wall:

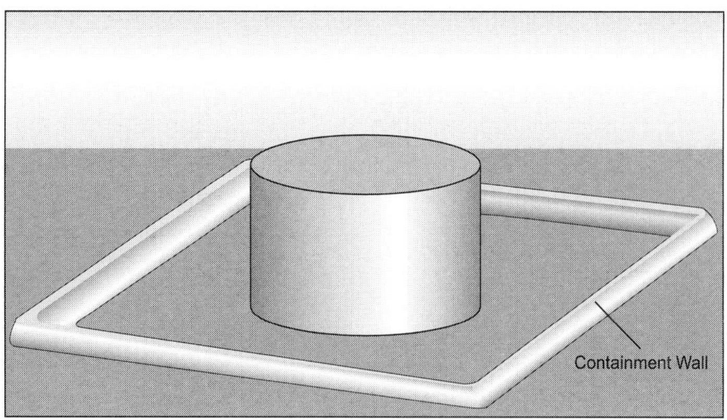

Figure 17-16: Containment wall

In the event of rainfall, technicians must open a valve and drain the liquid from within the containment wall to prevent the tank from floating.

MONITORING AND MAINTENANCE ACTIVITIES

When monitoring and maintaining vessels, technicians must always remember to look, listen, feel and smell for the following:

Look	Listen	Feel	Smell
■ Monitor levels. ■ Check firewalls, sumps, and drains. ■ Check auxiliary equipment associated with the tank. ■ Check to ensure the drain remains closed. ■ Visually inspect for leaks (especially if associated with abnormal odor). ■ Check sewer valves. ■ Use level gauges and sight glasses to monitor level. ■ Monitor leakage, level, and pressure. ■ Inspect for corrosion. ■ During heavy rainfall, open valves as needed to prevent the tank from floating.	■ Listen for abnormal noise.	■ Inspect for abnormal heat on vessels and piping. ■ Check for excessive vibration on pumps/mixers.	■ Be aware of abnormal odors that could indicate leakage. ■ Use sniffers to detect gas leaks and vapors.

Failure to perform proper maintenance and monitoring could impact the process and result in equipment damage.

HAZARDS ASSOCIATED WITH IMPROPER OPERATION

There are many hazards associated with vessels. Table 17-1 lists some of those hazards.

TABLE 17-1: Hazards associated with improper operation

IMPROPER OPERATION	POSSIBLE IMPACTS			
	Individual	**Equipment**	**Production**	**Environment**
Overfilling	Exposure to hazardous chemicals; possible injury	Damage to tank and equipment, especially floating roof tanks	Added cost for clean-up; lost product or raw material	Spill, possible fire, vapor release
Putting wrong or off-spec material in storage tank	Discipline		Added cost to remove material, clean tank, and re-run material	Possible spills when removing material and cleaning tank, or unwanted chemical reactions
Misalignment of blanket system		Loss of blanket; collapse of tank due to vacuum	Loss of production due to reduced storage	Possible vapor release
Misalignment of pump systems		Damaged pump	Contaminate other tanks	Possible spill
Pulling a vacuum on a tank while emptying		Collapse of tank due to vacuum	Loss of production due to reduced storage	Possible vapor release
Overpressure	Exposure to hazardous chemicals; possible injury	Possible rupture of vessel	Loss of production due to reduced storage	Possible vapor release

PURPOSE OF REACTORS

Reactors are vessels in which chemical reactions are initiated and sustained. Within a reactor, raw materials are combined at various flow rates, pressures and temperatures, and then reacted to form a product. These reactions can be batch reactions or continuous.

A **batch** reaction is a quantity of a product made in a single operation. In a batch reaction, raw materials are carefully measured and added to the reaction. They are then mixed and allowed to react. After a predetermined amount of time, the material is removed from the reactor, and a new batch is mixed.

A **continuous reaction** is a reaction in which products are continuously being formed as raw materials (reactants) are fed into the reactor. In other words, continuous reactors add raw materials and remove end products from the reactor.

The rate of chemical reactions can be altered (sped up or slowed down) through the application of heat, a catalyst, or an inhibitor. **Catalysts** (which can be either solid, liquid, or gas) are substances that affect the rate of a chemical reaction but are not part of the reaction. **Inhibitors** are substances which slow or stop a chemical reaction.

TYPES OF REACTORS

Reactors come in many shapes and sizes. The most common types of reactors, however, are stirred tank, fixed bed, fluidized bed, tubular, and reaction furnace. The factors that determine which type of reactor will be used are the type of catalyst and the properties of the reactants.

Stirred Tank Reactor

A **stirred tank reactor** is a reactor that contains a mixer or agitator mounted to the tank. The shell of this type of reactor may be heated or cooled, depending on the process and the design of the reactor.

Figure 17-17 shows an example of a stirred tank reactor.

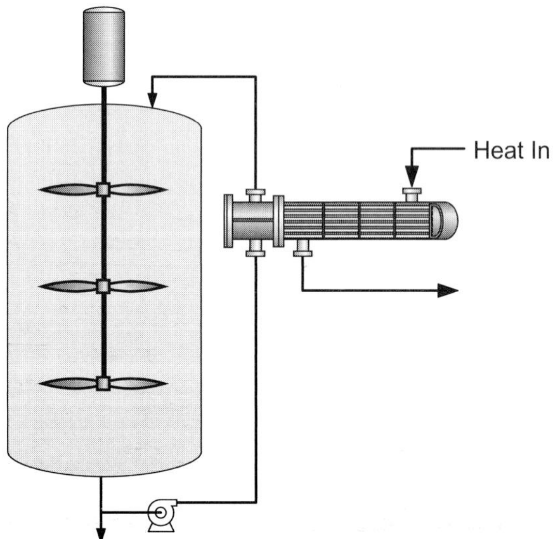

Figure 17-17: Stirred reactor

Fixed Bed Reactor

A **Fixed bed reactor** is a reactor in which the catalyst bed is stationary as the reactants are passed over it. In this type of reactor, the catalyst occupies a fixed position and is not designed to leave the reactor with the process.

Figure 17-18 shows an example of a fixed bed reactor.

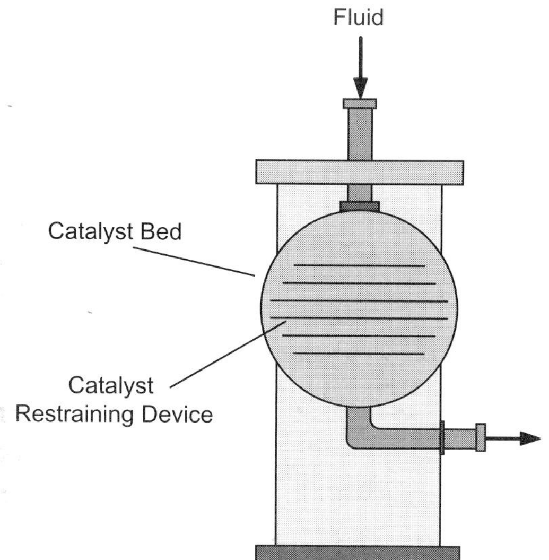

Figure 17-18: Fixed bed reactor

Fluidized Bed Reactor

A **fluidized bed reactor** is a reactor which uses countercurrent flows of gas to suspend and separate solids. In this type of reactor, gas is pumped into the bottom of the reactor, while solid particles are pumped in from the side. The countercurrent flow of gas suspends these particles, until the heavier components eventually fall to the bottom and the lighter components rise to the top.

Figure 17-19 shows an example of a fluidized bed reactor.

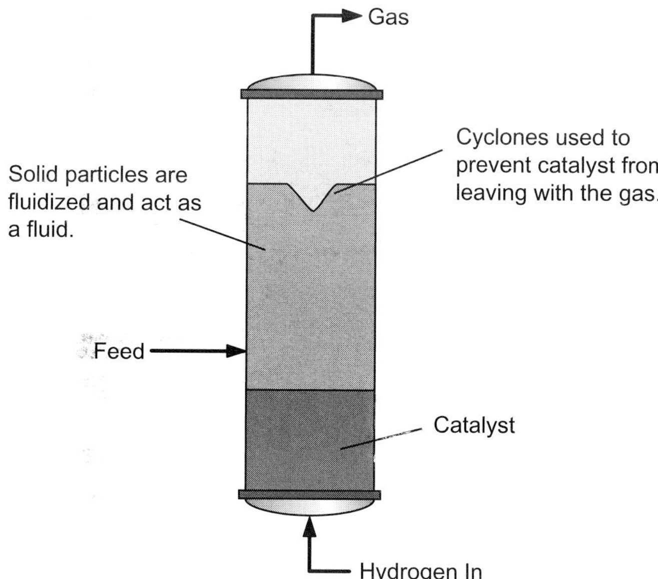

Figure 17-19: Fluidized bed reactor

Tubular Reactor

A **Tubular reactor** is a tubular heat exchanger used to contain a reaction. Based on process requirements (such as volume, surface area and pressure), the design of a tubular reactor can range from a simple jacketed tube to a multi-pass shell and tube exchanger.

Figure 17-20 shows an example of a tubular reactor.

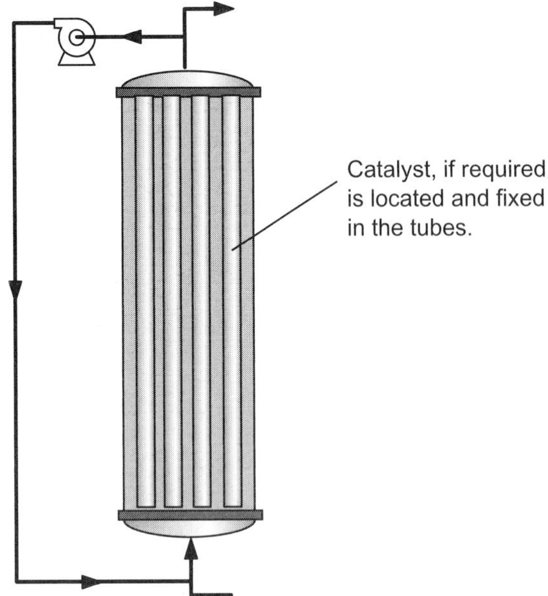
Catalyst, if required,
is located and fixed
in the tubes.

Figure 17-20: Tubular reactor

Reaction Furnaces

A Reaction furnace combines a firebox with tubing to provide heat for a reaction that occurs to the stream in the tubes. These furnaces can be used to crack hydrocarbons (decomposing the hydrocarbons using heat) or synthesize a product (such as carbon monoxide and hydrogen from steam and methane). In some processes, the furnace tubes are filled with catalyst.

SYMBOLS FOR VESSELS AND REACTORS

In order to accurately locate vessels (e.g. tanks, drums), and reactors on a piping and instrumentation diagram (P&ID) process technicians need to be familiar with the symbols that represent different types of tanks, drums, vessels, and reactors.

Figure 17-21 shows examples of vessel symbols.

Storage Symbols

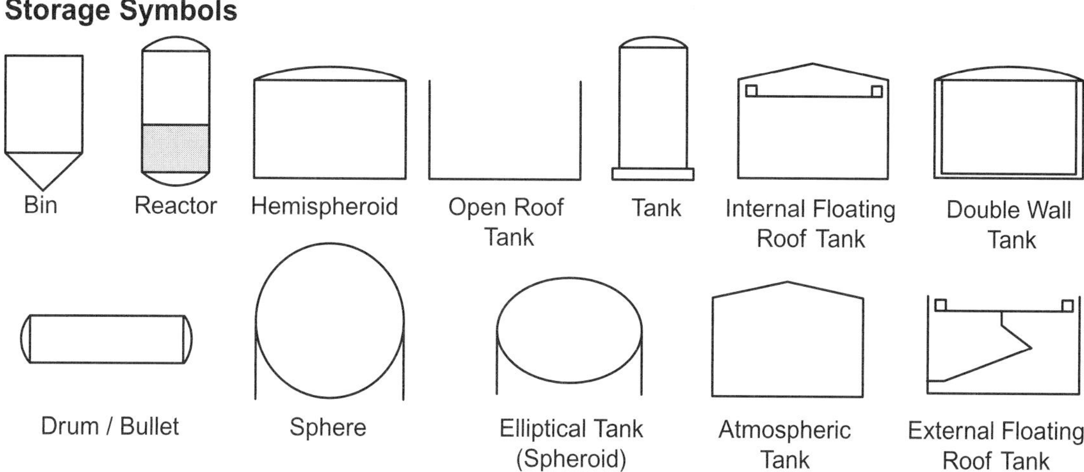

Figure 17-21: Vessel symbols

SUMMARY

Vessels are used to store raw materials and additives (process inputs), intermediate products (not yet finished), products (process outputs) and wastes (recoverable or non-recoverable off-spec products and byproducts).

Vessels include tanks, drums, cylinders, hoppers, bins and other similar containers that are used to store materials. Vessels vary greatly in design (such as size and shape) based on the requirements of the process. Factors that affect vessel design and manufacture can include pressure requirements (e.g. high and low), product storage (liquid, gas or solid), temperature requirements (e.g. insulation), corrosion factors and volume.

Reactors are specialized vessels that are used to contain a controlled chemical reaction, changing raw materials into finished products. Reaction variables include temperature, pressure, time, concentration, surface area and other factors. Like vessels, reactor designs also vary widely, based on the chemical reaction that must occur in the process.

Pressure and temperature requirements, and materials of construction are a key factors that affect~~s~~ the thickness of the wall and shape of the tank. There are two main types of storage tanks: atmospheric and pressurized.

Atmospheric tanks are enclosed vessels in which atmospheric pressure is maintained (i.e., they are neither pressurized, nor placed under a vacuum). Pressurized tanks are enclosed vessels in which a pressure greater than atmospheric is maintained. The most common types of pressurized tanks are: hemispheroid, bullet and spherical.

The process technician must recognize and understand vessel components, including floating roof, articulated drain, blanketing, spherical tank, foam chamber, sump, mixer, manway, vapor recovery system, vortex breaker, baffle, weir, boot, mist eliminator, and vane separators.

Some vessels have a containment system built around them for personal and environmental safety in case of leaks, spills or other accidents. A containment system, usually made from earth or concrete, can take the form of a dike, firewall, or containment wall.

Process technicians must always remember to look, listen, feel and smell when monitoring and maintaining vessels. They must also be aware of the hazards of improper operations of vessels.

Reactors come in many shapes and sizes. The most common types of reactors, however, are stirred tank, fixed bed, fluidized bed, tubular, and reaction furnace.

Process technicians need to be familiar with the symbols that represent different types of tanks, drums, vessels, and reactors and be able to accurately locate vessels on a piping and instrumentation diagram (P&ID).

CHECKING YOUR KNOWLEDGE

1. Define the following key terms:
   a. Articulated drain
   b. Baffle
   c. Blanket
   d. Boot
   e. Containment wall
   f. Floating roof
   g. Gauge hatch
   h. Manway
   i. Mist eliminator
   j. Mixer
   k. Sphere
   l. Sump
   m. Vane separators
   n. Vapor recovery
   o. Vortex breaker
   p. Weir

2. Which of the following tanks would the most appropriate choice for storing volatile substances under pressure?

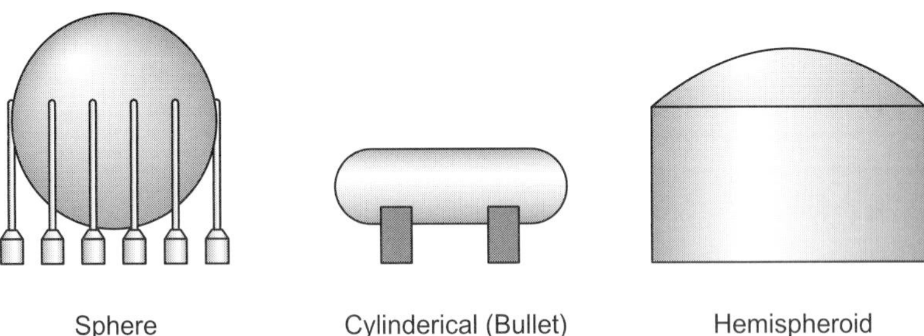

3. *(True or False)* Atmospheric tanks are good for storing substances with toxic vapors.

4. List three things dikes and containment walls are used for.

5. List at least three things a process technician should look, listen, and feel for during normal monitoring and maintenance

6. List at least three hazards/impacts associated with improper tank operation, and explain the impacts these hazards might have on individuals, equipment, production, and the environment.

ACTIVITIES

1. Given a picture of a tank, drum, or vessel, identify the following components:
   a. Articulated drain
   b. Baffle
   c. Blanket
   d. Boot
   e. Containment wall
   f. Floating roof
   g. Gauge hatch
   h. Manway
   i. Mist eliminator
   j. Mixer
   k. Sphere
   l. Sump
   m. Vane separators
   n. Vapor recovery
   o. Vortex breaker
   p. Weir

2. Describe the following types of reactors, including its design, purpose and how they work:
   a. Stirred Tank
   b. Fixed Bed
   c. Fluidized Bed
   d. Tubular
   e. Furnace

3. Given a Piping and Instrumentation Diagram (P&ID), identify vessels.

Chapter 18

Pumps

OBJECTIVES

Upon completion of this chapter you will be able to:

1. Describe the purpose or function of pumps in the process industries.
2. Explain the difference between the two common types of pumps used in the process industries: centrifugal (horizontal and vertical) and positive displacement (rotary and reciprocating).
3. Identify the primary parts of a typical centrifugal pump.
4. Describe the operations of a centrifugal pump.
5. Explain the difference between the rotary and reciprocating type of positive displacement pumps.
6. Identify the primary parts of a typical reciprocating type positive displacement pump.
7. Discuss the hazards associated with the improper operation of both the positive displacement and centrifugal pump.
8. Describe the monitoring and maintenance activities associated with pumps.
9. Identify the symbols used to represent the different types of pumps presented in this chapter.

KEY TERMS

- **Axial pump**—a dynamic pump that uses a propeller or row of blades to propel liquids axially along the shaft.
- **Cavitation**—a condition inside a pump wherein the liquid being pumped partly vaporizes due to factors such as temperature and pressure drop; occurs when the pressure on the eye of a pump impeller falls below the boiling pressure of the liquid being pumped; can be identified by noisy operation and erratic discharge pressure; can cause excessive wear on the impeller and case; often remedied by increasing the suction pressure on the pump, usually by raising the level of liquid in the suction line.
- **Centrifugal pump**—a pump that imparts velocity to liquid by centrifugal force and then converts some of the velocity to pressure.
- **Dynamic pump**—a non-positive displacement pump, classified as either centrifugal or axial, that converts centrifugal force to dynamic or flowing pressure to move liquids.
- **Positive displacement pump**—a pump that moves a constant amount of liquid through a system at a given pump speed.
- **Priming**—filling the liquid end of a pump with liquid to remove vapors present and eliminate the tendency to become vapor bound or lose suction.

- **Reciprocating pump**—a positive displacement pump that use the inward stroke of a piston or diaphragm to draw liquid into a chamber (intake) and then positively displaces (discharges) the liquid using an outward stroke.
- **Rotary pump**—a positive displacement pump that moves liquids by rotating a screw or a set of lobes, gears or vanes.
- **Volute**—a spiral casing for a centrifugal pump, designed so that speed will be converted to pressure without shock.

INTRODUCTION

Pumps are an important part of the process industries. Through pumps, technicians can move liquids from one location to another. Pumps come in many different sizes and can be either positive displacement or dynamic.

Positive displacement pumps use pistons, lobes, gears or vanes to move or "push" liquids, while centrifugal pumps use impellers to generate centrifugal force which is then converted to dynamic pressure to move liquids.

Centrifugal pumps tend to be used more often than positive displacement pumps because they are less expensive, require less maintenance, require less space, and are easier to operate.

When working with pumps, technicians should always conduct monitoring and maintenance activities according to manufacturer recommendations. These are done to ensure the pump is not over-pressurizing, overheating, leaking or cavitating, and is properly lubricated and sealed.

TYPES OF PUMPS

The two main categories of pumps are dynamic and positive displacement. Within each of these categories are subcategories (e.g., centrifugal, axial, rotary, and reciprocating).

Figure 18-1 shows a "family tree" diagram of the different types of pumps and shows how they are interrelated.

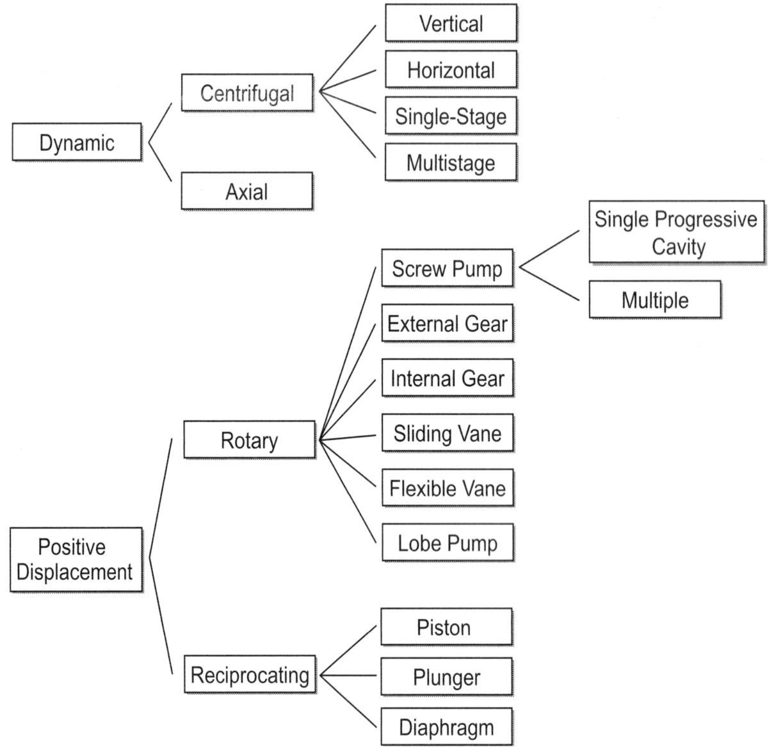

Figure 18-1: Pump "Family Tree"

POSITIVE DISPLACEMENT PUMPS

> **DID YOU KNOW?**
>
> The screw pump is one of the oldest pump designs.
>
> 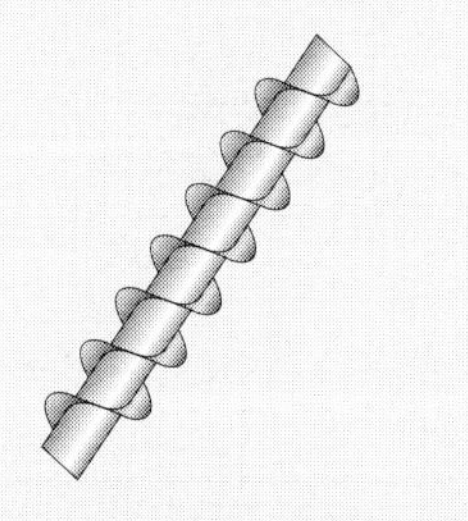
>
> It was created by Archimedes around 200 BC, and is still in use today.

Positive displacement pumps are piston, diaphragm, gear or screw pumps that deliver a constant volume with each stroke.

Unlike dynamic pumps, positive displacement pumps deliver the same amount of liquid regardless of the discharge pressure.

The two main types of positive displacement pumps are rotary and reciprocating.

Positive displacement pumps develop pressure by displacement of volume.

The pressure that is developed is not a function of pump speed. There is essentially no theoretical limit to the amount of pressure developed by a positive displacement pump. For this reason, a positive displacement pump must never be "dead headed" or "blocked in" while running.

The flow rate of a reciprocating positive displacement pump is equal to the piston size, times the stroke length, times the stroke rate.

By varying the stroke rate, the flow rate can be precisely controlled. These pumps are frequently used as "metering pumps."

Rotary Pumps

Rotary pumps move liquids by rotating a screw or a set of lobes, gears or vanes that physically push the liquid through the pump. As these screws, lobes, gears or vanes rotate, the liquid is drawn into the pump by lower pressure on one side and forced out of the pump (discharged) through higher pressure on the other.

Figure 18-2 contains three examples of rotary pumps.

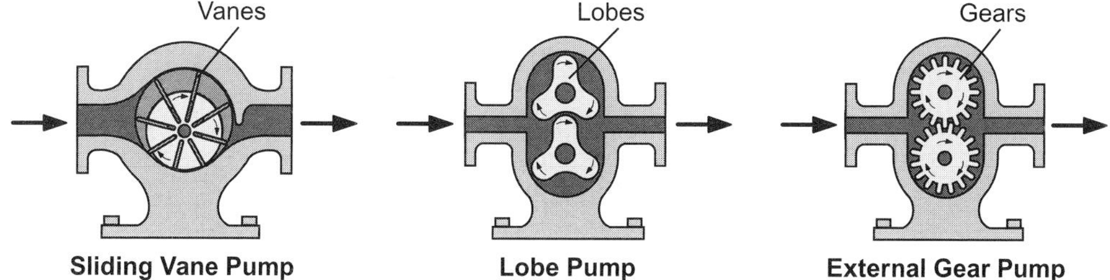

Figure 18-2: Examples of Rotary Pumps

The diagram in Figure 18-3 shows the main components of a rotary pump. These include:

- Housing
- Shaft
- Inlet valve
- Outlet valve
- Lobes, gears or vanes

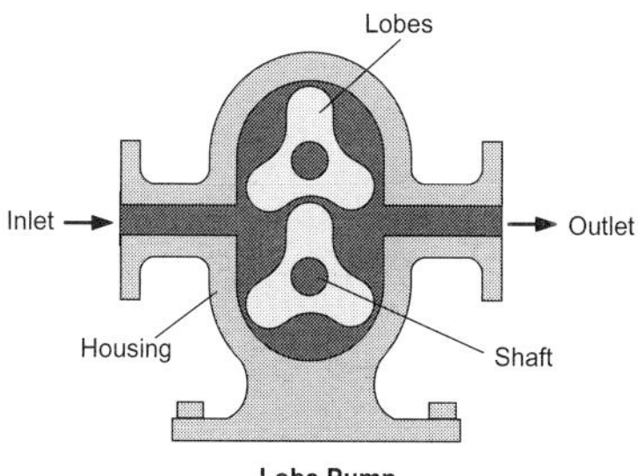

Figure 18-3: Rotary Pump Components

Reciprocating Pumps

Reciprocating pumps use the inward stroke of a piston or diaphragm to draw (intake) liquid into a chamber and then uses an outward stroke to positively displace (discharge) the liquid.

To better illustrate the actions of a reciprocating pump, think of a syringe. As a syringe plunger is pulled out of its housing, liquid is drawn in. As the plunger is pushed back in to the housing, liquid is forced out (discharged).

Figure 18-4 shows an example of a reciprocating pump and how liquid is drawn in and discharged.

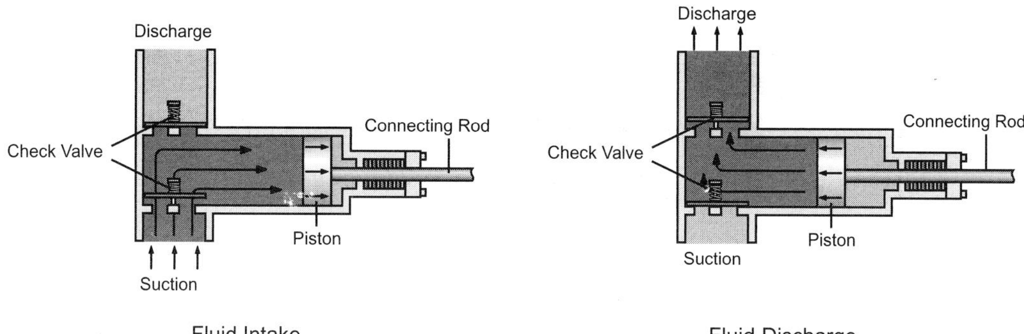

Figure 18-4: Piston-Type Reciprocating Pump

Figure 18-5 shows the main components of a reciprocating pump. These include:

- Housing
- Inlet check valve
- Outlet check valve
- Piston and connecting rod (shaft)
- Packing/Seal

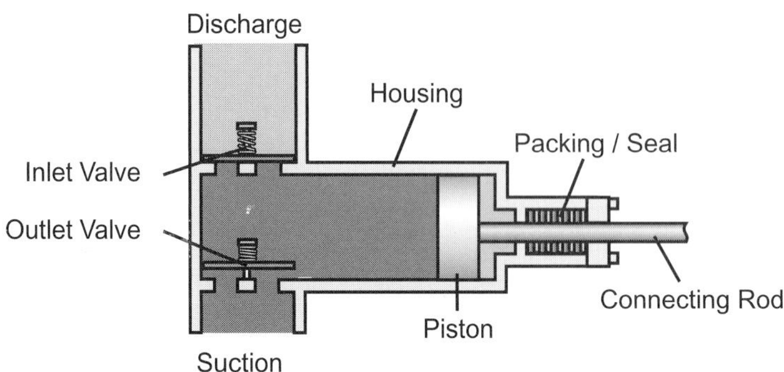

Figure 18-5: Piston pump components (shown in intake position)

DYNAMIC PUMPS

Dynamic pumps are non-positive displacement pumps that convert centrifugal force to dynamic pressure to move liquids (as opposed to positive displacement pumps which use a piston to "push" liquids). They are classified as either centrifugal or axial.

Centrifugal Pumps

Centrifugal pumps use an impeller on a rotating shaft to generate pressure and move liquids. In a centrifugal pump, a rotating impeller spins creating centrifugal force. This centrifugal force creates pressure in the liquid as it passes through a widening of the casing known as a volute.

Centrifugal pumps develop pressure by centrifugal force. The pressure that is developed is called "liquid head". The amount of liquid head that is developed is a function of the tip speed of the impeller (Impeller RPM & Impeller diameter).The higher the tip speed, the higher the liquid head.

Each centrifugal pump has a characteristic pump curve. The maximum liquid head occurs at zero flow, which is called the dead head pressure.

As the flow rate is increased, the liquid head decreases. This continues until the maximum pumping rate of the pump is reached.

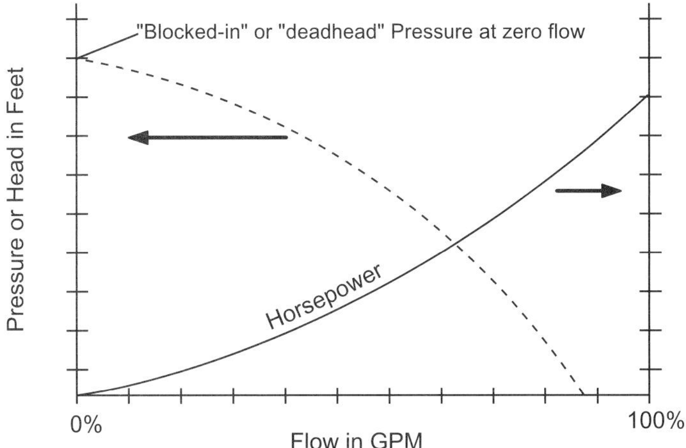

Figure 18-6: Typical centrifugal pump curve

Centrifugal pumps can be “dead headed” or “blocked-in” (zero flow) for short periods of time with no adverse consequences. This is often necessary when starting up or switching pumps. Some centrifugal pumps (especially very high speed and multi-stage pumps) may have excess vibration or undesired overheating if “dead headed.” “Dead heading” would not be allowed.

Figure 18-7 shows an example of a centrifugal pump.

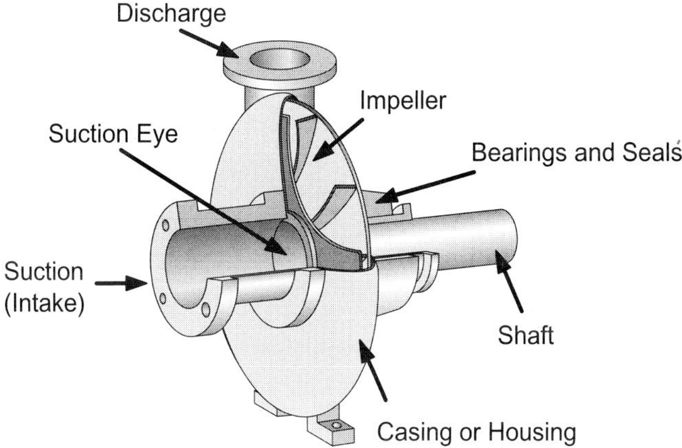

Figure 18-7: Centrifugal Pump

The main components of a centrifugal pump include:

- Housing (casing)
- Shaft
- Inlet (suction eye)
- Outlet (discharge)
- Impeller
- Bearings and seals

Centrifugal pumps differ from positive displacement pumps in that the amount of liquid they deliver is dependent on the discharge pressure, not the size of the chamber.

In centrifugal pumps there is a direct relationship between speed, velocity, pressure and flow. As the velocity decreases, the pressure increases

Axial Pumps

Axial pumps use a *propeller* or rows of blades to propel liquids axially along the shaft (as opposed to centrifugal pumps which use an *impeller* to force liquids to the outer wall of the chamber).

Figure 18-8 shows an example of an axial pump.

The main components of an axial pump include:

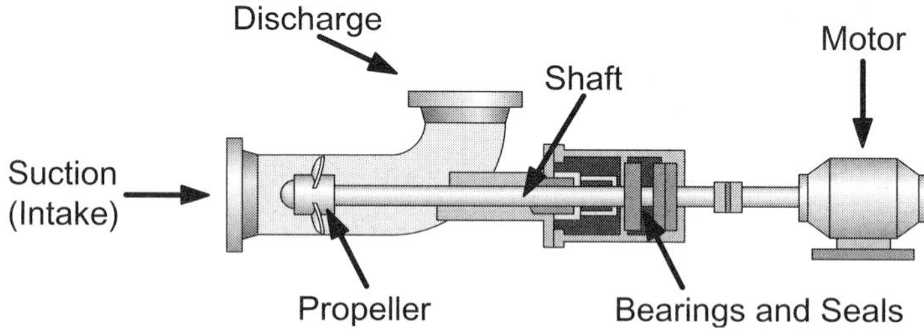

Figure 18-8: Axial Pump

- Shaft
- Inlet (intake)
- Outlet (discharge)
- Propeller
- Bearings and seals

HAZARDS ASSOCIATED WITH IMPROPER OPERATION

When working with pumps, process technicians should always be aware of potential hazards such as over-pressurization, overheating, cavitation, and leakage.

Positive Displacement Pumps

OVER-PRESSURIZATION

Pump over-pressurization can occur if the valves beyond the pump are incorrectly closed or blocked. Consider the example in Figure 18-9.

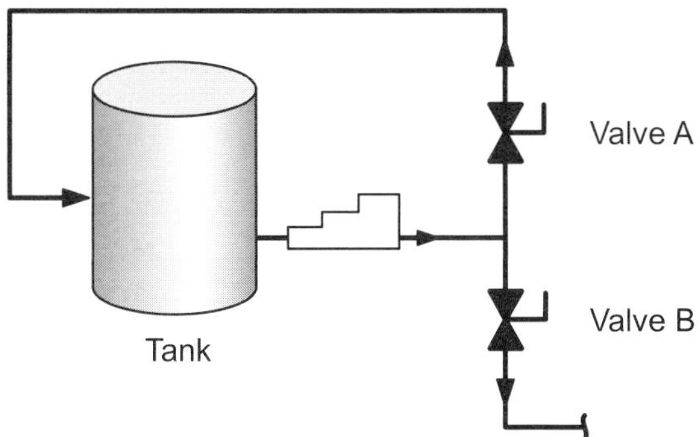

Figure 18-9: Improper valve operation ("Deadheading")

In Figure 18-9, valve A is open and valve B is closed. This means the liquid is flowing through valve A, back into the tank.

If a technician were to redirect the flow of the liquid through valve B, he or she must open valve B before closing valve A.

If both valves are closed at the same time, the liquid has no place to go. The result is back-pressure that can damage ("dead head") the pump or cause serious personal injury.

Centrifugal Pumps

OVERHEATING

Pump overheating is often caused by improper lubrication. Without lubrication, bearings fail and equipment surfaces rub together. As these surfaces rub against one another, friction is produced and heat is generated. This can cause mechanical failures, swelling, leakage, and decomposition of the process fluid.

If the pump is dead headed, the liquid in the pump will be heated by the mechanical energy of the motor. In many cases recycle loops will be added to allow flow through the pump even if the valves downstream are closed to prevent overheating of the liquid.

Process technicians should always monitor rotating equipment for unusual sounds and excessive heat, since operating equipment under these conditions can lead to permanent equipment damage or personal injury (e.g., burns).

CAVITATION

Cavitation is a condition inside a pump wherein the liquid being pumped partly vaporizes due to variables such as temperature and pressure drop. These variable changes cause vapor pockets (bubbles) to form and collapse (implode) inside a pump.

Cavitation occurs when the pressure on the eye of a pump impeller falls below the boiling pressure of the liquid being pumped. This is a very serious problem in dynamic pumps, especially centrifugal pumps. Cavitation is also a problem in vacuum operations since low pressure liquids boils at lower temperatures.

Key characteristics of cavitation include large pressure fluctuations, inconsistent flow rate, and severe vibration. Technicians who identify cavitation should always try to eliminate it as quickly as possible, since it can cause excessive wear on the pump seal, impeller, bearings, and case.

DID YOU KNOW?

When a pump cavitates it sounds as if rocks are being poured into the suction line.

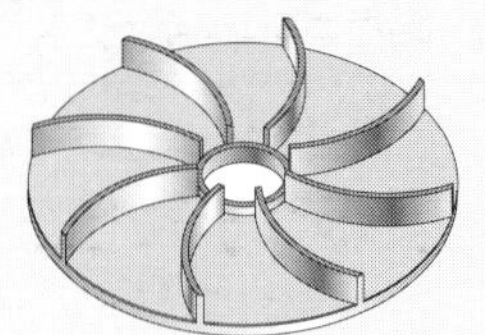

(Impeller)

These bursts of liquid and gas create a "liquid hammer" that can seriously damage the inside of the pump and the impeller.

Pumps that have been damage by cavitation frequently appear as though the internal workings of the pump have been sandblasted or pitted, particularly on the outer edges of the vanes.

To prevent cavitation, a pump should always be primed before it is started. **Priming** a pump involves filling the liquid end of a pump with liquid to remove any vapors that might be present. This eliminates the tendency to become vapor bound or lose suction.

Cavitation can still occur even if a pump was properly primed, however. For example, if the liquid becomes too hot, vapor bubbles can be created. The way to correct this is by cooling the liquid. Another way to stop cavitation is to raise the level of the liquid in the suction line to increase the suction pressure on the pump.

LEAKAGE

Process technicians should always check pumps for leaks, since leaks can introduce slipping hazards, exposure to harmful or hazardous substances, and process problems (e.g., inferior product produced as a result of improper feed supply).

Leaks most frequently occur where the pump shaft exits the pump housing. Packing (rope-like material) or a mechanical seal is normally used to prevent this leakage from occurring. Leaks occurring at this location are normally corrected by tightening or replacing the packing or replacing the mechanical seal.

A minimal amount of leakage is necessary on pump packing for pump lubrication and cooling.

MONITORING AND MAINTENANCE ACTIVITIES ASSOCIATED WITH PUMPS

When monitoring and maintaining pumps technicians must always remember to look, listen and feel for the following:

Look	Listen	Feel
■ Check oil levels to make sure they are satisfactory. ■ Check to make sure water is not collecting under the oil (water is not a lubricant, so it can cause bearing failure). ■ Check seals and flanges to make sure there are no leaks. ■ Check suction and discharge pressure gauges (note: technicians need to be aware that pressure gauges in vibrating areas can lose their calibration).	■ Listen for abnormal noises.	■ Feel for excessive vibration. ■ Feel for excessive heat.

Failure to perform proper maintenance and monitoring could impact the process and result in equipment damage.

PUMP SYMBOLS

In order to accurately locate pumps on a piping and instrumentation diagram (P&ID) process technicians need to be familiar with the different types of pumps and their symbols.

Figure 18-10 shows a few of the symbols for both centrifugal and positive displacement pumps.

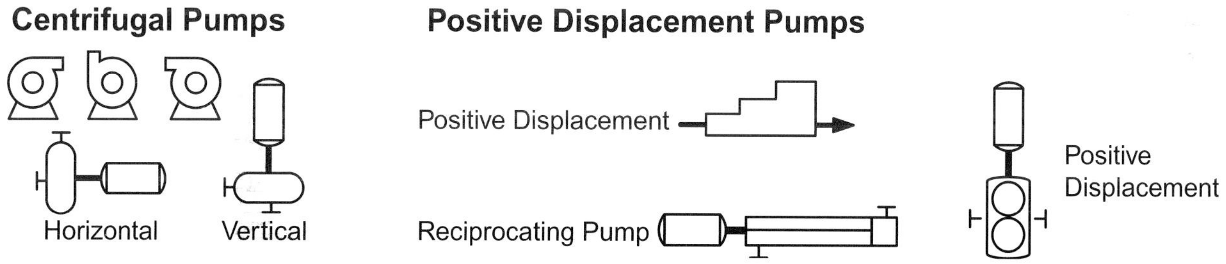

Figure 18-10: Common Pump Symbols

SUMMARY

Pumps are devices that are used to move liquids from one location to another. Pumps may be large or small, and may be dynamic (centrifugal) or positive displacement (reciprocating/rotary).

Positive displacement pumps draw a liquid into a chamber and then use pistons, lobes, gears or vanes to force the liquid out.

Centrifugal pumps use an impeller on a rotating shaft to generate pressure and force liquids out of the pump. As the impeller speed increases, velocity increases. As the volute size increases, velocity decreases, which causes the pressure to increase.

There are many hazards associated with improper pump operation. These include over-pressurization, overheating, cavitation, and leakage.

When monitoring and maintaining pumps, technicians must always remember to look, listen and feel to ensure oil levels are correct, that there are no leaks, that there are no abnormal noises, and that the equipment is not producing excessive heat or vibration.

CHECKING YOUR KNOWLEDGE

1. Explain the purpose of pumps in the process industries.

2. Is the diagram below an example of a centrifugal pump, or a positive displacement pump?

3. Which type of pump uses a piston to force liquids out of a chamber?
   a. Axial
   b. Centrifugal
   c. Positive Displacement

4. On the diagram below, identify the following parts of a centrifugal pump.
   - Bearings and seals
   - Casing
   - Discharge
   - Impeller
   - Shaft
   - Suction eye

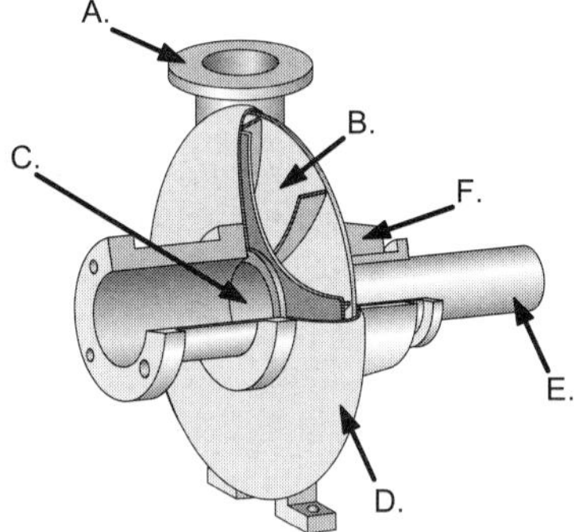

5. On the diagram below, identify the following parts of a reciprocating pump.
   - Casing
   - Connecting rods
   - Inlet valve
   - Outlet valve
   - Piston
   - Seals

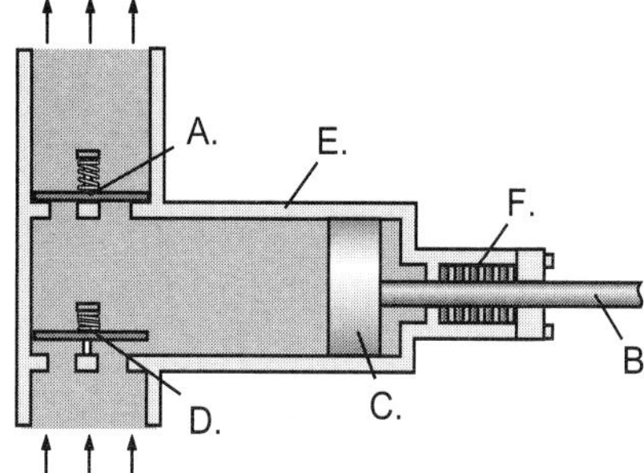

6. List four hazards associated with improper pump operation.

7. List five monitoring and maintenance activities that a technician should perform when working with pumps.

8. Draw piping and instrumenation diagram (P&ID) symbols for the following:
   a. Centrifugal pump
   b. Positive displacement pump
   c. Reciprocating pump

ACTIVITIES

1. Given a centrifugal pump, or a picture of a centrifugal pump, identify the following components:
   a. Casing
   b. Discharge
   c. Impeller
   d. Shaft
   e. Suction eye

2. Given a reciprocating pump, identify the following components:
   a. Casing
   b. Connecting rods
   c. Inlet valve
   d. Outlet valve
   e. Piston

3. Explain how you would perform the following maintenance and monitoring activities on a typical pump:
   a. Inspect for abnormal noise
   b. Inspect for excessive heat
   c. Check oil levels
   d. Check for leaks around seals and flanges
   e. Check for excessive vibration

4. Given a piping and instrumenation diagram (P&ID), identify all of the pumps and tell how many of them are centrifugal and how many are positive displacement.

5. Examine the concept of centrifugal force by doing the following:
   a. Obtain a small sand bucket or pail.
   b. Fill the pail with water until it is half full.
   c. Locate an open area away from other individuals or obstructions.
   d. In the open area, grasp the pail by its handle and swing it in a circular motion, arms fully extended, repeatedly rising over your head and back down to your knees.
   e. Examine what happens. Did the water stay in the pail or did it spill out?

6. Examine the actions of a reciprocating pump by doing the following:
   a. Obtain a syringe (without a needle) and a cup of water.
   b. Depress the syringe plunger all the way into the housing.
   c. Place the syringe in the cup of water and pull the plunger back until the syringe housing is full of water.
   d. Lift the syringe out of the water and then depress the plunger so the water is forced out of the syringe, into the cup.
   e. Repeat steps b-d several times. Each time, vary the amount of pressure applied to the plunger when forcing the water out of the syringe.
   f. Examine what happens. Did the amount of liquid discharged change as the pressure changed, or did it remain the same?

Chapter 19

Compressors

OBJECTIVES

Upon completion of this chapter you will be able to:

1. Describe the purpose or function of compressors in the process industries.
2. Explain the difference between a pump and compressor in terms of what function each performs.
3. Explain the difference between the two more common types of compressors used in the process industries: positive displacement and dynamic.
4. Explain the difference between the rotary and reciprocating type of positive displacement compressors.
5. Identify the primary parts of a typical reciprocating type positive displacement compressor.
6. Describe the operations of a positive displacement compressor.
7. Identify the primary parts of a typical centrifugal compressor.
8. Describe the operations of the centrifugal compressor.
9. Discuss the hazards associated with the improper operation of both the positive displacement and centrifugal compressor.
10. Describe the monitoring and maintenance activities associated with compressors.
11. Identify the symbols used to represent the different types of compressors.

KEY TERMS

- **Axial Compressor**—a dynamic-type compressor which uses a series of blades with a set of stator blades between each rotating wheel. In this type of compressor, the gas flow is axial, or straight through, parallel to the compressor shaft.
- **Blower**—a limited discharge compressor (usually below 100 PSI) that is used to move (airvey) powders or pellets from one point to another.
- **Centrifugal Compressor**—a dynamic type compressor using a series of impellers in which the gas flows from the inlet located near the shaft to the outer tip of the impeller blade. Flow is then routed from the outer edge of one stage back to the inlet port of the next stage.
- **Compression Ratio**—the ratio of discharge pressure (psia) to inlet pressure (psia).
- **Compressor**—mechanical device used to compress gases and vapors for use in a process system that requires a higher pressure.
- **Discharge**—normally refers to the outlet side of a pump, compressor, fan or jet.

- **Dynamic Compressor**—a compressor that uses centrifugal or rotational force to move gases (as opposed to positive displacement compressors which use a piston to compress the gas). Dynamic compressors are classified as either centrifugal or axial.
- **Positive Displacement Compressor**—compressors that use screws, sliding vanes, lobes, gears or diaphragms to deliver a set volume of gas; utilizes either reciprocating or rotary motion to trap a specific amount of gas and reduce its volume, thereby increasing the pressure at the discharge.
- **Reciprocating Compressor**—a type of positive displacement compressor that consists of a cylinder that contains a piston that travels back and forth (reciprocates) in a cylinder containing suction valves and discharge valves.
- **Rotary Compressor**—a type of positive displacement compressor that uses a rotating motion to move the gas. There are three basic compressor designs: screw, sliding vane, and lobe.
- **Suction**—normally refers to the inlet side of a pump, compressor, fan or jet.

INTRODUCTION

Compressors are an important part of the process industries. The primary function of a compressor is to compact or compresses gases and vapors.

Compressors can be used in a wide variety of applications. For example, they can be used to accelerate or compact gases (e.g. carbon dioxide, nitrogen, light hydrocarbons), or compress the air that is used to control instruments, or operate equipment power tools. Specialized compressors called blowers can also be used to transfer granular powders or pellets within a process.

The two most common compressor types are positive displacement and dynamic (centrifugal). Positive displacement compressors use pistons, lobes, gears or vanes to "push" gases. Dynamic, or centrifugal, compressors use impellers to generate centrifugal force.

Dynamic compressors are more commonly used than positive displacement compressors because they are less expensive, more efficient, have a larger capacity, and require less maintenance.

All compressors require a drive mechanism (e.g. electric motors, turbines) to operate, and are rated according to capacity (referencing discharge pressure in psi and flow rate in cubic feet per minute). Most compressors require auxiliary components for cooling, lubrication, filtering, instrumentation and control. Some compressors require a gearbox between the driver and compressor to increase the speed of the compressor.

DIFFERENCES BETWEEN COMPRESSORS AND PUMPS

A **compressor** is a mechanical device used to compress gases and vapors for use in a process system that requires a higher pressure. Compressors are similar to pumps in some ways, but different in others. Compressors and pumps are both commonly used in the process industries to move products in a process (pumps are used to move liquids from one location to another in a process system, and compressors are used to compress gases and vapors). In addition, they both use centrifugal or positive displacement action. They both have similar operation and maintenance tasks. Line-up (opening the necessary valves in a piping system prior to placing a device in service), suction and discharge are very important to both compressors and pumps.

Pumps are used to move liquids. Compressors are used in gas or vapor service and cannot tolerate any liquid, while pumps are used in liquid and slurry service and cannot tolerate any gas. The startup procedures are different for compressors and pumps.

POSITIVE DISPLACEMENT COMPRESSORS

Positive displacement compressors are compressors that use screws, sliding vanes, lobes, gears or diaphragms to deliver a set volume of gas with each stroke.

The following sections discuss the two main types of displacement compressors: rotary and reciprocating.

Rotary Compressors

Rotary compressors move gases by rotating a screw, a set of lobes, or a set of vanes. As these screws, lobes or vanes rotate, gas is drawn into the compressor by negative pressure on one side, and forced out of the compressor (discharged) through positive pressure on the other.

Figure 19-1 shows three examples of rotary compressors.

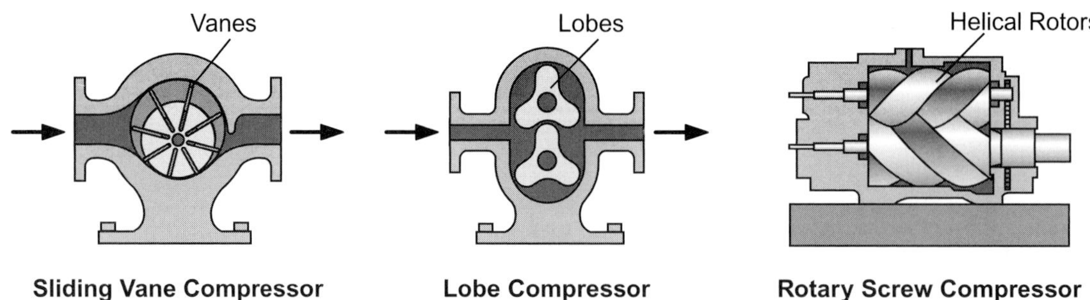

Figure 19-1: Examples of Rotary Compressors

The diagram in Figure 19-2 shows the main components of a rotary compressor:

- Housing (casing)
- Shaft
- Inlet (suction)
- Outlet (discharge)
- Lobes, gears or vanes

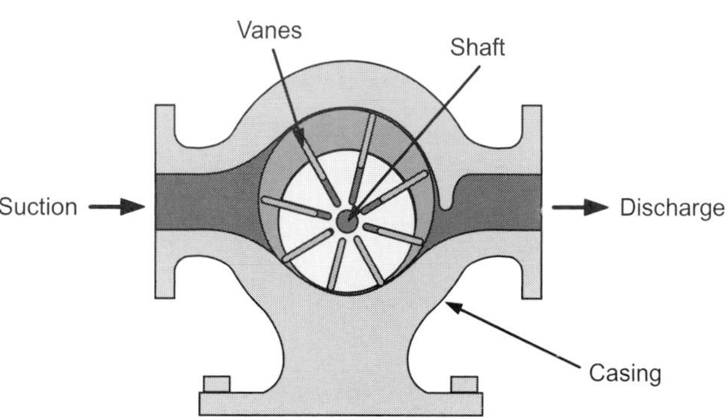

Figure 19-2: Rotary (Sliding Vane) Compressor Components

Reciprocating Compressors

Like a reciprocating pump, a **reciprocating compressor** uses the inward stroke of a piston or diaphragm to draw (intake) gas into a chamber and then uses an outward stroke to positively displace (discharge) the gas.

A good example of a reciprocating compressor would be a manually operated bicycle pump. In this type of compressor, air is drawn into the chamber when the handle (which is attached to a piston) is pulled up, and forced out when the handle is pushed down.

Figure 19-3 shows an example of a reciprocating compressor and how gas is drawn in and discharged.

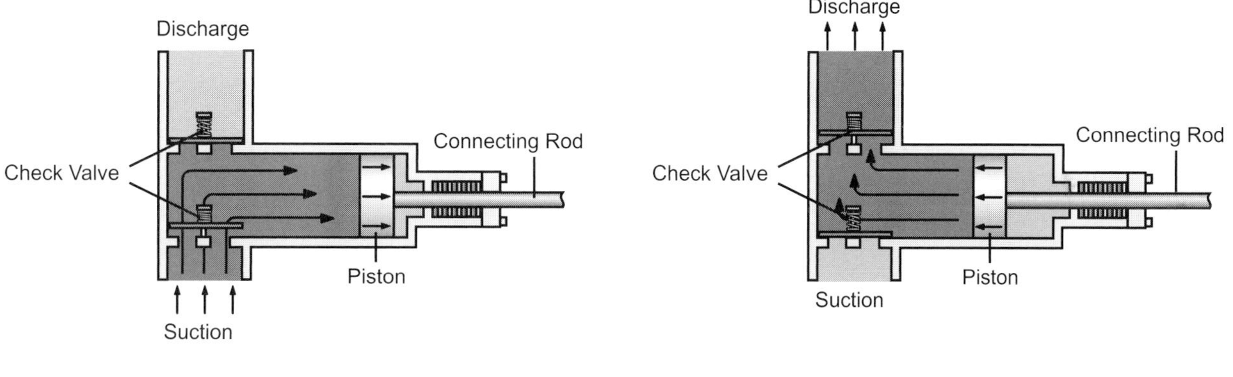

Figure 19-3: Piston-Type Reciprocating Compressor

Figure 19-4 shows the main components of a reciprocating compressor:

- Housing
- Inlet (suction)
- Outlet (discharge)
- Piston
- Connecting rod (shaft)
- Valves

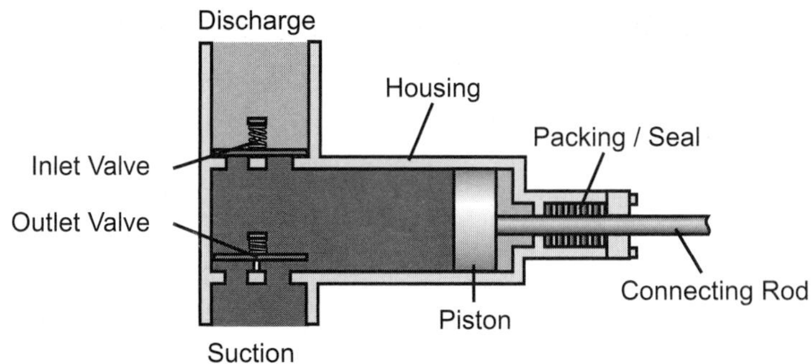

Figure 19-4: Piston compressor components

DYNAMIC COMPRESSORS

Dynamic compressors are non-positive displacement compressors that use centrifugal or rotational force to move gases (as opposed to positive displacement compressors which use a piston to compress gas). These types of compressors are classified as either centrifugal or axial, and they use principles such as centrifugal force or axial movement to increase the pressure of the gas.

Centrifugal Compressors

Centrifugal compressors use an impeller on a rotating shaft to generate pressure and move gases. In a centrifugal compressor, a rotating impeller spins creating centrifugal force. This centrifugal force creates pressure in the gas as it passes through a widening of the casing, known as a volute. As pressure increases, so does gas flow.

Centrifugal compressors may be single stage, or multiple stages, and the stages may be contained in one casing or several different cases.

Figure 19-5 shows an example of a centrifugal compressor.

The main components of a centrifugal compressor include:

- Housing (casing)
- Shaft
- Suction (inlet)
- Discharge (outlet)
- Impeller

Centrifugal compressors differ from positive displacement compressors in that the amount of gas they deliver is dependent on the discharge pressure, not the size of the chamber.

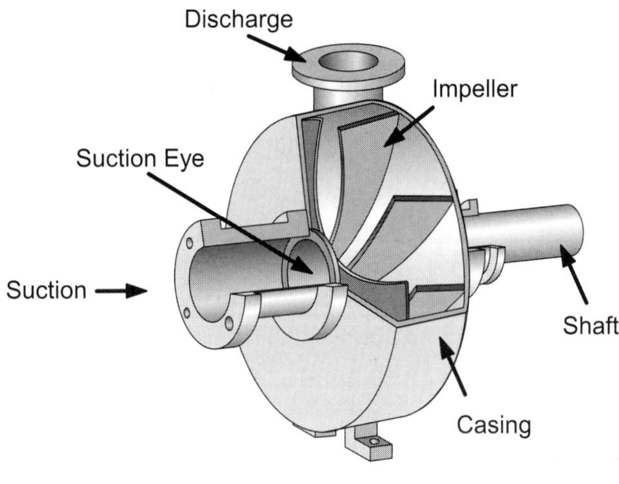

Figure 19-5: Centrifugal Compressor

In centrifugal compressors there is a direct relationship between impeller speed, velocity, pressure and flow. As the impeller speed increases, velocity increases. As velocity increases, pressure increases. As pressure increases, flow increases.

Centrifugal compressors are one of the most common types of compressors in the process industries because they are relatively economical, deliver much higher flow rates than positive displacement compressors, and take up less space.

Large multistage compressors can be extremely complex with many subsystems including bearing oil systems, seal oil systems, and extensive vibration detection systems.

Axial Compressors

Axial compressors use a series of rotor and stator blades to move gases axially along the shaft (as opposed to centrifugal compressors which use an impeller to force gases to the outer wall of the chamber).

Rotor blades are attached to the shaft and stator blades are attached to the internal walls of the compressor casing. These blades decrease in size as the casing size decreases. Rotation of the shaft causes flow to be directed axially along the shaft building higher pressure toward the discharge of the machine.

Figure 19-6 shows an example of an axial compressor.

The main components of an axial compressor include:

- Shaft
- Suction (inlet)
- Discharge (outlet)
- Inlet guide vanes
- Rotor blades
- Stator blades

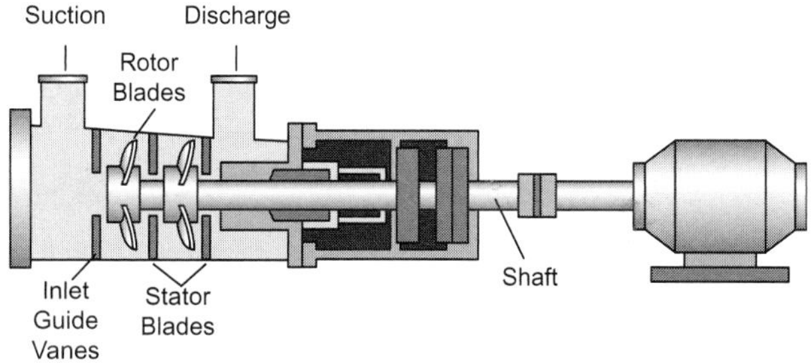

Figure 19-6: Axial Compressor

MULTI-STAGE COMPRESSORS

In multi-stage compressors, the temperature of a gas increases as it is compressed. The amount of temperature increase is a function of the gas and the compression ratio. The compression ratio in a compressor must usually be limited to around 3:1 to 5:1 to avoid extremely high discharge temperatures.

Frequently, the desired discharge pressure is over 10 times that of the inlet pressure. A single stage compressor could not be used because of the high temperature. A multi-stage compressor with cooling after each stage would be required. When the discharge gas from a compressor is cooled, liquids are frequently condensed. These liquids must be removed and not allowed to enter the compressor. Since the liquids are non-compressible, severe damage to the compressor will occur if they are allowed to enter the compressor.

HAZARDS ASSOCIATED WITH IMPROPER OPERATION

When working with compressors, process technicians should always be aware of potential hazards such as over-pressurization, overheating, surging, and leakage.

Over-Pressurization

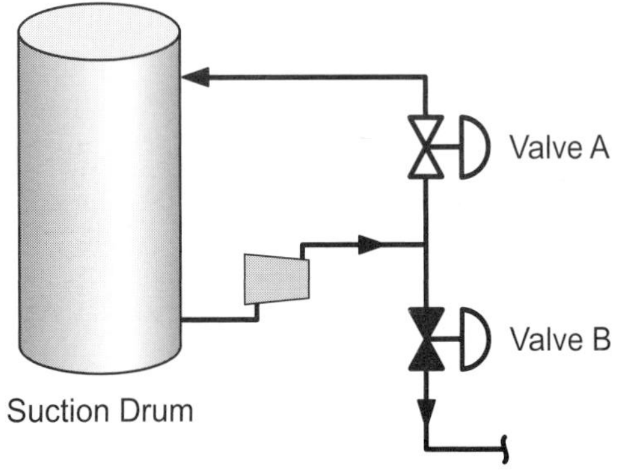

Figure 19-7: Valve operation

Compressor over-pressurization can occur if the valves associated with the compressor are incorrectly closed or blocked. Consider the example in Figure 19-7 (which is the same as the example presented in the chapter on pumps).

In Figure 19-7, valve A is open and valve B is closed. This means the gas is flowing through valve A, and then back into the tank. If a technician were to redirect the flow of the gas through valve B, he or she must open valve B before closing valve A. If both valves are closed at the same time, the gas has no place to go. This is referred to as "deadheading." The result is backflow pressure that can damage the compressor or cause serious personal injury.

Overheating

Compressor overheating is often caused by improper lubrication. Without lubrication, bearings fail and equipment surfaces rub together. As these surfaces rub against one another, friction is produced and heat is generated.

Overheating can also occur, especially on multi-stage compressors, when compressor valves malfunction. This can result in excessive compression ratios and very high gas temperatures.

Process technicians should always monitor rotating equipment for unusual sounds and excessive heat, since operating equipment under these conditions can lead to permanent equipment damage or personal injury (e.g., burns).

Surging

Another operational hazard is surging. Surging is a temporary loss of flow to one or more impellers or stages of the compressor. Surging, which is typically associated with centrifugal compressors, causes the compressor speed to fluctuate wildly and vibration to increase dramatically.

Leakage

Process technicians should always check compressors for leaks, since leaks can introduce harmful or hazardous substances into the atmosphere and create process problems (e.g., improper instrument and controller air pressure could cause erroneous readings). Check the cylinders or housing for liquid.

MONITORING AND MAINTENANCE ACTIVITIES ASSOCIATED WITH COMPRESSORS

When working with compressors, process technicians should always conduct monitoring and maintenance activities to ensure the compressor is not over-pressurizing, overheating or leaking.

On a daily basis process technicians should check vibration, oil flow, oil level, temperature and pressure. They should also make sure that connectors, hoses, and pipes are in proper condition.

On a periodic basis, technicians should also check overspeed trips, oil levels, and check for leaks at seals, packing and flanges.

While working with compressors, process technicians should, look, listen and feel for the following:

Look	Listen	Feel
■ Check oil levels to make sure they are satisfactory. ■ Check seals and flanges to make sure there are no leaks. ■ Check vibration monitors to ensure they are within operating range. ■ Check liquid level in suction drum to make sure liquid level is correct.	■ Listen for abnormal noises.	■ Feel for excessive vibration. ■ Feel for excessive heat.

Failure to perform proper maintenance and monitoring could impact the process and result in equipment damage.

COMPRESSOR SYMBOLS

In order to locate compressors on a piping and instrumentation diagram (P&ID), process technicians must be familiar with the different types of compressors and their symbols. Figure 19-8 shows examples of symbols for both centrifugal and positive displacement compressors. However, these symbols may be variable from facility to facility.

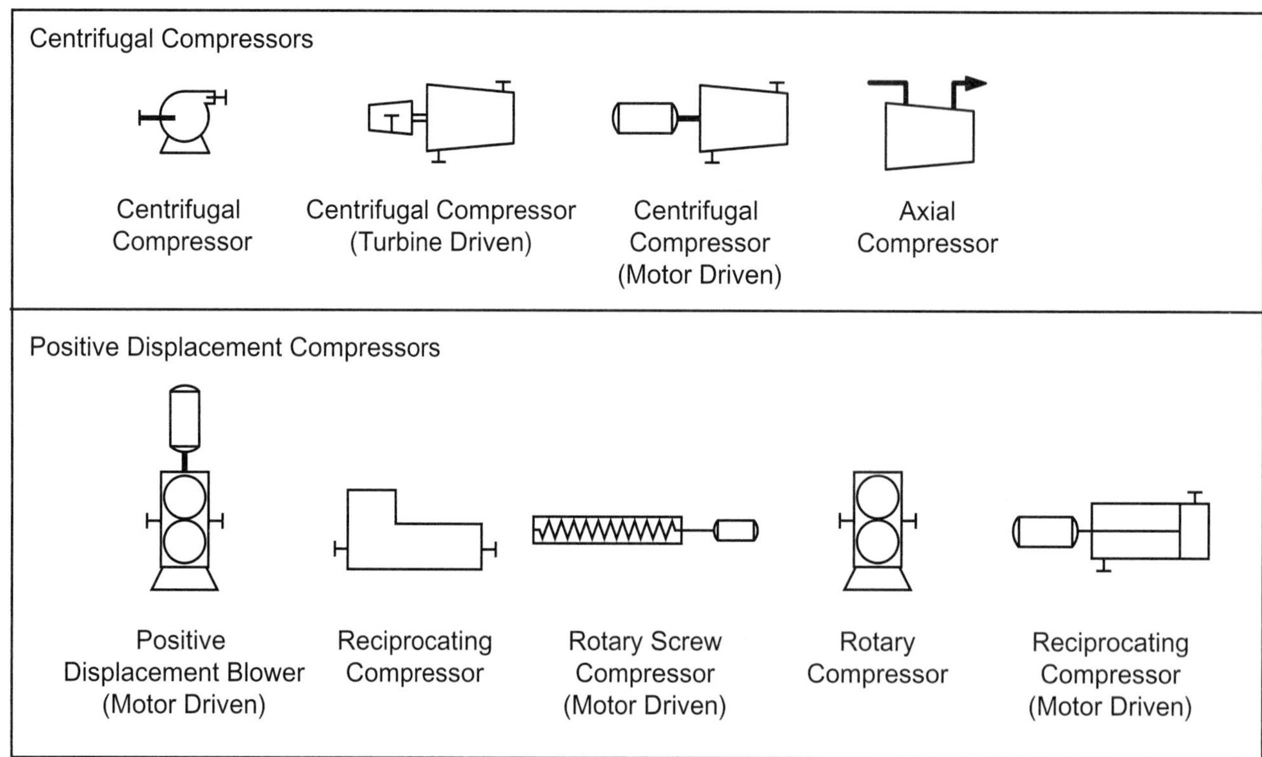

Figure 19-8: Compressor Symbols

SUMMARY

Compressors are devices that are used to compress gases from one location to another. Compressors may be single or multi-stage and are sized in accordance with the amount of gas to be compressed. They may also be dynamic (centrifugal) or positive displacement.

Positive displacement compressors draw gas into a chamber and then use pistons, lobes, gears or vanes to force the gas out.

Centrifugal compressors utilize impellers to increase the velocity of a gas and slow it down at the volute in order to increase pressure. Axial compressors utilize a set of rotor and stator blades to move the gas axially along the compressor shaft increasing pressure as the casing and blades decrease in size.

There are many hazards associated with improper compressor operation. These include over-pressurization, overheating, surging, and leakage.

When monitoring and maintaining compressors, technicians must always remember to look, listen and feel to ensure oil levels are correct, that there are no leaks, that there are no abnormal noises, and that the equipment is not producing excessive heat or vibration.

CHECKING YOUR KNOWLEDGE

1. Explain the purpose of compressors in the process industries.

2. Is the diagram below an example of a centrifugal compressor, or a positive displacement compressor? How do you know?

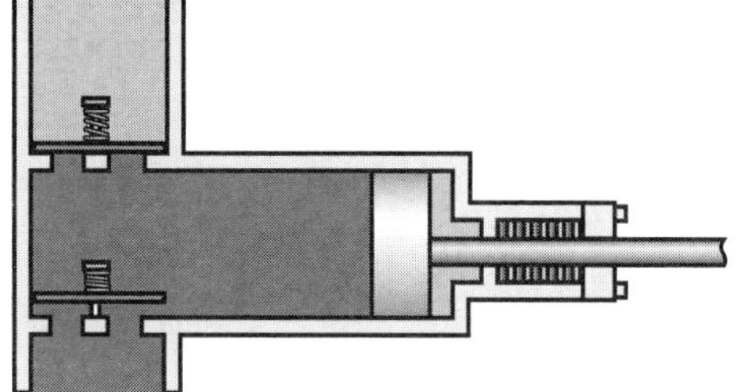

3. Which type of compressor uses vanes to force gases out of a chamber?
   a. Axial
   b. Centrifugal
   c. Positive Displacement

4. On the diagram below, identify the following parts of a centrifugal compressor.
   - Casing
   - Discharge
   - Impeller
   - Shaft
   - Suction

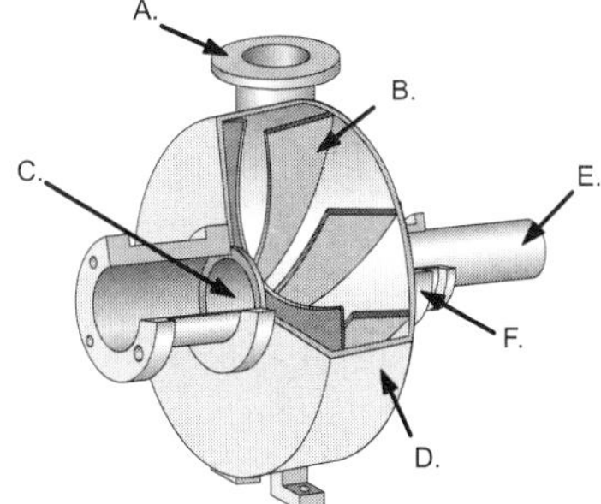

5. On the diagram below, identify the following parts of a reciprocating compressor.
   - Casing
   - Connecting rods
   - Suction
   - Discharge
   - Piston

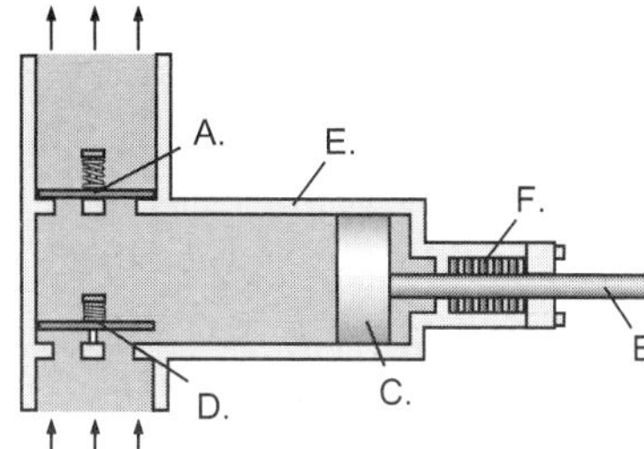

6. List three hazards associated with improper compressor operation.
7. List five monitoring and maintenance activities that a technician should perform when working with compressors.
8. Draw piping and instrumenation diagram (P&ID) symbols for the following:
   a. Centrifugal compressor
   b. Positive displacement compressor
   c. Reciprocating compressor
   d. Rotary compressor

ACTIVITIES

1. Given a cutaway of a centrifugal compressor, identify the following components:
   a. Casing
   b. Discharge
   c. Impeller
   d. Shaft
   e. Suction
2. Given a cutaway of a reciprocating compressor, identify the following components:
   a. Casing
   b. Connecting rods
   c. Suction valve
   d. Discharge valve
   e. Piston
3. Given a compressor, perform the following maintenance and monitoring activities:
   a. Inspect for abnormal noise
   b. Inspect for excessive heat
   c. Check oil levels
   d. Check for leaks around seals and flanges
   e. Check for excessive vibration
4. Given a piping and instrumenation diagram (P&ID), identify all of the compressors and tell how many of are centrifugal and how many are positive displacement.

Chapter 20

Turbines

OBJECTIVES

Upon completion of this chapter you will be able to:

1. List and describe different types of turbines and how they operate.
2. Describe the purpose or function of steam turbines in the process industries.
3. Identify the primary parts of a typical (non-condensing) steam turbine:
   - Casing
   - Shaft
   - Moving and fixed blades
   - Governor
   - Nozzle
   - Inlet
   - Outlet
   - Trip Valve
   - Throttle Valve
4. Discuss the hazards associated with the improper operation of a steam turbine.
5. Describe the monitoring and maintenance activities associated with a steam turbine.
6. Identify the symbols used to represent the steam turbine and associated equipment presented in this session.

KEY TERMS

- **Actuate**—to put into action.
- **Carbon Seal**—a sealing system that utilizes carbon rings surrounded by springs. As the shaft heats, these carbon seals help prevent the escape of steam from the turbine casing.
- **Casing**—the housing around the internal components of a turbine. The component of a steam turbine which holds all moving parts, including the rotor, bearings, seals, and is connected to the driven end equipment.
- **Fixed Blades**—blades inside a steam turbine that remain stationary when steam is applied.
- **Gas Turbine**—a device that consists of an air compressor, combustion chamber, and turbine. Hot gases produced in the combustion chamber are directed towards the turbine blades causing the rotor to move. The rotation of the connecting shaft can be used to operate other equipment.

- **Governor**—a device used to control the speed of a piece of equipment such as a turbine.
- **Hunting**—a term used to describe the condition when a turbine's speed fluctuates while the governor/controller is searching for the correct operating speed.
- **Hydraulic Turbine**—a turbine that is moved, operated, or effected by liquid.
- **Inlet**—the point where something enters.
- **Kinetic Energy**—energy associated with mass in motion.
- **Mechanical Energy**—the energy of motion that is used to perform work.
- **Nozzle**—a small spout or extension on a hose or pipe that directs the flow of steam.
- **Outlet**—the point where something exits.
- **Rotor**—The rotating member of a motor or turbine.
- **Set Point**—the point or place where the control index of a controller is set.
- **Shaft**—a metal rod, attached to an impeller, which is suspended by bearings.
- **Steam Turbine**—a turbine that is driven by the pressure of steam discharged at high velocity against the turbine vanes.
- **Throttle Valve**—a valve that can be opened or closed to quickly (tripping) or slowly (throttling) to control flow. This valve works in conjunction with the governor to control the speed of the turbine.
- **Trip Valve**—a safety valve that can be opened or closed quickly.
- **Turbine**—a machine for producing power. Activated by the expansion of a fluid (e.g., steam, gas, air) on a series of curved vanes on an impeller attached to a central shaft.
- **Wind Turbine**—a device that converts wind energy into mechanical energy.

INTRODUCTION

Turbines are machines that are used to produce power and rotate shaft-driven equipment such as pumps, compressors, and electricity generators. Turbines are activated by the expansion of a fluid on a series of curved impeller vanes attached to a central shaft.

Turbines can be powered by a variety of different fluids including, steam, gas, liquid or air. Steam powered turbines are frequently used as backups for electric motors.

DIDYOU KNOW?

Steam can be a very powerful locomotive (moving) force.

So powerful, in fact, that it has been used to drive many types of large industrial equipment (e.g. trains, dredges, and shovels).

TYPES OF TURBINES

There are four main types of turbines: steam, gas, hydraulic, and wind. Each of these turbine types are classified according to how they operate and the fluid that turns them (i.e., steam, gas, water or air).

While each of these turbine types is discussed, the most common turbine type used in industry is the steam turbine. For this reason, steam turbines will be the main focus of this chapter.

Steam Turbines

A **steam turbine** is a turbine that is driven by the pressure of steam discharged at high velocity against the turbine vanes. Within steam turbines there is a direct relationship between the boiling point of the water and the pressure of the steam, since steam turbines use the temperature and pressure of steam to turn a rotor and produce mechanical energy.

Mechanical energy is the energy of motion that is used to perform work (as opposed to **Kinetic Energy**, which is energy in motion).

Figure 20-1 shows an example of a steam turbine.

Steam turbines have many advantages over electrical equipment. For example, they are free from spark hazards, so they are good in areas where volatile substances are produced. In addition, they do not require electricity to run, so they are good during power outages. They are also suitable for damp environments that might cause electrical equipment to fail.

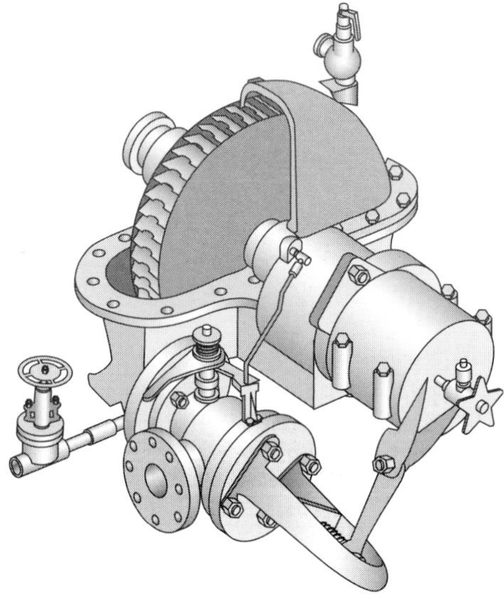

Figure 20-1: Steam Turbine

Gas Turbines

Gas turbines are devices that consist of an air compressor, combustion chamber, and a turbine. In a gas turbine, hot gases produced in the combustion chamber are directed towards the turbine blades causing the rotor to move. As the rotor moves, the shaft turns. The rotation of the connecting shaft can be used to operate other equipment. Gas turbines are often coupled to a smaller steam turbine. The steam turbine spins the gas turbine/compressor to provide compressed air for startup.

Figure 20-2 shows an example of a gas turbine.

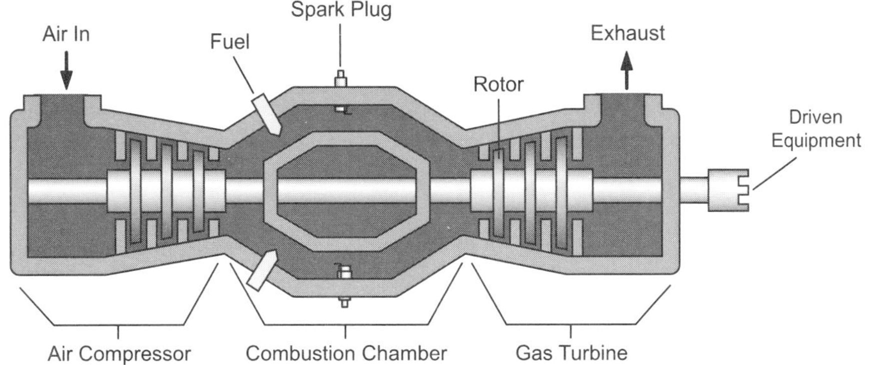

Figure 20-2: Gas Turbine

DID YOU KNOW?

Jet engines used in airplanes are actually gas turbines.

In a gas turbine an electric motor is used to rotate the shaft and get the air compressor up to speed and start the machine. Once the turbine is started, it provides enough power to drive the air compressor to keep the gas turbine running and drive the other connected equipment (e.g., pumps, compressors and generators).

Hydraulic Turbines

Hydraulic turbines are turbines that are moved, operated or affected by a fluid.

In a hydraulic turbine, a fluid flows across the rotor blades, forcing them to move. The faster the fluid flows, the faster the wheel turns.

An example of a simple hydraulic turbine is a water wheel like the one shown in Figure 20-3.

Figure 20-3: Water Wheel (Hydraulic Turbine)

Wind Turbines

Wind turbines are mechanical devices that convert wind energy into mechanical energy. These types of turbines use air pressure to move a rotor. A windmill is an example of a wind turbine.

In a wind turbine, air currents move across fan-like blades, causing them to turn. As they turn, a shaft is rotated. The rotation of the shaft drives devices such as pumps or electrical generators.

Figure 20-4 shows an example of a simple wind turbine and how it might be used to **actuate** (put into action) a water pump.

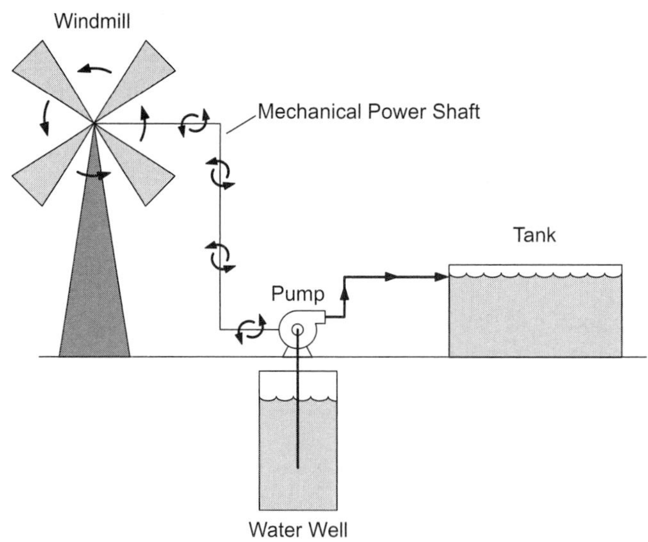

Figure 20-4: Wind turbine actuating a pump

STEAM TURBINE COMPONENTS

Turbine components vary depending on the type of turbine and its application. However, the main components of a steam turbine include:

- Casing
- Shaft
- Carbon Seals
- Moving and fixed blades
- Governor
- Valve
- Nozzle
- Inlet
- Outlet
- Trip Valve
- Throttle

Figure 20-5 shows an example of a steam turbine and its components.

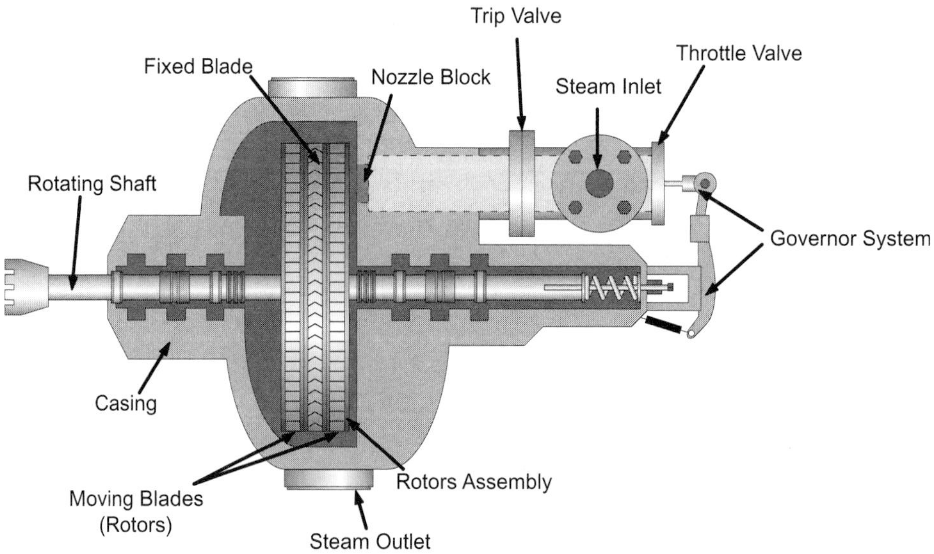

Figure 20-5: Steam Turbine Components

The **casing** is the housing around the internal components of a turbine. Within the casing are a shaft and a set of blades.

The **shaft** is a metal rod, attached to an impeller, which is suspended by bearings. The **moving blades** are rotor blades that move or rotate when steam is applied. The **fixed blades** are blades that remain stationary when steam is applied.

The **governor** is a device used to control the speed of the turbine as steam is channeled through the nozzle.

The **nozzle** is a small spout or extension on a hose or pipe that directs the flow of steam. Steam enters the nozzle from the steam **inlet**, and exits the turbine through the steam **outlet**.

The **trip/throttle valve** is a valve that can be opened or closed to quickly (tripping) or slowly (throttling) to control flow. This valve works in conjunction with the governor to control the speed of the turbine.

PRINCIPLES OF OPERATION

Turbines are devices that create rotational movement that can be used to drive other equipment (e.g., pumps, compressors, and blowers). While each of the different turbine types varies in how they work, all of them use the same basic principles. That is, they use some kind of force (e.g., water pressure, air pressure, steam pressure, or the pressure of combustion) to turn a rotor. As the rotor (the driver) is turned, another device (the mover) is moved or actuated. A pump is an example of a mover.

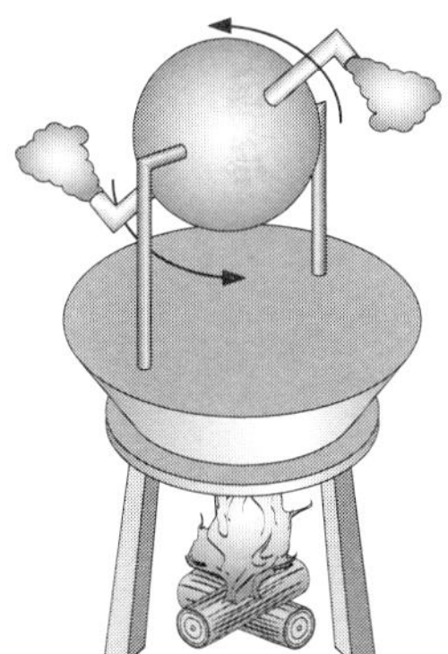

Figure 20-6: Simple Reactive Turbine

One of the earliest types of turbines is the simple **reactive turbine** like the one shown in Figure 20-6.

In a simple reactive turbine, water is placed in a globe that contains two opposing nozzles. As the water is heated, steam and pressure are produced which force the steam out of the nozzles. As the steam exits, propulsive (rotational) force is created. This propulsive force causes the globe to spin.

Another type of turbine is an impulse turbine. Like a reactive turbine, impulse turbines use steam to move a rotor. However, instead of generating their own steam the way a reactive turbine does, **impulse turbines** are acted upon by an external steam source.

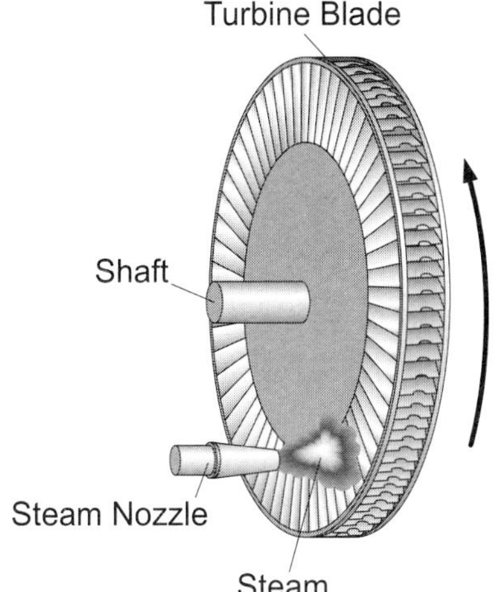

Figure 20-7: Impulse turbines are acted upon by an external steam source

Figure 20-7 shows an example of an impulse turbine.

In an impulse turbine, steam is channeled through a steam nozzle onto the turbine blades. As the steam passes through the nozzle it is converted to velocity. This force causes the wheel to turn, thereby rotating the shaft and any equipment coupled to it on the driven end.

HAZARDS ASSOCIATED WITH STEAM TURBINES

There are many hazards associated with steam turbines. Table 20-1 lists some of these hazards and their impacts.

TABLE 20-1: Hazards associated with improper steam turbine operation

IMPROPER OPERATION	POSSIBLE IMPACTS			
	Individual	**Equipment**	**Production**	**Environment**
Touching the external housing of a steam turbine that is full of steam	Burns caused by exposure to heat or steam			
Allowing hot steam to enter the turbine without following the proper warm up procedure	Burn injury from steam	Equipment damage due to thermal shock	Downtime due to repair	
Running a turbine at a speed above or below normal operating **setpoint** (the point or place where the control index of a controller is set).	If the overspeed trip does not activate, the turbine could fly apart causing injury.	**Hunting** may occur (when a turbine's speed fluctuates while the controller is searching for the correct operating speed)	Downtime due to repair. Driven equipment (e.g., a pump or compressor) may not pump enough if the turbine is running too slowly.	
Turbine may shut down.			Improper pump speed might cause product to go off-spec.	
Failure to lubricate the linkage between the governor and the governor valve.		May cause hunting as a result of valve sticking or binding.		
Failure to maintain sufficient inlet steam pressure.		Hunting may occur due to insufficient steam pressure.	Insufficient power to run the process.	
Allowing "wet steam" or condensate into the turbine.		Can cause total failure of the turbine.		

TABLE 20-1: Continued

IMPROPER OPERATION	POSSIBLE IMPACTS			
	Individual	Equipment	Production	Environment
Failure to maintain critical speeds.		Turbine could be damaged as a result of harmonic vibrations if the turbine is not moved through critical speeds as quickly as possible.		
High discharge steam pressure.		Turbine slowdown and relief valve popping.	Possible loss of production.	Noise pollution.
Steam Leaks.	Burn injuries.			Noise hazard.

MONITORING AND MAINTENANCE ACTIVITIES ASSOCIATED WITH STEAM TURBINES

When monitoring and maintaining steam turbines, technicians must always remember to look, listen and feel for the following:

Look	Listen	Feel
■ Check oil levels to make sure they are satisfactory. ■ Check seals and flanges to make sure there are no leaks. ■ Check for water in the oil reservoir ■ Check steam pressure and temperature ■ Observe governor operation for proper operation	n Listen for abnormal noises.	■ Feel for excessive vibration. ■ Feel for excessive heat.

Failure to perform proper maintenance and monitoring could impact the process and result in equipment damage.

STEAM TURBINE SYMBOLS

In order to accurately locate steam turbines on a piping and instrumentation diagram (P&ID), process technicians must be familiar with turbine symbols.

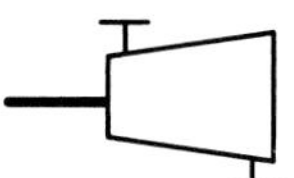

Figure 20-8: Steam Turbine Symbols

Figure 20-8 provides an example of a steam turbine symbol. However, this symbol may vary from facility to facility.

SUMMARY

Steam turbines are machines that are used to produce power. They are activated by the expansion of steam on a series of curved impeller vanes attached to a central shaft.

Turbines are used to rotate shaft-driven equipment such as pumps, compressors, and electrical generators. Steam powered turbines are good backups for electric motors.

There are four main types of turbines: steam, gas, hydraulic, and wind. Each of these turbine types are classified according to how they operate and the fluid that turns them (i.e., steam, gas, water or air).

Improperly operating a steam turbine, or any other type of turbine, can lead to safety and process issues, including burns, and equipment damage.

Technicians should always monitor turbines for abnormal noises, leaks, excessive heat, pressure drops, temperature drops, or other abnormal conditions, and conduct preventive maintenance as needed. Lubrication needs special attention.

CHECKING YOUR KNOWLEDGE

1. Define the following key terms:
   a. Casing
   b. Governor
   c. Hunting
   d. Inlet
   e. Moving and fixed blades
   f. Nozzle
   g. Outlet
   h. Shaft
   i. Trip valve
   j. Throttle valve

2. Which type of turbine uses high pressure steam to turn a rotor?
   a. Gas turbine
   b. Hydraulic turbine
   c. Steam turbine
   d. Wind turbine

3. A jet engine is an example of which type of turbine?
   a. Gas turbine
   b. Hydraulic turbined
   Steam turbine
   Wind turbine

4. A water wheel is an example of which type of turbine?
   a. Gas turbine
   b. Hydraulic turbine
   c. Steam turbine
   d. Wind turbine

5. List three things a process technician should look, listen and feel for when working with steam turbines.

ACTIVITIES

1. List three hazards associated with steam turbines, and identify any impacts these hazards might have on individuals, equipment, production, and the environment.

2. Given a piping and instrumentation diagram (P&ID), identify all of the turbines.

3. Given a diagram of a steam turbine, identify the following components and explain the function of each:
   a. Casing
   b. Shaft
   c. Moving and fixed blades
   d. Governor
   e. Nozzle
   f. Inlet
   g. Outlet
   h. Trip valve
   i. Throttle valve

Chapter 21
Electricity and Motors

OBJECTIVES

Upon completion of this chapter you will be able to:

1. Explain the difference between Alternating Current (AC) and Direct Current (DC).
2. Identify what current (AC or DC) is most commonly used in the process industries.
3. Describe the purpose or function of the electric motors in process industries.
4. Identify the primary parts of a typical electric motor:
   - Frame
   - Fan Shroud
   - Stator
   - Rotor
   - Fan
   - Bearings
   - Lubrication System
5. Discuss the hazards associated with the improper inspection and operation of an AC motor.
6. Describe the monitoring and maintenance activities associated with an electric motor.
7. Identify the symbols used to represent electric motors and associated equipment presented in this session.

KEY TERMS

- **Alternating Current (AC)**—electric current that reverses direction periodically, usually sixty times per second; the primary type of electrical current used in processing plants.
- **Ammeter**—device used to measure current.
- **Amperes (AMPs)**—a unit of measure of the electrical current flow in a wire; similar to "gallons of water" flow in a pipe.
- **Circuit**—a system of one or many electrical components connected together to accomplish a specified purpose.
- **Conductor**—a substance or body that allows a current of electricity to pass continuously along it.
- **Direct Current (DC)**—electric current that always travels in the same direction.
- **Electricity**—electricity is a flow of electrons from one point to another along a pathway, called a conductor.
- **Electromotive Force**—The force that causes the movement of electrons through an electrical circuit.
- **Generator**—a device that converts mechanical energy into electrical energy.
- **Grounding**—the process of using the earth as a return conductor in a circuit.

- **Ground Fault Circuit Interrupter (GFCI)**—a safety device that detects the flow of current to ground and opens the circuit to interrupt the flow.
- **Insulator**—a device made from a material that will not conduct electricity; the device is normally used to give mechanical support to electrical wire or electronic components.
- **Motor**—a mechanical driver with rotational output; usually electrically operated.
- **Ohm**—a measurement of resistance in electrical circuits.
- **Resistance**—Opposition to the flow of electrons, measured in ohms.
- **Rotor**—The rotating member of a motor, with shaft. Commonly called the armature in DC motors.
- **Semiconductor**—materials that are conductors at one time, and insulators at another, depending on voltage drop across the material.
- **Shaft**—a round metal tube that holds the rotor.
- **Stator**—a stationary magnet in an AC or DC motor.
- **Transformer**—a device that will raise or lower the voltage of alternating current of the original source. Transformers are used to step up or step down AC voltages. Transformers do not convert AC to DC or DC to AC.
- **Volt**—one volt is the electromotive force, which will establish a current of one amp through a resistance of one ohm.
- **Voltmeter**—device used to measure voltage.
- **Watt**—work is being done at the rate of one watt when an equivalent current of one amp is maintained through a resistance by an electromotive force of one volt.

INTRODUCTION

Electricity and motors are important components in the process industry. Electricity allows us to power up equipment and run electric lights. Motors allow us to operate rotating equipment like pumps and compressors.

The electricity used by motors and other equipment can be either Alternating (AC) or Direct Current (DC). AC, however, is the most common type used in the process industries.

When working with electricity, process technicians should always wear proper protective gear, follow safety procedures, and be aware of potential hazards.

Process technicians should also be able to recognize the symbols for various pieces of electrical equipment on process drawings.

WHAT IS ELECTRICITY?

Electricity is a flow of electrons from one point to another along a pathway, called a conductor. Process technicians should have a good understanding of electricity, since it is integral to the functioning of a plant and its systems.

The chapter on chemistry described atoms as the smallest particle of an element that can combine with other elements. Atoms contain positive, negative, and neutrally charged particles. The negatively charged particles are called electrons.

When electrons flow from one atom to another, this is called electrical current. Free electrons flow along a path like a river. This pathway is called a conductor.

DID YOU KNOW?

Ben Franklin's famous experiment with flying a kite in a thunderstorm demonstrated the true nature of lightning—that it is electricity.

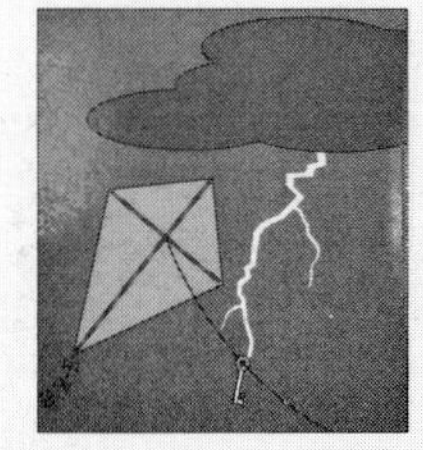

Conductors are materials that have electrons that can break free and flow more easily than other materials. Metals, as well as some types of hot gasses (plasmas), and liquids, are good conductors.

Materials that do not give up their electrons as easily are called **insulators**. Insulators are poor conductors. Air, rubber, and glass are examples of insulators.

Materials that are neither conductors nor insulators are called **semiconductors**. Semiconductors are most commonly used in the electronics industry and are being used increasingly in power applications.

Another type of electricity is static electricity. **Static electricity** occurs when a number of electrons build up on the surface of a material, but have no positive charge nearby to attract them and cause them to flow. When the negatively charged surface comes into contact with (or comes near) a positively charged surface, current flows until the charges on each surface become equalized, sometimes creating a spark. Lightning, and the shock that occurs from touching a doorknob after shuffling across the carpet, are both good examples of static electricity.

To better understand electricity, process technicians need to be familiar with its principles and watts (horsepower) and know that it is measured in Volts, Ohms and Amps.

Volts

Using the water analogy again, electric current flows like a river down a slope (the path of least resistance). The greater the angle of a slope, the faster the water flows. Electric current behaves in a similar way. If the difference between positive and negative charges is low, electrons flow with little force. Increase the difference, and the electrons will flow with greater force. The force that makes electrons flow is called voltage or electromotive force (EMF), and it is measured in units called volts (or V).

Voltage is a measurement of the potential energy required to push electrons from one point to another. A **volt** is the electromotive force that will establish a current of one amp through a resistance of one ohm.

A **voltmeter**, which is a device used to measure voltage, can be connected to a circuit to determine the actual voltage present.

Resistance (Ohms)

The force that opposes the push of electrons is called **resistance** (or R). Resistance in electrical equipment is measured in ohms. One **ohm** is the amount of resistance where one volt produces a flow of one amp. Conductors provide little resistance to electricity, while insulators provide high resistance.

Amps

Electrical current is measured using amperes (amps). **Amps** (or I) are a unit of measure for electrical current flow in a wire (similar to "gallons of water" flow in a pipe). Amps describe how many electrons are flowing at a given time.

Amps are one indication of how much work a circuit can do. In other words, amps show the capacity of a battery or other source of electricity to produce electrons. An **ammeter**, which is a device used to measure amperage, must be connected in series to an electrical circuit to display the actual amps.

Ohm's Law

Ohm's Law describes how volts, ohms and amps act upon each other. This law states that the amount of steady current through a conductor is proportional to the voltage across that conductor. This means a conductor with one ohm of resistance has a current of one amp under the potential of one volt. Simply put, volts equals amps times ohms (**V = IR**).

VOLTAGE (OHM'S LAW)

The formula for Ohm's Law is:

$$V = IR$$

(V = volts, I = current flow in amps, R = resistance to flow in ohms)

When working with Ohm's law, if you know the value of two units, you can always figure out the third using one of the following calculations:

I = V/R (Current = Volts ÷ Resistance) or

R = V/I (Resistance = Volts ÷ Current)

Watts

Electrical power is measured in watts. A **watt** is the work being done when a constant current of one amp is maintained through a resistance by an electromotive force of one volt.

If you know the volts and amps of a circuit, you can figure how many watts a circuit uses by employing the formula **W=VI** (watts = volts x amps).

WATTAGE

The formula for determining how many watts a circuit uses, if you know the volts and amps, is:

$$W = VI$$

(W = watts, V = potential difference in volts, I = current flow in amps)

The formula for determining how many watts a circuit uses, if you know the amps and resistance, is:

$$W = I^2R$$

(W = watts, I = current flow in amps squared, R =resistance to current flow in ohms)

Circuits

A **circuit** is a system of one or many electrical components that accomplish a specified purpose. Circuits combine conductors with a power supply and usually some kind of electrical component (such as a switch or light) in a continuously conducting path. In a circuit, electrons flow along the path, uninterrupted, and return to the power supply to complete the circuit.

Circuits fall into one of two types: series and parallel.

- **Series**—Electrons only have one path to flow along. An example is a string of Christmas lights; if one bulb burns out, then the circuit (path) is interrupted and none of the bulbs will light.

Figure 21-1: Series Circuit Example

- **Parallel**—Electrons are given a choice of paths to flow along. Some strings of Christmas lights use this type of circuit, so if one bulb burns out then the others continue to light. Houses are wired using parallel circuits.

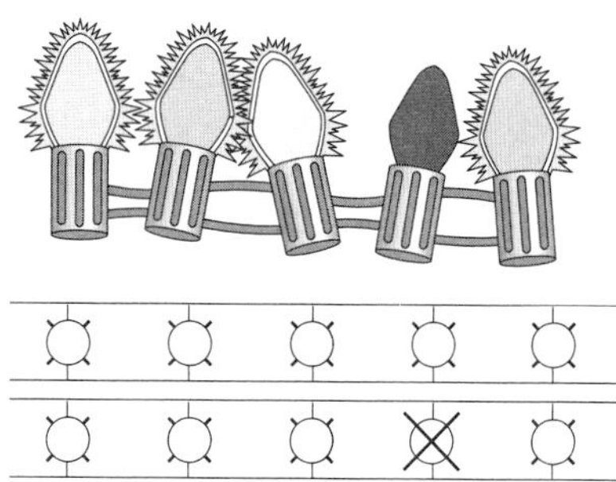

Figure 21-2: Parallel Circuit Example

Grounding

All energized conductors supplying current to equipment are kept insulted from each other, from ground (earth) and from the equipment user. Many types of equipment have exposed conductive parts, such as metal covers, that are routinely touched during normal operation. If these surfaces become energized, a difference in electrical potential will exist between the equipment and ground (earth). The process technician can complete the circuit, since he or she can be grounded. An energized equipment case can present a shock hazard.

For that reason, non-current carrying conductive materials enclosing electrical conductors or equipment are required to be connected to earth so as to limit the voltage to ground on these materials. This is typically accomplished by a separate conductor specifically designed for the purpose. With this conductor in place, if the equipment case becomes energized, a low resistance path for the flow of ground current back to the source is already in place that reduces or eliminates the shock hazard to the operator.

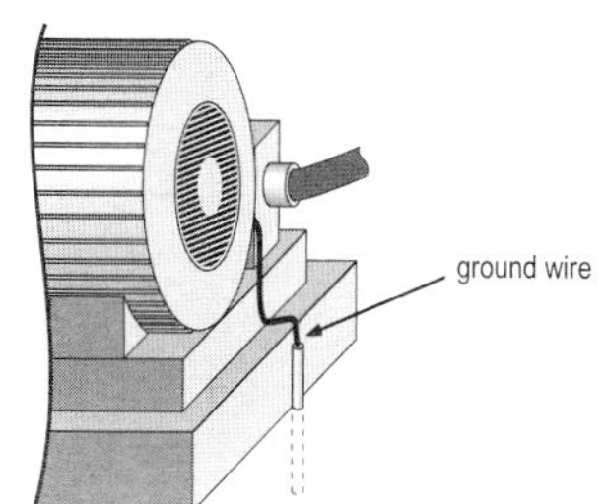

Figure 21-3: Grounding Wire

ELECTRICAL TRANSMISSION

Electricity must come from a power source, such as from batteries or generators in a power station. Although electricity is a form of energy, it is not an energy source and must be manufactured. This section describes how electricity is moved from a power station to homes and businesses.

Generators are devices that convert mechanical energy into electrical energy. For a power station, the most common way to manufacture electricity is by burning fuels that make turbines rotate magnetic fields inside generators to create electric current. Other methods include hydroelectric and nuclear power generation.

Electricity flows in a continuous current from a high potential point (the power source) to a point of lower potential (like your home or plant), through a conductor such as a wire. High voltage electricity is transmitted from the plant to the power grid through a system of wires.

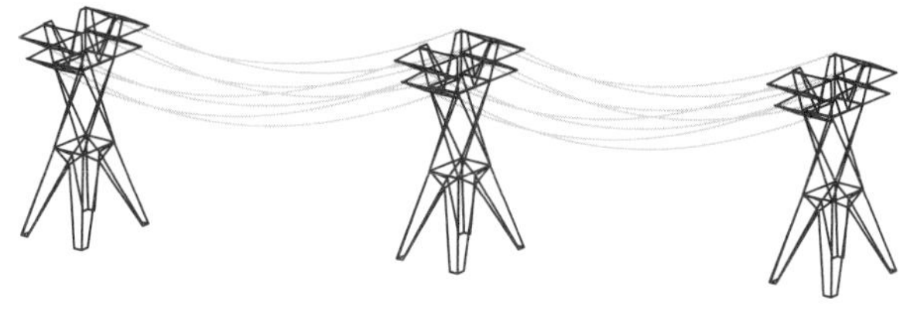

Figure 21-4: Electrical Power Lines

High voltage electricity is routed to a substation that steps the electricity down to a lower, safer voltage. The substation distributes the electricity through feeder wires to a **transformer**, a device that steps down the voltage again. The electricity must be stepped down for use by residential or commercial customers.

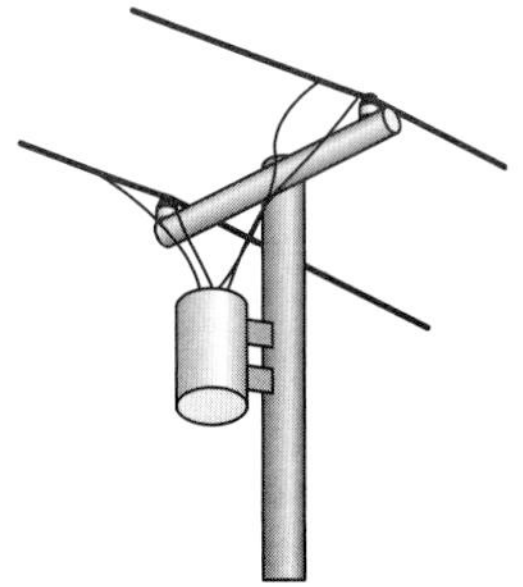

Figure 21-5: Power Transformer

The electricity is then used as energy to do work, such as lighting a bulb or operating a motor to activate a pump. Safety devices such as fuses, protective relays, ground fault detectors and others are used throughout the power transmission process to make the system as safe as possible.

UNDERSTANDING ALTERNATING (AC) AND DIRECT CURRENT (DC)

The two types of electrical currents are Alternating Current (AC) and Direct Current (DC):

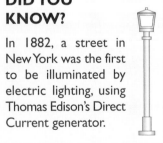

- **AC**—electrical current that uses a back and forth movement of electrons in a conductor. The movement of alternating current is similar to water sloshing backwards and forwards in a pipe. When a negative charge is at one end of a conductor and a positive charge is at the other end, the electrons move away from the negative charge. But if the charges (polarity) at the end of the conductors are reversed, the electrons switch directions. In the U.S., the AC power supply changes direction 60 times per second; this is called frequency or cycles per second.

DID YOU KNOW?

In 1882, a street in New York was the first to be illuminated by electric lighting, using Thomas Edison's Direct Current generator.

- **DC**—electrical current that results in a direct flow of electrons through a conductor. Direct current flows like water moving in one direction through a pipe. With a battery as a power source, electricity flows in only one direction through a circuit.

In the late 1800s, Edison power stations (created by Thomas Edison) supplied direct current (DC) electricity to customers scattered across the United States. DC had a limiting factor though: it could only be sent economically a short distance (about a mile) before the electricity began to lose power.

To remedy these distance limitations, George Westinghouse introduced an Alternating Current (AC) power systems, designed by Nikola Tesla, as an option to DC power. AC has an advantage over DC because AC voltages can be easily increased or decreased using transformers. By employing transformers to boost voltage levels, AC systems economically distribute electricity for hundreds of miles.

In addition to traveling more efficiently over longer distances, AC has another major advantage over DC. AC can easily be "stepped up" or "stepped down" using a transformer.

Transformers takes low-voltage current and make it high-voltage, and vice versa. Power stations send out high voltage electricity that is stepped down using a transformer so homes and businesses can use the electricity.

AC and DC produce the same amount of heat, which is proportional to the product of the current (amps) squared times the resistance (ohms) and written using the formula I^2R. However, it is more difficult to measure a circuit using AC than DC, since AC voltage cycles from zero, to positive, to zero, to negative, to zero during every cycle.

To summarize, the following are the similarities and differences between alternating (AC) and Direct Current (DC):

AC	DC
Polarity is switched constantly	Polarity is fixed
Voltage varies during cycles	Voltage remains constant
Can be varied for power distribution (transformers can amplify or reduce AC)	Cannot be varied (a steady value is produced that transformers do not affect)
Measurements are harder than DC	Measurements are easier than AC
Voltage can be stepped up or down by a transformer.	Voltage cannot be stepped up or down by a transformer.
Heating effect is the same as DC	Heating effect is the same as AC

WHICH TYPE OF CURRENT IS USED MOST IN THE PROCESS INDUSTRIES?

Alternating Current (AC) is the most common type of electrical current used in the process industries. AC provides most of a plant's electrical power requirements.

DID YOU KNOW?

An English scientist, Michael Faraday, is credited for generating electric current on the first practical scale.

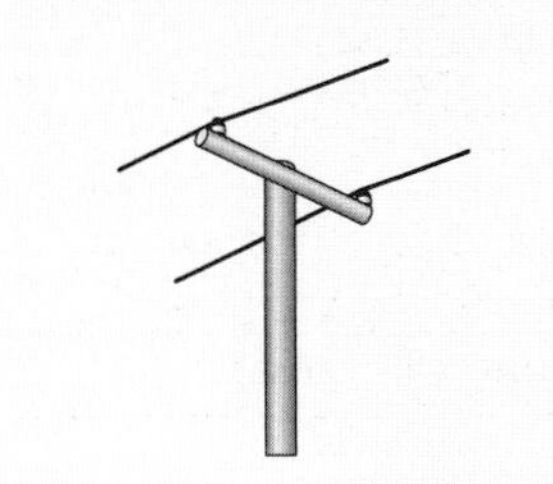

Plants typically receive high voltage electricity from the electric company. Transformers then step this electricity down so it can be distributed throughout the plant.

When AC power reaches a unit, it is stepped down to 4160V for use with large motors and is then stepped down again to 480V (sometimes referred to as 440V), which is a common motor voltage used around a plant.

Finally, the power can be stepped down once again to somewhere between 120V and 240V for use with air conditioning, heating, lighting and other applications.

Direct Current (DC) is used primarily in the form of batteries, which can provide electrical power to portable equipment or emergency uses. For example, DC can power flashlights and other types of lighting, basic power tools, carts, and other small motors designed for DC current.

To maintain power to critical equipment and instrumentation during a power outage or emergency, a device called an Uninterruptible Power Supply (UPS) is used. A typical UPS first converts AC to DC and then remanufactures AC from DC for use by plant instrumentation and other critical loads. A battery bank is connected to, and kept charged by, the DC section of the UPS. If the plant AC power should fail, the UPS uses DC from the battery to continue manufacturing AC for the critical loads.

Plants can also generate their own electricity in some cases. For example, if power fails, emergency diesel generators can provide electricity to critical equipment and instrumentation.

However, this textbook focuses primarily on AC, since it is currently most common in the industry.

USES OF ELECTRIC MOTORS IN THE PROCESS INDUSTRIES

Motors are mechanical drivers (usually electrically operated) with rotational output. Electric motors turn electrical energy into useful mechanical energy and provide power for a variety of rotating equipment including:

- Pumps
- Compressors
- Mixers
- Fans
- Feeders
- Valves
- Blowers

The rotation of a motor can affect the speed of rotating equipment, so different types of motors are used depending on the application (e.g. a three-phase motor). Of the many types of motors used in industrial applications, induction motors are the simplest, most rugged and most common. Induction motors "induce" or draw in current into the rotor by the stator. The rotor current generates magnetic fields in the rotor which then interact with those produced by the stator to cause the rotor to turn.

DID YOU KNOW?

Nikola Tesla, a Serbian engineer who invented the first AC motor, discovered a way to broadcast radio waves before Guglielmo Marconi, the person who is generally credited with inventing radio.

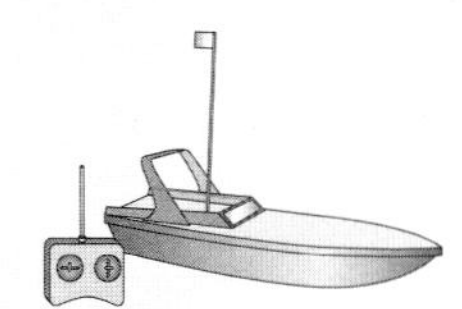

He also created the world's first radio-controlled boat.

THE PRIMARY PARTS OF A TYPICAL AC INDUCTION MOTOR

The primary parts of a typical AC motor include a: frame, shroud, rotor, stator, fan, bearings, and an AC power source.

The following is a description of each of these components.

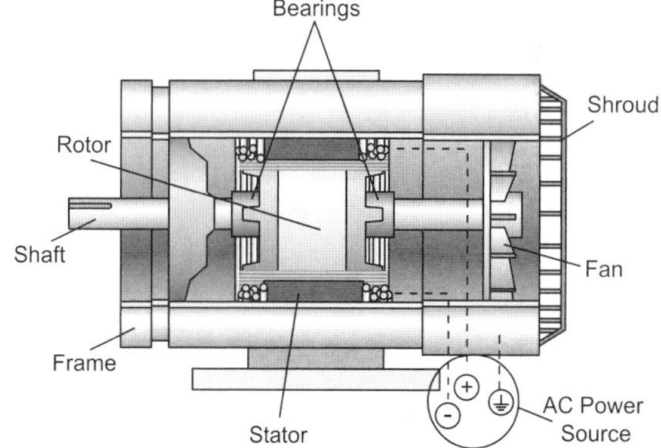

Figure 21-6: AC Motor components

- **Frame**—a structure that holds the internal components of a motor and motor mounts.
- **Shroud**—a casing over the motor that allows air to flow into and around the motor. The air keeps the temperature of the motor cool.
- **Rotor**—an iron core with copper bars attached to it; when the stator creates an electric current in the rotor, this creates a second magnetic field in the rotor. The magnetic fields from the stator and rotor interact, causing the rotor to turn.
- **Shaft**—a round metal tube that holds the rotor.
- **Stator**—a stationary part of the motor where the alternating current flows in and a large magnetic field is created using magnets and coiled wire.
- **Fan**—rotating blades that cool the motor by pulling air in through the shroud.
- **Bearings**—a machine part that supports another part, which rotates, slides, or oscillates in or on it. Bearings reduce friction between the motor's rotating and stationary parts.
- **AC power source**—a device that supplies current to the stator.

HOW AN AC ELECTRIC MOTOR WORKS

Early pioneers in electricity discovered the principle of electromagnetism. When current is run through a coil of insulated wire that is wrapped around a soft iron bar, a strong magnetic field is created. When the current is removed, the magnetic field diminishes. This is electromagnetism.

Electromagnetism plays an important role in electric motors. Electric motors, whether AC or DC, operate on the same three electromagnetic principles:

- Electric current generates a magnetic field.
- Like magnetic poles repel each other (positive to positive or negative to negative), while opposite poles attract each other (positive to negative).
- The direction of the electrical current determines the magnetic polarity.

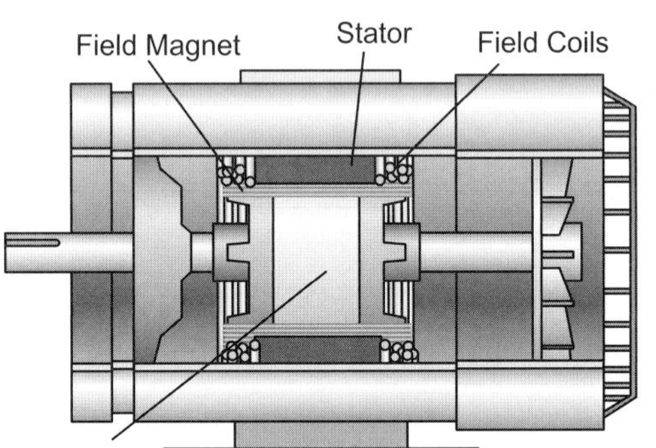

Figure 21-7: AC Motor field magnet, stator, field coils and rotor

A motor consists of two main parts: a stationary magnet, called a **stator**, and a rotating conductor, called a **rotor**. The stator includes field magnets and field coils, through which AC current runs. This generates a rotating magnetic field. The rotor features an iron core and copper bars (two highly conductive metals).

The magnetic field of the stator creates an electric current in the rotor, generating a second magnetic field. When the two magnetic fields interact, this causes the rotor to turn. The rotor is then used to supply mechanical energy to another device, such as a pump or compressor. So, motors convert electrical energy into mechanical energy that can be used to power other equipment.

HAZARDS OF ELECTRICITY

All personnel working with electrical equipment MUST be properly trained and authorized for the particular type and level of work being performed.

The most challenging electrical control and maintenance activities are likely to be reserved for personnel with specialized training in this area. However, process technicians might be required to perform some electrically related tasks or assist an electrician.

Electricity can be dangerous. Electrical current passing through a body causes a shock, which can result in:

- Serious bodily injuries
- Burns
- Death

DID YOU KNOW?

Did you know? Approximately 15% of the OSHA general industry citations relate to electrical hazards.

Shock can make your muscles tighten (e.g., your chest muscles and diaphragm), which can restrict breathing. Shock can also interrupt the rhythm of the heart by interfering with the natural electrical impulses that control it.

Shock and other electrical hazards occur when a person contacts a conductor carrying electricity while also touching the ground or an object that has a conductive path to the ground. The person completes the circuit as the current passes through his or her body.

The amount of current and the contact point or path determines the amount of damage to the body. For example, current passing from a finger to an elbow causes less damage than current passing from a hand to a foot.

Electrical shocks can occur when a person:

- Comes in contact with a bare wire (either bare on purpose or as a result of cracked or worn insulation.
- Uses improperly grounded electrical equipment.
- Works with electrical equipment in a wet or damp environment, or when they are sweating heavily.
- Works on electrical equipment without checking that the power source has been turned off.
- Uses long metal equipment, such as cranes or ladders, which can come into contact with a power source.

Static electricity discharges and lightning strikes are two other ways a person can be shocked.

Short circuits are a common cause of electrical hazards. A short circuit occurs when electrons in a current find a path of least resistance, which is outside of the normal circuit, and flow to it. For example, if the insulation is cracked on two wires that are near each other, the electrons jump between the wires and create a "short" circuit (a short cut outside the intended circuit).

Water is another potential hazard when dealing with electricity. Water decreases the resistivity of materials to electricity. For example, dry skin can resist an average of up to 100,000 Ohms. Wet skin, however, reduces resistivity to as low as 450 Ohms. Even sweat on the skin can decrease resistivity.

When energized conductors are close together or touching, electricity can arc through the air from one to the other and complete the circuit. Also, static electricity can discharge, causing a spark. Arcs or sparks can ignite nearby hydrocarbons or other flammable materials, causing a fire or explosion. Fires or explosions can endanger workers, the unit or plant, and nearby communities.

Along with electrical hazards, motors present mechanical hazards to workers. Motors have moving parts that can pull, tear or rip clothing or skin if not properly de-energized before working on them. Motors can also impact production (slowing or halting it) if they become inefficient or seize up.

All motors should have controllers that incorporate protection from instantaneous overload, such as a short circuit, as well as thermal overload from working beyond design limits. Motors will also have a disconnecting means to isolate the circuitry for maintenance work. Typically, the motor's overload, starting and stopping means, and disconnecting device are built together in one motor controller. A Motor Control Center (MCC) is a grouping of these motor controllers.

Many accidents are caused by carelessness. Electrically related accidents are no different. If proper safety procedures are followed when working on electrical circuits and/or electrical devices, then the risk of any hazards occurring are minimized.

The following are some general, yet essential, tips for reducing electrical hazards:

Tips for reducing the risk of electrical hazards

- Consider all equipment and electrical systems energized until it is verified that they are not.
- Always inspect and test safety equipment.
- Plan your work before you start.
- Familiarize yourself with the electrical equipment you are using and the electrical circuit on which you will work.
- Do not wear conductive metal (such as jewelry).
- Understand lockout/tagout procedures (methods of controlling hazardous energy).
- Disconnect and lock the electrical circuit or system yourself; do not depend on someone else.
- After a circuit or system is disconnected, test the circuit to ensure it was properly disconnected and is still not energized or does not contain residual energy.
- Only use equipment and tools as they are intended.
- Use a buddy system when working on electrical circuits.
- Never distract or startle a co-worker.
- For electrical fires, only use an extinguisher approved for such a fire; never use water.
- Make sure you are trained in CPR and first aid.

Here are some ways to minimize the risks associated with specific electrical hazards:

- **Wear Proper Personal Protective Equipment (PPE)** when engaging breakers that are 480v or above. Failure to do this could cause serious bodily injury or death. It could also lead to an equipment breaker explosion and flash fire.
- **Use properly grounded 110V lighting** with a Ground Fault Circuit Interrupter (GFCI) when working inside a vessel. Unlike circuit breakers that only sense excessive current, GFCIs detect the flow of current to ground and open the circuit to interrupt the flow of current. Otherwise, seriously bodily injury, burns or death could occur. This could also lead to a short in electrical equipment, or energized vessel floors and walls.
- **Use Ground Fault Circuit Interrupter (GFCI).** A GFCI works on the principle that the amount of current fed from the source to a device, such as a hand tool, equals the amount of current returned from the device to the source. Therefore a simultaneous measurement of the current going to and coming from any device should always be equal in a properly working system. The GFCI continuously monitors and verifies that the two currents are, in fact, equal. The GFCI assumes that any imbalance is most likely due to the current returning to the source through a human body & ground, rather than through the circuit conductors. If the GFCI measures an imbalance of 5 milliamps (ma) (.005 ampere) or greater, the GFCI will automatically shut off the circuit to prevent possible electrocution. 5 ma is the typical threshold of feeling for most people, but well below the lethal level.

- **Use properly grounded electrical tools** (with a GFCI). Also, make sure electrical tools are working properly (e.g., housings are not accidentally energized) and do not have frayed cords. Otherwise, serious bodily injury, burns or death could occur.
- **Turn off and lock out breakers using the proper methods** (called lockout/tagout). Otherwise, you or coworkers working on the equipment could be injured.
- **Inspect wiring for corrosion or fraying** and insulation for cracks, burns and degradation.

MONITORING AND MAINTAINING ACTIVITIES

Scheduled maintenance activities are an essential part of keeping equipment in good condition and preventing equipment failure.

The following are typical, basic maintenance routines necessary for the upkeep of motors:

Look	Listen	Feel
■ Inspect wires to ensure they are insulated and the insulation is not cracked or worn. ■ Check for loose covers and shrouds. ■ Visually inspect bearings for wear. ■ Occasionally measure bearings and bushing to check for excessive wear. ■ Look for signs of corrosion. ■ Lubricate bearings when scheduled or noisy. ■ Check the motor shaft for wear and verify that it is not bent. ■ Inspect oil passages to make sure they are not plugged. ■ Ensure oil wells or holes are clean.	■ Listen for abnormal noise. ■ Listen and check the cooling fan. ■ Listen to the bearings.	■ Feel for excessive heat. ■ Check for excessive vibration. ■ Check the bearings for heat.

SYMBOLS FOR ELECTRICAL EQUIPMENT AND MOTORS

There are several types of equipment associated with electricity. Each type has a unique symbol that is used to identify it on process drawings. The following table provides some examples of these symbols:

Symbol	Name	Description
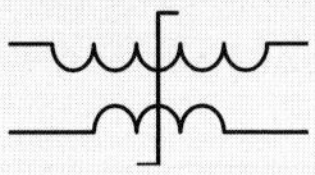	Transducer	A device that converts one type of energy to another, such as electrical to pneumatic.
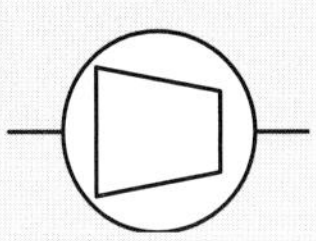	Motor Driven	A symbol that indicates a piece of equipment is motor driven (either AC or DC).
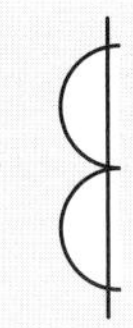	Current Transformer	A device that can provide circuit control and current measurement.
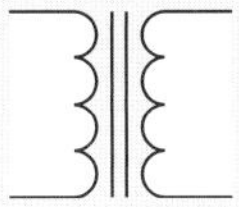	Transformer	A device that can either step up or step down the voltage of AC electricity.

Symbol	Name	Description
- - - - - - - - -	Electrical Signal	A signal that indicates voltage or current.
	Potential Transformer	A device that monitors power line voltages for power metering.
	Inductor	An electronic component consisting of a coil of wire.
Motor	Motor	A motor (either AC or DC) that converts electrical energy into mechanical energy.
	Outdoor Meter Device	A meter used to monitor electricity, such as a Voltmeter (volt measurement) or Ammeter (current measurement)

SUMMARY

Electricity is a flow of electrons from one point to another along a pathway, called a conductor. **Conductors** are a substance or body that allows a current of electricity to pass continuously along it. An **insulator** is a device made from a material that will not conduct electricity; the device is normally used to give mechanical support to electrical wire or electronic components.

Electricity travels through **circuits**, a system of one or many electrical components connected together to accomplish a specified purpose. An **ohm** is a measurement of resistance in electrical circuits. An **amp** is a unit of measure of the electrical current flow in a wire; similar to "gallons of water" flow in a pipe. A **volt** is the electromotive force, which will establish a current of one amp through a resistance of one ohm. A **watt** is work being done at the rate of one watt when an equivalent current of one amp is maintained through a resistance by an electromotive force of one volt.

Electrical energy can either be Alternating Current (AC) or Direct Current (DC). Alternating Current (**AC**) uses a back and forth movement of electrons in a conductor (similar to water sloshing backwards and forwards in a pipe). Direct Current (**DC**) results in a direct flow of electrons through a conductor (like water moving in one direction through a pipe). DC is most commonly used in batteries. AC is the most commonly used power type in industry.

AC and DC power can be used to turn motors. **Motors** are mechanical drivers that turn electrical energy into useful mechanical energy and provide power for a variety of rotating equipment (e.g., pumps, compressors and fans). The main components of an electric motor include a frame, shroud, rotor, stator, fan, and bearings.

Both AC and DC Motors work off the same three principles of electromagnetism: (1) Electric current generates a magnetic field; (2) Like magnetic poles repel each other (positive to positive or negative to negative), while opposite poles attract each other (positive to negative); (3) The direction of the electrical current determines the magnetic polarity.

Because of the hazards associated with electric shock, all personnel working with electrical equipment MUST be properly trained and authorized for the particular type and level of work being performed.

Electricity can be dangerous. Electrical current passing through a body causes a shock, which can result in serious bodily injuries, burns or death. For this reason, process technicians should always wear proper protective gear, follow safety procedures, and be aware of potential hazards (e.g., frayed cords and cracked insulation).

Process technicians should also inspect and perform scheduled maintenance on equipment in order to keep it in good condition and prevent equipment failure (e.g., lubricate bearings, inspect wires to make sure they are not cracked or worn, listen for abnormal noises, and check for excessive heat or vibration).

CHECKING YOUR KNOWLEDGE

1. Define the following key terms:
   a. Alternating Current (AC)
   b. Direct Current (DC)
   c. Circuit
   d. Electricity
   e. Grounding
   f. Insulator
   g. Motor
   h. Generator
   i. Amp
   j. GFCI
   k. Ohm
   l. Transformers
   m. Watts
   n. Volt

2. *(True or False)* DC current is the most common current used in the process industries.

3. What type of equipment converts mechanical energy to electrical energy?

4. What type of equipment converts electrical energy to mechanical energy?

5. The __________ is a casing over an electric motor that allows air to flow into and around the motor.
   a. Stator
   b. Shell
   c. Shroud
   d. Frame

6. What is the stator on a motor?

7. Which of the following is NOT a component of a motor?
   a. Stator
   b. Bearings
   c. Fan
   d. Shield

8. Which of the following hazards can result in death?
   a. Engaging 480v and above breakers without PPE
   b. Using faulty electrical equipment in the field without a GFCI
   c. Using 110v lighting inside a vessel without a GFCI
   d. All of the above

9. Most accidents involving electric motors are:
   a. Caused by carelessness
   b. Cannot be avoided
   c. Someone else's fault
   d. Minor enough to be ignored

10. Electrical shock occurs when your _______ becomes part of the circuit.

11. What is the resistivity of wet skin versus dry skin?

12. What is a GFCI and how does it work?

13. Fill in the blanks using look, listen or touch:
    a. _________ to ensure wires are insulated.
    b. _________ to check for excessive vibration.
    c. _________ to check the cooling fan.

14. Name three items to visually inspect on an electric motor.

15. What does the following symbol represent? What does this device do?

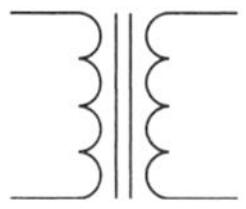

ACTIVITIES

1. Describe the basic principles of electricity, including the difference between AC and DC current and which type is commonly used in the process industry.

2. Write a one-page paper explaining how an electric motor works.

3. Given a picture or a cutaway of an AC motor, identify the following components:
   a. Frame
   b. Shroud
   c. Rotor
   d. Stator
   e. Fan
   f. Bearings
   g. Power supply

4. List five causes of electrical shock and explain what you could do to prevent them.

5. Given two strings of lights (one wired in series and the other wired parallel), examine what happens when a "good" bulb (one that lights properly) is replaced with a "bad" bulb (one that no longer lights). Did the entire string stay lit? If not, explain what happened.

6. Given two magnets with both poles exposed (as opposed to a refrigerator magnet which may be covered on one side), observe what happens when like poles are placed together (positive to positive, or negative to negative). Then observe what happens when opposite poles are placed together (positive to negative). Think of how these forces could be used to turn a rotor in an AC motor.

7. Using the formula for Ohm's law:
   a. Determine the voltage (V) if current (I) = 0.2A and resistance (R) = 1000 ohms.
   b. Determine the current (I) if voltage (V) = 110V and resistance (R) = 22000 ohms.
   c. Determine the resistance (R) if voltage (V) = 220V and current (I) = 5A.

8. Determine the watts required if the voltage is 480 and the amps are 5.

Chapter 22

Heat Exchangers

OBJECTIVES

Upon completion of this chapter you will be able to:

1. Describe the purpose or function of heat exchangers in the process industries.
2. Recall the three (3) methods of heat transfer.
3. Identify the primary parts of a typical heat exchanger.
4. Describe the operations of a typical tube heat exchanger.
5. Describe the different applications of typical heat exchangers.
6. Discuss the hazards associated with the improper operation of a heat exchanger.
7. Describe the monitoring and maintenance activities associated with a heat exchanger.
8. Identify the symbols used to represent the heat exchanger and associated equipment.

KEY TERMS

- **Aftercoolers**—a heat exchanger located on the discharge side of a compressor with the function of removing excess heat from the system created during compression.
- **Back Flush**—to wash by reversing the normal flow.
- **Baffles**—partitions located inside a shell and tube heat exchanger that increase turbulent flow and reduce hot spots.
- **Chiller**—a device used to cool a fluid to a temperature below ambient temperatures; chillers generally use a refrigerant as a coolant.
- **Condenser**—a heat exchanger that is used to condense vapor to a liquid.
- **Conduction**—the transfer of heat through matter via vibrational motion.
- **Convection**—the transfer of heat through the circulation or movement of a liquid or gas.
- **Distillation**—the separation of the constituents of a liquid mixture by partial vaporization of the mixture and separate recovery of vapor and residue.
- **Exchanger Head** (also called the channel head)—a device at the end of a heat exchanger that directs the flow of the fluids into and out of the tubes.
- **Heat Exchanger**—a device used to exchange heat from one substance to another.
- **Interchanger** (also called a cross exchanger)—One of the process-to-process heat exchangers.
- **Laminar Flow**—a condition in which fluid flow is smooth and unbroken; viewed as a series of laminations or thin cylinders of fluid slipping past one another inside a tube.

- **Pre-heater**—a heat exchanger used to warm liquids before they enter a distillation tower or other part of the process.
- **Radiation**—the transfer of heat energy through electromagnetic waves.
- **Reboiler**—a tubular heat exchanger placed at the bottom of a distillation column or stripper to supply the necessary column heat.
- **Shell**—the outer housing of a heat exchanger that covers the tube bundle.
- **Shell Inlets and Outlets**—the openings that allow process fluids to flow into and out of the shell side of a shell and tube heat exchanger.
- **Spacer Rods**—the rods that space the tubes in a tube bundle apart so they do not touch one another.
- **Tube Bundle**—a group of fixed or parallel tubes, such as is used in a heat exchanger; the tube bundle includes the tube sheets with the tubes, the baffles and the spacer rods.
- **Tube Inlets and Outlets**—the openings that allow process fluids to flow into and out of the tube bundle in a shell and tube heat exchanger.
- **Tube Sheet**—a flat plate to which the tubes in a heat exchanger are fixed.
- **Turbulent Flow**—a condition in which the fluid flow pattern is disturbed so there is considerable mixing.

INTRODUCTION

Heat exchangers facilitate the transfer of heat from one process fluid to another without physically contacting each other. Without this heat exchange many process temperature changes could not efficiently occur.

There are many different types of heat exchangers in use today. While their designs differ, all have similar components and use the same principles of heat transfer.

When working with heat exchangers, process technicians should be aware of operational aspects of the exchanger and the factors that could impact heat exchange and the operation of the exchangers themselves.

PURPOSE OF HEAT EXCHANGERS

A **heat exchanger** is a device used to exchange heat from one substance to another without physically contacting the other substance. Without heat exchangers, many processes could not occur properly.

Heat exchangers come in a variety of types (e.g., air cooled, plate and frame, double-pipe, and shell and tube). The most common type, however, is the shell and tube exchanger.

PARTS OF A TYPICAL HEAT EXCHANGER

The components of a heat exchanger may vary based on the design and purpose of the exchanger. However, there are some commonalities between exchangers.

The components of a typical shell and tube heat exchanger include:

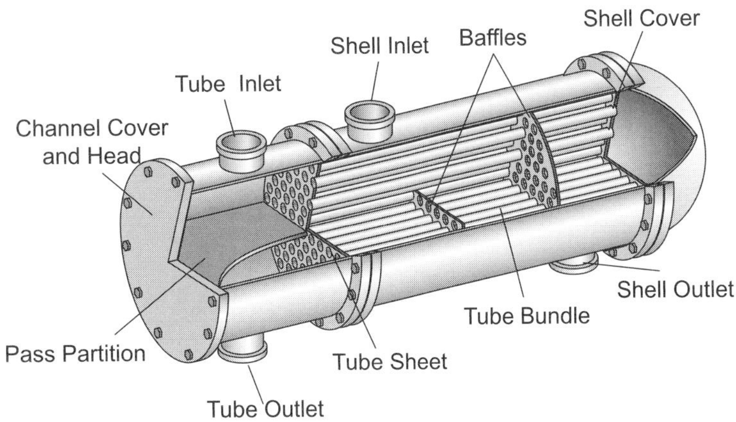

Figure 22-1: Common heat exchanger components

- Tube bundle
- Tube sheet
- Baffles
- Tube inlet and outlet
- Shell
- Shell inlet and outlet
- Exchanger (channel) head

The **tube bundle** is a group of fixed or parallel tubes through which process fluids are circulated. Tube bundles include the **tube sheet,** a flat plate to which the tubes in a heat exchanger are fixed, **baffles,** partitions located inside a shell and tube heat exchanger that increase turbulent flow and reduce hot spots, and **spacer rods** (not shown in Figure 22-1), the rods that space the tubes in a tube bundle apart so they do not touch one another.

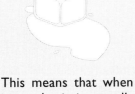

DID YOU KNOW?

Heat always travels from hot to cold.

This means that when ice melts it is actually absorbing heat, not giving up cold.

Fluids move into the tube bundle through the **tube inlet**, and out of the tube bundle through the **tube outlet**. Covering the tube bundle is an outer housing called a **shell**. Fluid moves into the shell through the **shell inlet**, and out of the shell through the **shell outlet**.

The **exchanger head** (also called the channel head) is located on the end of a heat exchanger and it directs the flow of fluids into and out of the tubes.

HOW A HEAT EXCHANGER WORKS

Heat exchangers facilitate the transfer of heat through one or more of the three heat transfer methods: conduction, convection, and radiation.

If you recall from the *Physics* chapter, **Conduction** is the flow of heat through a solid (e.g., a frying pan transferring heat to an egg). **Convection** is the transfer of heat through a fluid medium (e.g., warm air circulated by a hair dryer); and **radiation** is the transfer of heat through space (e.g., warmth emitted from the sun).

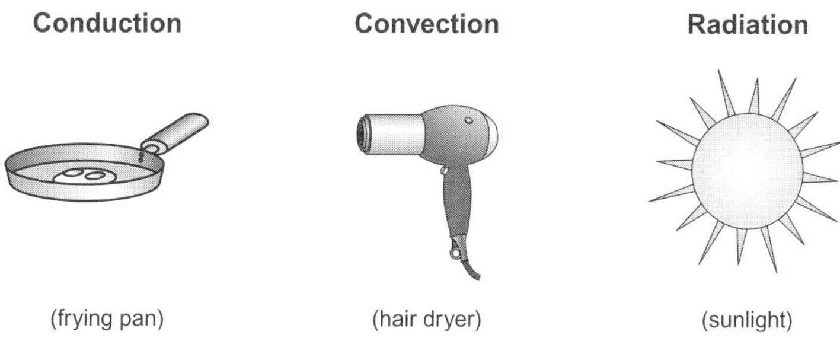

Figure 22-2: Examples of convection, conduction and radiation

The main methods of heat transfer in a shell and tube heat exchanger are conduction (through the tube wall and tube surface, to shell fluid) and convection (fluid movement within the shell and the tubes).

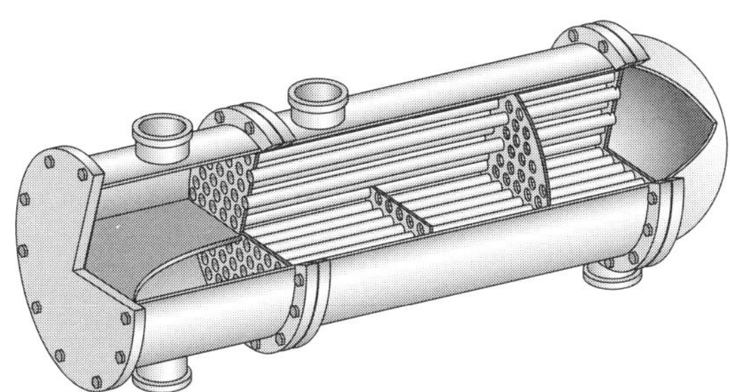

Figure 22-3: Shell and tube heat exchanger

In a shell and tube exchanger, liquids or gases of varying temperatures are pumped into the shell and tubes in order to facilitate heat exchange.

For example, hot process fluids can be pumped into the tubes while cool water is pumped into the shell. As the cool water circulates around the tubes, heat is transferred from the tubes to the water in the shell.

A practical example of a heat exchanger is a car radiator. In this type of exchanger, hot fluids flow through the radiator tubes. As the car moves forward, cool air is drawn into the radiator grill and over the radiator tubes. As the air passes over the tubes, the heat from the fluid is transferred to the air via conduction and convection.

As fluids move through a pipe or a heat exchanger, the flow can be either **laminar** (a condition in which the fluid flow is smooth and unbroken, like as a series of laminations or thin cylinders of fluid slipping past one another inside a tube) or **turbulent** (a condition in which the fluid flow pattern is disturbed so there is considerable mixing).

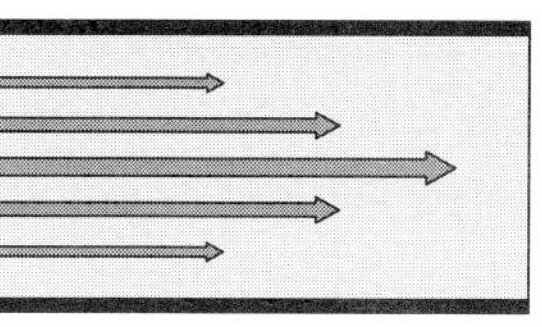

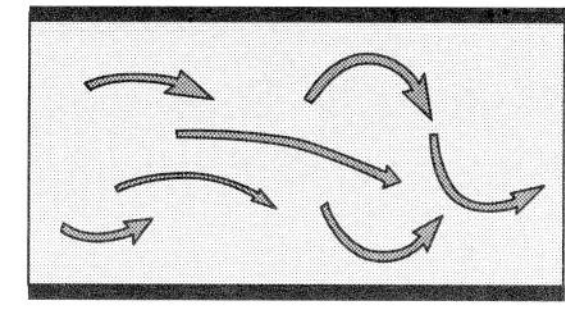

Figure 22-4: Examples of laminar and turbulent flow

The ideal flow type in a heat exchanger is turbulent flow because it provides more mixing and better heat transfer for heating and cooling. That is why heat exchangers contain baffles. Baffles support the tubes, increase turbulent flow, increase heat transfer rates, and reduce hot spots.

DID YOU KNOW?

Turbulent flow provides more fluid mixing than laminar flow, so it is the desired flow type in most heat exchangers.

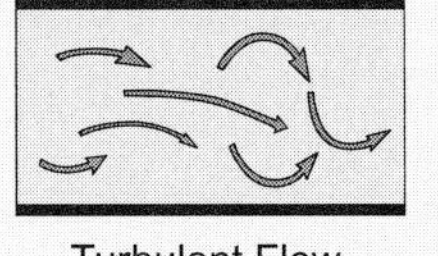

Turbulent Flow

HEAT EXCHANGER APPLICATIONS

Heat exchangers are used to heat or cool process fluids. They can be used alone or coupled with other heat exchanging devices (e.g., cooling towers). The most common applications for heat exchangers are: reboilers, pre-heaters, aftercoolers, condensers, chillers, and interchangers.

A **reboiler** is a tubular heat exchanger, placed at the bottom of a distillation column or stripper, which is used to supply the necessary column heat. The main purpose of a reboiler is to convert a liquid to a vapor, and to control temperature, pressure, or product quality. The heating medium for a reboiler can be steam or hot fluids from other parts of the plant.

A **pre-heater** is a heat exchanger that adds heat to some substance prior to a process operation.

An **aftercooler** is a shell and tube heat exchanger, located on the discharge side of a compressor, which is used to remove excess heat created during compression.

A **Condenser** is a heat exchanger that is used to condense vapor to a liquid. The design of a condenser can be the same as a pre-heater (i.e., a typical shell and tube exchanger). The difference, however, is the temperature of the fluid being used for heat exchange (i.e., warm vs. cool).

Chillers are devices used to cool a fluid to a temperature below ambient temperature. Chillers generally use a refrigerant as a coolant. Chillers are frequently found in industry.

An **interchanger** (cross exchanger) is a process-to-process heat exchanger. Interchangers use hot process fluids on the tube side and cooler process fluids on the shell side. Interchangers are used on columns or towers where the hot bottom fluids are pumped through the tube side of an exchanger, and the feed to the column is put through the shell side. This action preheats the feed to the column and cools the bottom material.

RELATIONSHIP BETWEEN DIFFERENT TYPES OF HEAT EXCHANGERS

To illustrate the various applications of heat exchangers, consider the distillation process.

Distillation (which is discussed in more detail in the *Distillation* chapter) is the separation of the constituents of a liquid mixture by partial vaporization. In order for vaporization to occur, heat must be added to the process fluids. This can be accomplished through pre-heaters and reboilers.

Pre-heaters warm the fluid before it enters the tower. Reboilers add additional heat to fluids that are already in the tower so they will vaporize.

Once vapors have formed, a condenser cools the vapors and returns them to their liquid state. Once liquefied, these vapors are stored or fed to other parts of the process.

Figure 22-5 illustrates the relationship between a pre-heater, a reboiler, and a condenser in the distillation process.

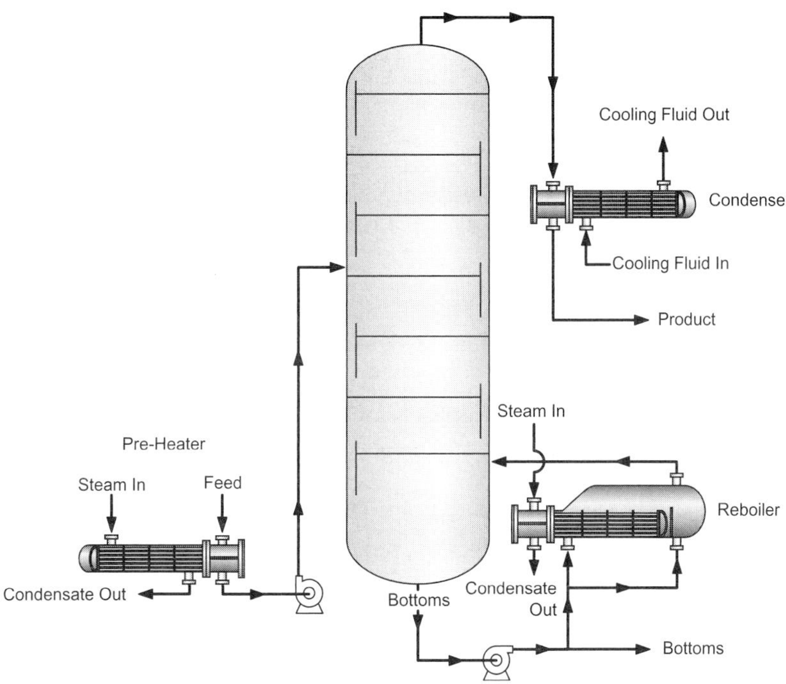

Figure 22-5: Heat exchangers attached to a distillation column

HAZARDS ASSOCIATED WITH IMPROPER HEAT EXCHANGER OPERATION

There are many hazards associated with improper heat exchanger operation. Table 22-1 lists some of these hazards and their impacts.

TABLE 22-1: Hazards associated with improper heat exchanger operation

IMPROPER OPERATION	POSSIBLE IMPACTS			
	Individual	Equipment	Production	Environment
Putting the exchanger online with the bleeder valve open on process side	Exposure to chemicals or steam			Spill to the environment
Applying heat to the exchanger without following the proper warm up procedure	Exposure to chemicals or steam if the tubes rupture	Tube rupture due to thermal shock	Downtime due to repair	Spill to the environment if the tubes rupture
Misalignment: cooling, heating or not opening inlet and outlet		Overheating, tube fouling or melting	Downtime due to repair	
Operating a heat exchanger with ruptured tubes	Exposure to chemicals or steam	Exchanger could blow up or be damaged if the safety valve is not working properly	Product ruined by contamination; downtime due to repair	Spill to the environment
Opening a cold fluid to a hot heat exchanger		Over pressurizing the heat exchanger; potential for thermal shock		

HEAT EXCHANGER MONITORING AND MAINTENANCE ACTIVITIES

When working with heat exchangers, process technicians need to be aware of problems that can impact the exchanger, and be able to perform preventive maintenance.

Table 22-2 lists some of the things a process technician should monitor when working with heat exchangers:

TABLE 22-2: Heat exchanger monitoring and maintenance activities

Look	Listen	Feel
■ Check for external leaks. ■ Check for internal tube leaks by collecting and analyzing samples. ■ Look for abnormal pressure changes (could indicate tube plugging). ■ Look for temperature changes. ■ Look for hot spots or uneven temperatures. ■ Inspect for leaky gaskets. ■ Inspect insulation. ■ Monitor inlet and outlet temperature gauges. ■ Monitor inlet and outlet pressure gauges. ■ Monitor inlet and outlet flow rate. ■ Monitor inlet and outlet samples. ■ When the heat exchanger is disassembled. ▪ Inspect for leaky tubes. ▪ Inspect baffles for proper fit. ▪ Inspect for corrosion. ▪ Inspect for distorted tubes. ■ Inspect vents to ensure they are working properly.	■ Listen for abnormal noises (e.g., rattling or whistling). Be extremely careful around high pressure leaks, they are often invisible but very audible.	■ Check for excessive vibration (vibration can loosen the tubes in the tube sheet). ■ Check inlet and outlet for heat and coolness.

In addition to the tasks listed above, process technicians should also back flush the water side of the heat exchanger to reduce fouling. **Back flushing (backwashing)** is washing by reversing the normal flow.

During back flushing, water is run backwards through a heat exchanger to remove deposits and reduce fouling. Heat exchangers are equipped with special valves so this back flushing can be performed.

Failure to perform proper maintenance and monitoring could impact the process and result in personnel or equipment damage.

HEAT EXCHANGER SYMBOLS

The following are examples of heat exchanger symbols a process technician might encounter on a Piping and Instrumentation Diagram (P&ID).

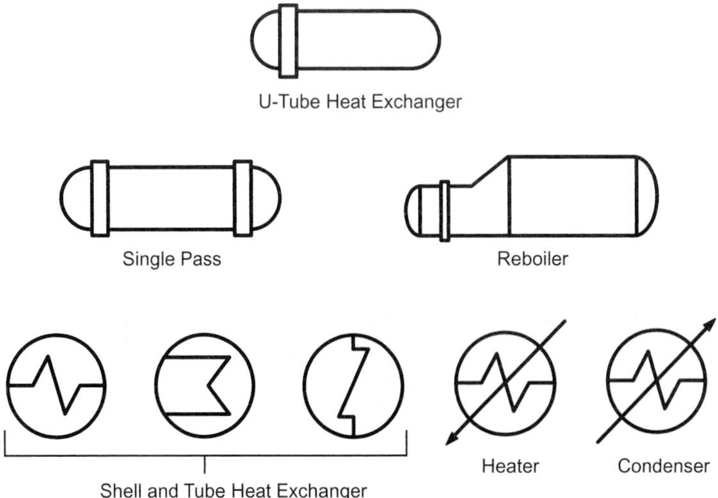

Figure 22-6: Heat exchanger symbols

SUMMARY

Heat exchangers are devices that use conduction, convection, and radiation to heat or cool process fluids.

The most common type of heat exchanger is the shell and tube design. The primary components of this type of exchanger are: a tube bundle, a tube sheet, baffles, tube inlet and outlet, shell, shell inlet and outlet, and an exchanger (channel) head.

In a shell and tube exchanger, liquids or gases of varying temperatures are pumped into the shell and tubes. As the fluids move through the shell, baffles cause turbulence (mixing) in the fluid. Turbulence is desired because it facilitates heat transfer and creates more surface area for heating and cooling.

Heat exchangers can be used in a variety of applications, including reboilers, pre-heaters, aftercoolers, condensers, chillers, or interchangers.

Improperly operating a heat exchanger can lead to safety and process issues, including possible exposure to chemicals or hot fluids, or equipment damage due to thermal shock.

Technicians should always monitor heat exchangers for abnormal noises, leaks, excessive heat, pressure drops, temperature drops, or other abnormal conditions, and conduct preventive maintenance as needed.

CHECKING YOUR KNOWLEDGE

1. Define the following key terms:
   a. Heat exchanger
   b. Shell
   c. Tube bundle
   d. Baffle
   e. Conduction
   f. Convection
   g. Turbulent flow
   h. Laminar flow
   i. Distillation
   j. Pre-heater
   k. Reboiler
   l. Condenser

2. Which of the following heat exchanger components is used to create turbulence in the process fluids and facilitate heat exchange?
   a. Shell
   b. Baffle
   c. Exchanger head
   d. Tube inlet

3. *(True or False)* Heat always moves from cold to hot.

4. The transfer of heat through a fluid medium is called:
   a. Convection
   b. Conduction
   c. Radiation

5. List three applications of heat exchangers (e.g., reboiler) and explain what each one is used for.

6. List at least three things a process technician should look, listen, and feel for when monitoring a heat exchanger.

ACTIVITIES

1. Given a cutaway picture of a heat exchanger, identify the following components:
   a. Tube bundle
   b. Tube sheet
   c. Baffles
   d. Tube inlet and outlet
   e. Shell
   f. Shell inlet and outlet
   g. Exchanger (channel) head

2. Sketch a simple diagram of distillation column and identify where the following components might be found:
   a. Pre-heater
   b. Condenser
   c. Reboiler

3. List two hazards associated with improper heat exchanger operations, and identify the possible impacts these hazards could have on individuals, equipment, production, and/or the environment.

4. Given a Piping and Instrumentation Diagram (P&ID), identify all of the heat exchangers.

Chapter 23
Cooling Towers

OBJECTIVES

Upon completion of this chapter you will be able to:

1. Describe the purpose or function of a cooling tower in the process industries.
2. Identify the primary parts and support systems of a typical cooling tower.
3. Describe the operation of a cooling tower.
4. Describe the different applications or use of water from a cooling tower.
5. Discuss the hazards associated with the improper operation of a cooling tower.
6. Describe the monitoring and maintenance activities associated with a cooling tower.
7. Identify the symbols used to represent cooling towers and fin fans, and associated equipment.

KEY TERMS

- **Basin**—a compartment, located at the base of a cooling tower, which is used to store water until it is pumped back into the process.
- **Blowdown**—taking basin water out of a cooling tower to reduce the level or impurity concentration.
- **Chemical Treatment**—chemicals added to cooling water that are used to control algae, sludge, and fouling of exchangers and cooling equipment.
- **Condenser**—a heat exchanger that is used to condense vapor to a liquid.
- **Cooler**—a heat exchanger that uses a cooling medium to lower temperature of a process material.
- **Cooling Tower**—a structure designed to lower the temperature of a water stream by evaporating part of the stream (latent heat of evaporation); these towers are usually made of plastic and wood and are designed to promote maximum contact of the water with the air.
- **Drift Eliminators**—devices that prevent water from being blown out of the cooling tower; the main purpose of a drift eliminator is to minimize water loss.
- **Evaporation**—an endothermic process in which a liquid is changed into a gas.
- **Fan**—a device used to force or draw air through a furnace or cooling tower.
- **Fill**—the material that breaks water into smaller droplets as it falls inside the cooling tower.
- **Forced Draft Tower**—cooling towers that have fans or blowers at the bottom of the tower that force air through the equipment.

- **Induced Draft Tower**—cooling towers that have fans at the top of the tower that pull air through the tower.
- **Makeup Water**—water that is used to replace the water lost during blowdown and evaporation.
- **Natural Draft Tower**—cooling towers that use temperature differences inside and outside the stack to facilitate air movement.
- **Suction Screens**—usually a cone-shaped or flat metal strainer used to remove debris.
- **Water Distribution Header**—a device that evenly distributes the water on top of the cooling tower through the use of distribution valves, and allows the water to evenly flow into the tower fill.

INTRODUCTION

Cooling towers play an important role in many processes. Through cooling towers, heat exchangers are able to change the temperature of process fluids.

There are many different types of cooling towers in use today. While their designs differ, all have similar components and use the same principles of heat transfer.

When working with cooling towers, process technicians should be aware of operational and safety aspects of the tower and the factors that could impact heat exchange and the operation of the exchangers.

PURPOSE OF COOLING TOWERS

Cooling towers are structures designed to lower the temperature of a water stream by evaporating part of the stream (latent heat of evaporation). These towers are usually made of plastic and wood and contain various types of fill or packing, which is designed to break the water into tiny droplets and promote maximum water to air contact.

The main purpose of cooling towers is to remove heat from process cooling water so it can be recycled and re-circulated through the process.

DID YOU KNOW?

Many homes use evaporative cooling units to cool the air. These units lower the temperature of the air by filtering it through water.

Because the process of evaporation is impacted by humidity, this type of unit works best in drier climates.

TYPES OF COOLING TOWERS

Cooling towers come in many shapes, sizes, and air flow types (i.e., atmospheric, natural draft, induced or forced draft).

- **Natural draft towers**—use temperature differences inside and outside the stack to facilitate air movement.
- **Forced draft towers**—have fans or blowers at the bottom of the tower that force air through the equipment.
- **Induced draft towers**—have fans at the top of the tower that pull air through the tower.

Figure 23-1 shows examples of the different types of cooling towers.

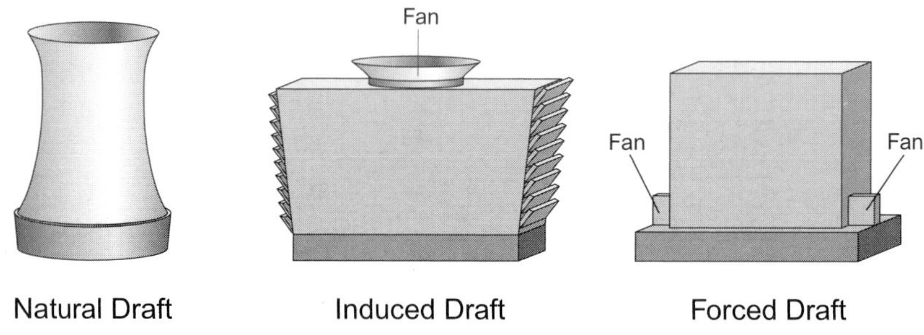

Figure 23-1: Types of cooling towers

PARTS OF A COOLING TOWER

While cooling tower designs may vary, most cooling towers are made of wood and have similar components. These components include:

- Water distribution header and cells
- Fill (splash bars)
- Basin
- Makeup water
- Suction screens
- Drift eliminators
- Fan with a chimney or stack (induced or forced draft only)

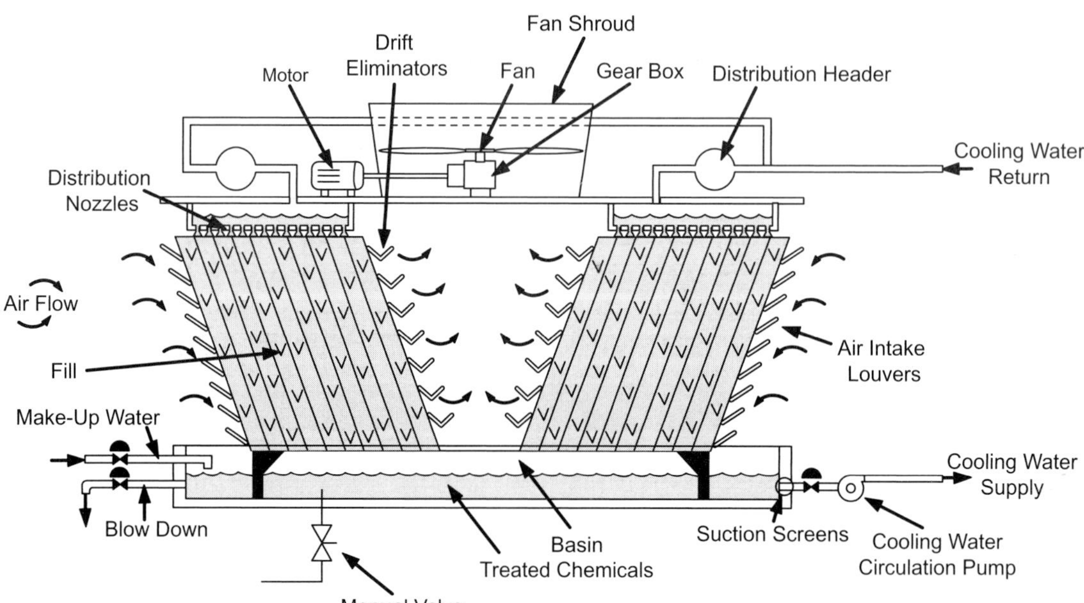

Figure 23-2: Parts of an induced draft, cross-flow cooling tower

In a cooling tower, the **water distribution header** evenly distributes the water on top of the cooling tower for more effective cooling. As the water leaves the header, it falls down on splash bars called "fill." **Fill** is the material that breaks water into smaller droplets as it falls inside the cooling tower.

In induced or forced draft cooling towers **fans** are used to force or induce air flow. As the water falls down, the air contacts the water, thereby reducing the temperature of the water by evaporation.

Drift Eliminators minimize water loss and prevent water from being blown out of the cooling tower. However, some water is always lost despite these measures. In order to compensate for this water loss, makeup water is added to the basin.

Makeup water is water that is used to replace the water lost during blowdown and evaporation. The **basin** is a compartment, located at the base of a cooling tower, which is used to store water until it is pumped back to the process.

Debris can damage or destroy expensive pumps. In order to prevent this damage, **suction screens** (which are usually cone-shaped or flat) are used to filter out debris (e.g., wood fibers and trash).

HOW A COOLING TOWER WORKS

Cooling towers work based on the process of **evaporation**, a process in which a liquid is changed into a vapor through the latent heat process. Through the process of evaporation, hot water from the heat exchangers is cooled. This cool water is then sent back to the heat exchangers for reuse.

The following is an overview of the cooling process:

1. The cooling tower sends cool water to the heat exchangers.
2. In the heat exchangers, heat is transferred to the cool water.
3. Hot exchanger water is returned to the cooling tower distribution header.
4. The distribution header sprays the hot water downward onto the fill (splash bars).
5. The fill breaks the water into smaller droplets, thereby increasing the surface area and facilitating heat exchange.
6. As water falls through the tower it is exposed to air that removes the heat through latent heat of evaporation, making it cool again.
7. When the cool water reaches the bottom of the tower it is collected in the basin.

8. Pumps take water out of the basin and return it to the heat exchangers for reuse.
9. Steps 2-8 are repeated (this is a continuous process).

During the cooling process some water is lost due to evaporation, splashing, leakage, or blowdown (approximately 80% of cooling water loss is due to evaporation). This water is replaced with makeup water, often using a simple level control (e.g., a toilet bowl float) to maintain a certain water level in the basin.

All cooling water, including makeup water, must undergo **chemical treatment** in order to prevent algae growth, sludge, scale buildup and fouling of the exchangers and cooling equipment.

In addition to chemical treatment, cooling water must also go through blowdown. **Blowdown** is a process in which water is taken out of a cooling tower basin to reduce the level of impurity concentration. The blowdown process can occur continuously or periodically.

FACTORS THAT IMPACT COOLING TOWER PERFORMANCE

There are many factors that affect cooling tower performance. These include:

- Temperature and relative humidity of ambient air
- Wind velocity and direction
- Water contamination
- Cooling tower design

Table 23-1 lists each of these variables and explains how they impact the performance of the cooling tower.

TABLE 23-1: Factors that impact cooling tower performance

Variable	How it impacts cooling tower performance
Temperature	↑ ambient temperature = ↑ cooling tower demands
Humidity	↑ humidity = ↑ water vapor in the air ↑ water vapor = ↓ cooling (Note: If humidity reaches 100% evaporation will not occur because the air is completely saturated with water vapor).
Wind velocity and direction	↑ wind velocity = ↑ air flow ↑ air flow = ↑ cooling Wind direction opposite of cooling tower orientation = decreased evaporation
Water contamination	↑ contamination (e.g., algae buildup) = ↓ cooling (e.g., due to fouling)
Tower design	↑ air flow = ↑ cooling (Note: induced or forced draft towers have greater air flow than natural draft towers, and are not dependent on wind currents).

COOLING TOWER APPLICATIONS

Cooling tower water can be used in a variety of applications including process condensers and equipment coolers.

Condensers are heat exchangers that cool vapors and convert them to a liquid. In a condenser, cool water flows through the tubes and vapor flows around the tubes. As the vapors flow, the cool water is heated and the vapors are cooled. As the vapors cool, they condense on the tubes, drip off, and collect in the bottom of the exchanger. This product is then pumped out for storage or returned to the process.

Coolers are heat exchangers that use a cooling medium to lower temperature of a process material. In lubricating systems (found in rotating equipment) coolers are used to cool lubrication oils on steam turbines and other large pieces of equipment.

HAZARDS ASSOCIATED WITH IMPROPER COOLING TOWER OPERATION

There are several hazards associated with improper cooling tower operation. Table 23-2 lists some of these hazards and their possible impacts.

TABLE 23-2: Hazards associated with improper cooling tower operation

IMPROPER OPERATION	POSSIBLE IMPACTS			
	Individual	**Equipment**	**Production**	**Environment**
Failure to chemically treat cooling water	Exposure to harmful microorganisms (e.g., Legionella bacteria, which causes Legionnaires' disease)	Algae growth or ludge buildup scan foul the exchanger	Products created are off spec due to improper heat exchange	
Improper valve operation	Scalding or burns from exposure to hot process cooling water	Pump damage due to "dead heading" or improper feedwater supply	Products created are off spec due to improper heat exchange	
Circulating cooling water through a heat exchanger with ruptured tubes	Exposure to chemicals, steam, or bacteria	Tower contamination; depletion of treatment chemicals	Product ruined by contamination	Hazardous chemicals spilled to the environment
Failure to use caution on wet, icy, or slippery surfaces	Injury due to slipping and falling			
Improper operation of the Chlorine system	Severe injury or death due to chlorine exposure (e.g., don't climb a tower when shock chlorinating)	Possible algae growth or sludge buildup that foul the exchanger (if chlorine levels are not high enough)	Products created are off spec due to improper heat exchange (e.g., if the exchanger fouls because chlorine levels are too low)	
Ingesting, or exposing skin to chemically treated cooling water.	Injury or death due to water treatment chemical exposure			

COOLING TOWER MONITORING AND MAINTENANCE ACTIVITIES

When working with cooling towers, process technicians should to be aware of problems that can impact exchangers and be able to perform preventive maintenance.

Table 23-3 lists some of the things a process technician should monitor when working with heat exchangers:

TABLE 23-3: Cooling tower monitoring and maintenance activities

Look	Listen	Feel
■ Check for leaks. ■ Check basin water levels to make sure they are adequate. ■ Check chemical balance (pH and hardness). ■ Check filter screens for plugging. ■ Check temperature differentials. ■ Look for broken fill materials to fix at next turnaround. ■ Look for ice build-up in cold climates. ■ Check for proper water distribution on top of the tower.	■ Listen for abnormal noises (e.g., grinding sounds associated with pump cavitation, or high pitched sounds associate with improperly lubricated fan bearings).	■ Feel for excessive heat or vibration in fans and pumps

In addition to the tasks listed above, process technicians should only run the number of fans required to produce the proper cooling water supply temperature.

Failure to perform proper maintenance and monitoring could impact the process and result in personal or equipment damage.

COOLING TOWER SYMBOLS

In order to accurately locate cooling towers on a piping and instrumentation diagram (P&ID) process technicians need to be familiar with the different cooling tower symbols.

Figure 23-3 shows some of the symbols used to indicate cooling towers.

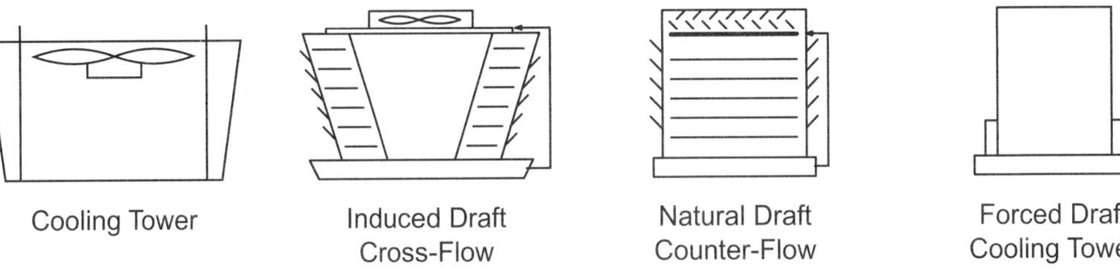

Figure 23-3: Cooling tower symbols

SUMMARY

Cooling towers are structures designed to lower the temperature of a water stream by evaporating part of the stream (latent heat of evaporation). The main purpose of cooling towers is to remove heat from process cooling water so it can be recycled and re-circulated through the process.

Cooling towers are used in conjunction with heat exchangers. Heat exchangers remove heat from process fluids. Cooling towers remove heat from heat exchanger cooling water. The heating and cooling relationship between exchangers and towers is a continuous process.

Cooling towers come in many shapes, sizes, and air flow types. The three main air flow types are **atmospheric** (rely on wind for air movement), **natural draft** (rely on temperature differences), and **mechanical** (use fans to move air).

The main components of a cooling tower include a **water distribution header** (distributes the water), **fill** (redirects the flow of air and water), a **basin** (stores water), **makeup water** (replaces lost fluids), **drift eliminators** (minimize water loss), **suction screens** (filter out debris), and a **fan** with a chimney or stack (used to force or induce air flow in induced or forced draft towers).

There are many factors that affect cooling tower performance. These include temperature, humidity, wind velocity, water contamination, and tower design

Cooling tower water can be used in a variety of applications including process condensers and equipment coolers.

When monitoring and maintaining cooling towers, technicians must remember to check pumps for excessive vibrations, noise, or heating. They must also ensure that water and chemical levels area adequate, verify that there are no leaks and that filter screens are not plugged, and that proper heat exchange is occurring. In cold climates they should check for excessive ice build-up inside the tower, since heavy ice layers can damage the internal structures of the tower.

CHECKING YOUR KNOWLEDGE

1. Define the following key terms:
   a. Cooling tower
   b. Condenser
   c. Cooler
   d. Water Distribution Header
   e. Drift Eliminators
   f. Basin
   g. Fill
   h. Blowdown
   i. Evaporation
   j. Makeup water
   k. Suction Screens

2. What is the purpose of a cooling tower?

3. Explain the difference between an atmospheric tower, a natural draft tower, and an induced or forced draft tower.

4. List five factors that affect cooling tower performance.

5. Explain why cooling water must be chemically treated.

ACTIVITIES

1. Given a model or diagram of a cooling tower, identify the following components and explain the purpose of each:
   a. Water distribution header
   b. Fill (splash bars)
   c. Basin
   d. Makeup water
   e. Suction screens
   f. Drift eliminators
   g. Fan

2. List the steps associated with the cooling process (i.e., explain how a cooling tower works).

3. List at least three monitoring and maintenance activities associate with cooling towers, and explain why those activities are important.

4. Given a Piping and Instrumentation Diagram (P&ID), identify all of the cooling towers.

Chapter 24
Furnaces

OBJECTIVES

Upon completion of this chapter you will be able to:

1. Describe the purpose or function of furnaces in the process industries.
2. Describe the operation of a furnace.
3. Describe the types of fuel used in a furnace.
4. Identify the primary parts of a typical furnace.
5. Describe the different types of furnaces by draft.
6. Describe the different furnace designs.
7. Describe the monitoring and maintenance activities associated with furnaces.
8. Discuss the hazards associated with the improper operation of a furnace.

KEY TERMS

- **Air Register**—devices, located on a burner, which are used to adjust the primary and secondary airflow to the burner; air registers are the main source of air to the furnace.
- **Balanced Draft Furnace**—a furnace that uses two fans to facilitate airflow, one inducing flow out of the firebox (induced draft) and one providing positive pressure to the burners (forced draft).
- **Burner**—a device used to introduce, distribute, and mix air, fuel and flame in the firebox.
- **Convection Section**—the upper portion of a furnace where heat transfer is primarily through convection.
- **Convection Tubes**—furnace tubes, located above the shock bank, that receive heat through convection.
- **Damper**—a valve, movable plate, or adjustable louvers used to regulate the flow of air or draft in a furnace.
- **Draft Gauges**—gauges, calibrated in "inches of water," that measure the firebox pressure, airflow, and differential pressure between the outside of the furnace and the flue gas inside.
- **Firebox**—the portion of a boiler/furnace where burners are located and radiant heat transfer occurs.
- **Flame Impingement**—a condition in which the flames from a burner touch tubes in a furnace.
- **Forced Draft Furnace**—a furnace that use fans or blowers to force air into the air registers.
- **Fuel Gas Valve**—a valve that controls fuel gas flow and pressure to burners.
- **Furnace**—an apparatus in which heat is liberated and transferred directly or indirectly to a fluid mass for the purpose of increasing the temperature of a process fluid.

- **Furnace Purge**—a method of removing combustibles from a furnace firebox in preparation for lighting burners before startup.
- **Hot Spot**—a furnace tube or area within a furnace that gets too hot.
- **Induced Draft Furnace**—a furnace that uses fans, located in the stack, to induce airflow from the firebox.
- **Natural Draft Furnace**—a furnace that has no mechanical draft or fans. Instead, the heat in the furnace causes draft.
- **Pilot**—an initiating device used to ignite the burner fuel.
- **Radiant Section**—the lower portion of a furnace (firebox) where heat transfer is primarily through radiation.
- **Radiant Tubes**—tubes located in the firebox that receive heat primarily through radiant heat transfer.
- **Refractory**—a form of insulation used inside high temperature boilers, incinerators, heaters, reactors, and furnaces.
- **Shock Bank**—tubes located directly above the firebox in a furnace that receive both radiant and convective heat.
- **Stack**—a cylindrical outlet, located on the top of a furnace, which remove flue gas from the furnace.

INTRODUCTION

Furnaces play an important role in many processes, and they can be used for many applications. The primary application of a furnace is to increase the temperature of process streams.

There are many different types of furnaces in use today. While their designs differ, all have similar components and use the same principles of combustion and heat transfer.

When working with furnaces, process technicians should be aware of operational aspects of the furnace and the factors that could impact furnace performance and safety.

DID YOU KNOW?

The Titanic had 159 coal-fired furnaces that used as much as 825 tons of coal per day!

These furnaces were used to generate the steam required to turn the propellers.

PURPOSE OF FURNACES

A **furnace**, also referred to as a process heater, is an apparatus in which heat is liberated and transferred directly or indirectly to a fluid mass for the purpose of increasing the temperature of the process stream.

Furnaces can be used for a variety of applications. For example, they can be used to heat water, superheat steam, incinerate waste products, remove metals from ore (smelting), and facilitate chemical reactions. For example, in the refining and petrochemical industries, they are used to break heavier hydrocarbon molecules into lighter hydrocarbon molecules (a process called "cracking") so products like gasoline, fuel oil, kerosene, plastic, feedstocks, and asphalt can be formed.

HOW A FURNACE WORKS

Furnaces burn fuel inside a containment area (firebox) to produce heat. While the firebox is being heated, process fluids are pumped through tubes inside the firebox. Heat is then transferred through the walls of the tubes to the process fluid via **conduction**. The product of combustion (flue gas) flows from the firebox, up through a stack, and to the atmosphere where it is released.

PARTS OF A FURNACE

Furnaces contain many parts. The most common components include:

- Firebox
- Stack
- Damper
- Air Register
- Burner (Gas or Oil)
- Pilot
- Refractory Lining
- Convection Section Tubes
- Radiant Tubes
- Fuel Gas Valve
- Draft Gauges
- Furnace Purge (Steam)

Figure 24-1 shows an example a cabin furnace and its components. Cabin furnaces are called "cabin" because they resemble a crude cabin with a chimney.

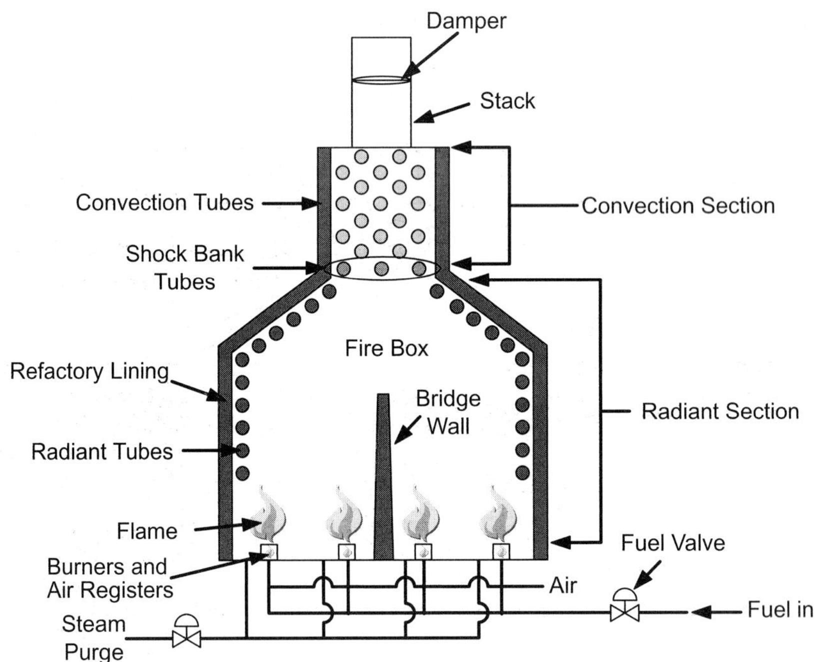

Figure 24-1: Cabin Furnace Components

Furnaces are divided into two sections: radiant and convection.

Radiant Section

The **radiant section** is the lower portion of a furnace (firebox) where heat transfer is primarily through radiation. Within the radiant section are the firebox, radiant tubes, and burners.

The **firebox** is the portion of a furnace where burners are located and radiant heat transfer occurs. Because so much heat is generated in this section, the firebox must be lined with a special refractory lining.

The **refractory** lining is a brick-like form of insulation used inside high temperature applications. The purpose of this lining is to reflect heat back into the firebox.

Radiant tubes are tubes contained within the firebox area that receive radiant heat from the furnace burners.

Burners are devices used to introduce air and fuel, distribute, mix, and produce a safe flame. Common furnace fuels include: natural gas, fuel oil, process oil, process gas, and fuel gas.

The **fuel gas valve** is a valve that controls fuel gas flow and pressure to burners. The **pilot** is an initiating device used to ignite the burner fuel.

Air registers are the main source of air to the furnace. They are used to adjust the primary and secondary airflow to the burner.

Draft gauges are devices, calibrated in "inches of water," that measure the firebox pressure, airflow, and differential pressure between the outside of the furnace and the flue gas inside.

Furnace purging is when steam or air is used to remove combustibles from a furnace firebox prior to burner lighting and startup.

Convection Section

The **convection section** is the upper portion of a furnace where heat is transferred by convection. Within this section are convection tubes, the stack and a damper.

Convection tubes are tubes that receive heat through the process of convection. Below the convection tubes, and above the firebox, is a row of tubes called the **shock bank**. Shock bank tubes receive both radiant and convective heat and protect the convection section from exposure to the radiant heat of the firebox.

Above the convection tubes is the stack. The **stack** is a cylindrical outlet at the top of a furnace, which removes flue (combustion) gas from the furnace. Within the stack is a valve or movable plate (louver) called the **damper**. The purpose of the damper is to regulate the flow of air or draft in the furnace.

FURNACE DESIGNS AND DRAFT TYPES

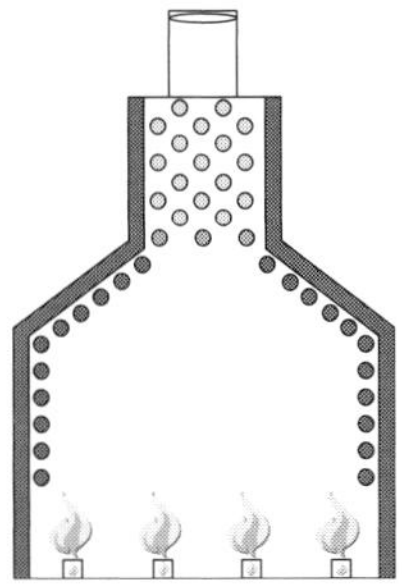

Furnaces come in a variety of styles (e.g., cabin, box, cylindrical, and A-Frame).

Figure 24-2 shows an example of a cabin furnace.

Figure 24-2: Cabin furnace

Furnaces also come in a variety of draft types including: natural draft, forced draft, induced draft, and balanced draft.

Natural Draft

Natural draft furnaces have no mechanical draft or fans. Instead, the heat in the furnace causes draft. As hot air rises through the stack, pressure is created inside the firebox burner air registers.

Figure 24-3 shows an example of a natural draft furnace.

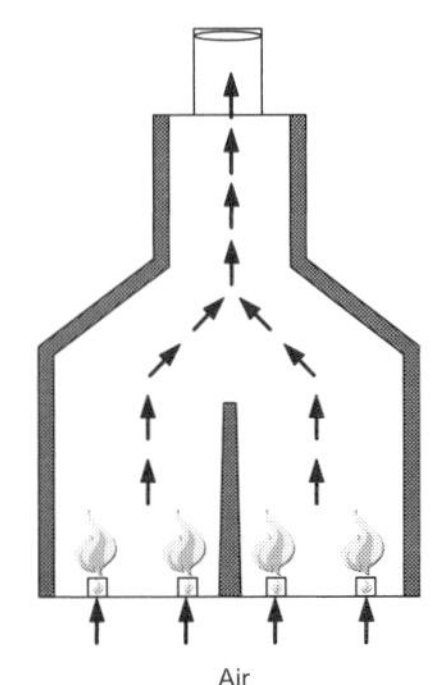

Figure 24-3: Natural draft furnace (cabin-type)

In order to achieve proper draft, the stacks on natural draft furnaces must be taller than furnaces that use other draft methods.

FORCED DRAFT

Forced draft furnaces use fans or blowers, located at the base of the furnace, to force airflow.

Figure 24-4 shows an example of a forced draft furnace.

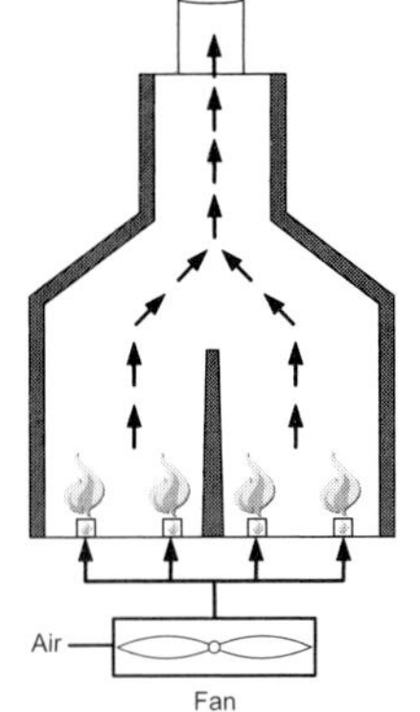

Figure 24-4: Forced draft furnace (cabin-type)

During combustion in a forced draft furnace, positive pressure is created by the fan or blower, which forces air into the burner air registers as flue gases leave the firebox.

Induced Draft

Induced draft furnaces use fans, located in the stack, to induce airflow by creating a lower pressure in the firebox.

Figure 24-5 shows an example of an induced draft furnace.

In this type of furnace, fans create a low-pressure area in the firebox that facilitates (induces) airflow.

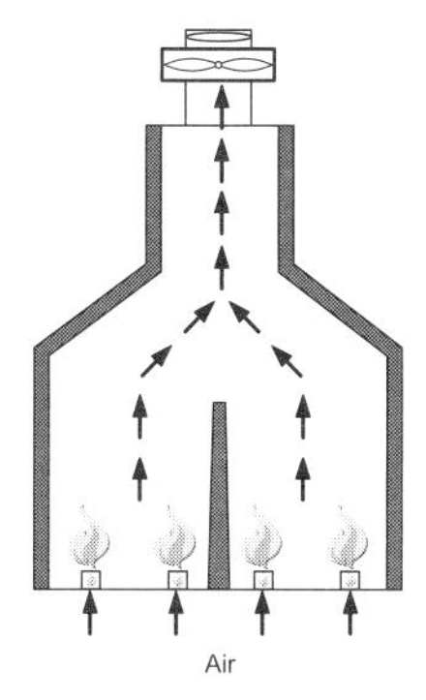

Figure 24-5: Induced draft furnace (cabin-type)

Balanced Draft

Balanced draft furnaces use two fans to facilitate airflow: one inducing flow out of the stack and one providing positive pressure to the burner.

Figure 24-6 shows an example of a balanced draft furnace.

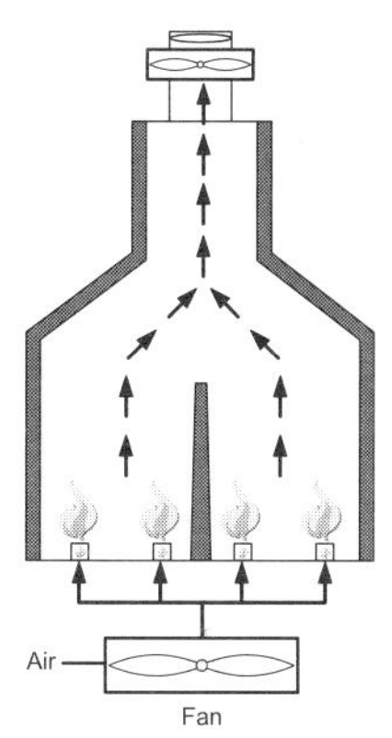

Figure 24-6: Balanced draft furnace (cabin-type)

MONITORING AND MAINTENANCE ACTIVITIES

When monitoring and maintaining furnaces technicians must always remember to look, listen and feel for the following:

Look	Listen	Feel
■ Check for secondary combustion in the convection section. ■ Check for **flame impingement** (when the flames from a burner touch tubes in a furnace) on wall and roof tubes. ■ Check for external wall **hotspots** (an area that has been overheated due to flame impingement or some other cause). ■ Check draft balance (pressures). ■ Check temperature gradient. ■ Check firing efficiency (CO_2 and O_2 in stack). ■ Check burner balance (fuel to air) ratios. ■ Check controlling instruments (fuel flow, feed, flow and pressure, temperature). ■ Check for proper flame pattern.	■ Listen for abnormal noise (fans, burners and leaks).	■ Feel for excessive vibration (fans and burners). WARNING: Process technicians should be very careful feeling for vibrations, since surfaces can be very hot!

Failure to perform proper maintenance and monitoring could impact the process and result in equipment damage.

HAZARDS ASSOCIATED WITH IMPROPER OPERATION

When monitoring and maintaining furnaces technicians must always remember to look, listen and feel for the following:

IMPROPER OPERATION	POSSIBLE IMPACTS			
	Individual	Equipment	Production	Environment
Opening inspection ports when the firebox has a positive pressure	Burns and eye injuries			
Failure to follow flame safety procedure (e.g., lighting off burners without purging firebox)	Burns, injuries or death	Explosion in firebox or flashbacks	Lost production due to down time for repairs	Exceeding EPA opacity limits
Poor control of excess air and draft control	Burns, injuries or death	Flame impingement and/or tube rupture and explosion	Lost production due to down time for repairs	Exceeding EPA opacity limits
Bypassing safety interlocks	Burns, injuries or death	Explosion in firebox	Lost production due to down time for repairs	
Opening furnace inspection ports when firebox pressure is greater than atmospheric pressure	Burns, injuries or death caused as hot flames or combustion gases are forced out of the inspection port			
Failure to wear proper protective equipment (e.g., gloves and face shield) when opening furnace inspection ports	Burns, injuries or death if the pressure in the firebox is greater than atmospheric pressure			

FURNACE SYMBOLS

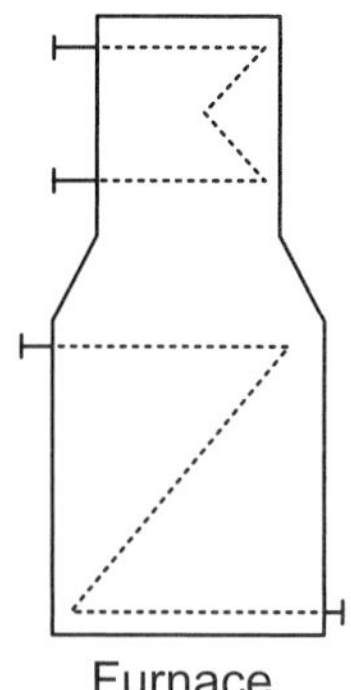

Figure 24-7: Furnace symbols

In order to accurately locate furnaces on a piping and instrumentation diagram (P&ID) process technicians need to be familiar with the different types of furnaces and their symbols.

Figure 24-7 shows an example of a furnace symbol. However, these symbols may vary from facility to facility.

SUMMARY

Furnaces are devices used to heat and increase the temperature of various process streams. Furnaces are divided into two sections: the **radiant section**, and the **convection section**. The radiant section includes components such as: the firebox, burners, radiant tubes, valves, and gauges. The convection section includes components such as: the stack and convection tubes.

Radiant furnaces work by burning fuel inside a containment area called the **firebox**. Within the firebox are tubes that receive heat and transfer it to process fluids through the process of conduction.

Furnaces come in many styles and draft types. **Natural draft** furnaces have no mechanical draft or fans. Instead, they use the difference in density between hot combustion gases and cold outside air to cause draft. **Forced draft** furnaces use fans or blowers, located at the bottom of the furnace, to force airflow through the furnace. **Induced draft** furnaces use fans, located in the stack, to induce airflow. **Balanced draft** furnaces use both forced and induced draft fans to facilitate airflow: one at the top to induce airflow, and one at the bottom to force airflow.

When working with furnaces, process technicians should always wear proper protective equipment and look, listen, and feel for potential problems or hazards. They should also be aware of the hazards associated with improper operation, and the impact those hazards could have on individuals, equipment, production, and the environment.

CHECKING YOUR KNOWLEDGE

1. Define the following key terms:
   a. Firebox
   b. Convection section
   c. Radiant section
   d. Refractory
   e. Pilot
   f. Burner
   g. Stack
   h. Furnace purge
   i. Induced draft furnace
   j. Natural draft furnace
   k. Balanced draft furnace
   l. Induced draft furnace
2. List at least three things a process technician should look, listen, and feel for during normal monitoring and maintenance
3. List at least three hazards associated with improper furnace operation, and explain the impacts these hazards might have on individuals, equipment, production, and the environment.

ACTIVITIES

1. Given a picture of a furnace, identify the following components:
   a. Firebox
   b. Convection section
   c. Radiant section
   d. Burner
   e. Stack
   f. Fans (if any are present)
   g. Refractory Lining
   h. Radiant tubes
   i. Convection tubes
   j. Shock bank

2. Given a picture of a furnace, tell whether the furnace is natural draft, forced draft, induced draft, or balanced draft.

3. Compare and contrast the differences between natural draft, forced draft, induced draft, and balanced draft furnaces.

4. Given a Piping and Instrumentation Diagram (P&ID), identify all of the furnaces.

Chapter 25
Boilers

OBJECTIVES

Upon completion of this chapter you will be able to:

1. Describe the fundamental principles of boiler operation.
2. Describe the operation of boilers in the process industries.
3. Identify the primary parts and support systems of a typical fuel-fired boiler.
4. Describe the types of fuels used in a boiler.
5. Describe the different types of boilers.
6. Describe the monitoring and maintenance activities associated with operating boilers.
7. Discuss the hazards associated with improper operation of a boiler.

KEY TERMS

- **Air Register**—air intake devices, located at the burner, which are used to adjust the primary and secondary airflow to the burner. Air registers are the main source of air to the furnace.
- **Boiler**—a closed vessel in which water is boiled and converted into steam under controlled conditions.
- **Burner**—a device used to introduce, distribute, mix, and burn a fuel. Usually associated with devices such as boilers, incinerators, heaters, furnaces and flares.
- **Blowdown**—removing water from a cooling tower basin or boiler drum in order to reduce the level of impurity concentration and control the chemical ratio
- **Damper**—a movable plate or adjustable louvers used to regulate the flow of air or draft in a furnace.
- **Downcomer**—a conduit or pipe which allows liquid to pass from one distillation tray to another.
- **Economizer**—the section of a boiler used to preheat feed water before it enters the main boiler system.
- **Fan**—a device used to force or draw air through the boiler firebox.
- **Firebox**—the portion of a boiler/furnace where burners are located and radiant heat transfer occurs.
- **Mud Drum**—(1) the lower drum of a boiler that is used as a junction area for boiler tubes. (2) a low place in a boiler where heavy particles in the water will settle out and can be blown down.
- **Pilot**—an igniting device used to light the primary burners in a furnace.
- **Radiant Tubes**—tubes located in the firebox that receive heat primarily through radiant heat transfer.
- **Riser**—tubes that allow liquids or vapors to move upward in a vessel or a distillation column.

- **Stack**—cylindrical outlet at top of a furnace which remove flue gas from the furnace.
- **Steam Drum**—the top drum of a boiler where all of the generated steam gathers before entering the separating equipment.
- **Superheater**—tubes, located toward the boiler outlet, which increase (superheat) the temperature of the steam flow.

INTRODUCTION

Boilers are vessels that are used to create steam. Steam has many applications in the process industries. For example, it is used to heat and cool process fluids, drive equipment, fight fires and purge equipment. Steam provides the energy required to drive many types of equipment and reactions. Without boilers, many processes could not occur.

PURPOSE OF BOILERS

Boilers (also referred to as steam generators) are vessels in which water is boiled and converted into steam under controlled conditions.

The steam produced by boilers provides mechanical energy to drive equipment such as turbines, compressors, and pumps. It is also used to provide the heat energy required to facilitate distillation and induce other physical and chemical reactions.

HOW A BOILER WORKS

Boilers use a combination of heat and pressure to convert water to steam. To illustrate how boilers works, consider a simple boiler like the one shown in Figure 25-1.

This simple boiler consists of a heat source, a water drum, a water inlet, and a steam outlet. In this type of boiler the water drum is partially filled with water and then heat is applied. Once the water is sufficiently heated, steam forms. As the steam leaves the vessel, it is captured and sent to other parts of the process (e.g., it is used to turn a steam turbine, or is sent to a heat exchanger to heat a process fluid). Makeup water is then added to the drum to compensate for the fluid lost during the formation and removal of steam.

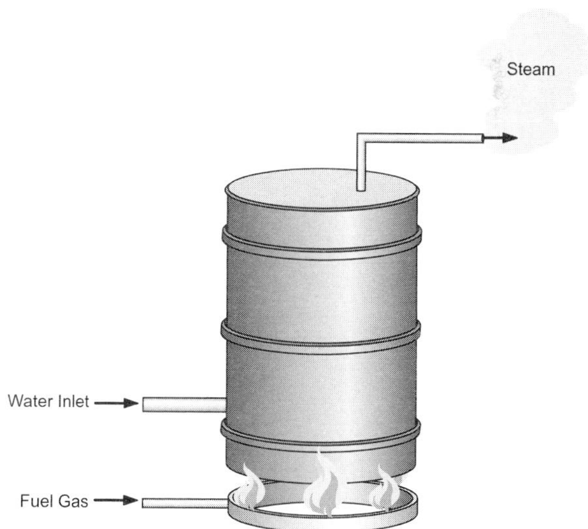

Figure 25-1: Simple boiler

Boilers use the principle of differential density when it comes to fluid circulation. In order for boilers to work properly, they must have adequate amounts of heat and water flow. Factors that affect boiler operation include pressure, temperature, water level, and differences in water density.

As fluid is heated, the molecules expand and it becomes less dense. When cooler, denser water is added to hot water, convective currents are created that facilitate water circulation and mixing.

PARTS OF A BOILER

Furnaces contain many parts. The most common components of a water tube boiler include:

- Firebox
- Stack
- Damper
- Air register
- Burner (gas or oil)
- Raw gas burners
- Pre-mix burners
- Combination burners
- Pilot
- Downcomer and risers
- Fan
- Lower (mud) drum
- Economizer
- Superheater
- Steam drum
- Water wall tubes

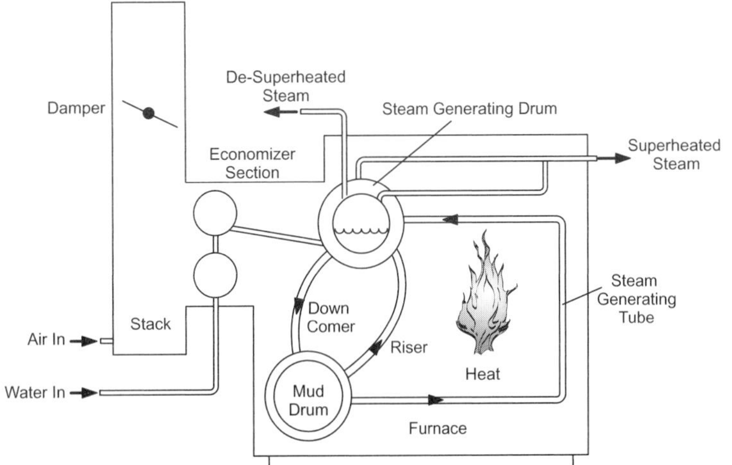

Figure 25-2: Water tube boiler Components

Figure 25-2 shows an example a water tube boiler and its components.

The heating portion of a boiler is very similar to the devices discussed in the *Furnaces* chapter. Like furnaces, boilers contain a firebox. The **firebox** is where burners are located and radiant heat transfer occurs. In order to contain and redirect heat back into the firebox, a special refractory lining (a brick-like form of insulation) is used.

Within the firebox are radiant tubes. **Radiant tubes** are located in the firebox area of a boiler, which receive radiant heat from the burners. These tubes contain water (called boiler feed water) that is heated and turned in to steam.

Burners are devices used to introduce, distribute, mix, and burn a fuel (e.g., natural gas or fuel oil). They can be either be raw gas, premix, or combination. A **pilot** is a device use to light a burner.

Associated with the burners are air registers. **Air registers** control the flow of air to the firebox section of a boiler. They are used to adjust the primary and secondary airflow to the burner. Air registers are used on burners to maintain the flame in a correct burn so as not to create smoke or soot. **Fans** provide the main source of air into the boiler firebox. They are used to force air flow through the firebox.

At the top of the boiler is a stack. A **stack** is the outlet at top of a boiler, which removes flue gas from the boiler. Contained within the stack is a **damper**, a movable plate used to regulate the flow of air or draft.

At the top of the boiler is a steam drum. The **steam drum** is the component where all generated steam gathers before exiting the boiler. Water enters the steam drum from the **economizer**, the section of a boiler used to preheat feed water before it enters the main boiler system.

The steam drum is connected to a lower drum, called the **mud drum**. The mud drum is used as a junction area for the boiler tubes. It is also a low place in a boiler where heavy particles in the water will settle out and can be blown down. The mud drum is connected to the steam drum through a set of downcomer and riser tubes.

Downcomers are tubes located in the firebox that transfer water from the steam drum to the mud drum. The tubes contain cooler water descending from the steam drum. As water flows through the downcomers, it picks up heat from the firebox and replenishes the water supply to the mud drum.

Another type of tube contained within a boiler is a riser. **Risers** are tubes that go into the side or top of the steam drum, and which allow heated water to flow up from the mud drum.

> **DID YOU KNOW?**
>
> American inventors George Herman Babcock and Stephen Wilcox patented the water-tube boiler in 1867.
>
> 
>
> *Courtesy of The Babcock & Wilcox Company*
>
> This type of boiler became the standard for all large boilers because it allowed for higher pressures than earlier boilers.

The **superheater** is a set of tubes located toward the boiler outlet that increase (superheats) the temperature of the steam flow. The steam drum is usually connected to the superheater through a coil or pipe.

A **de-superheater** is a temperature control point at the outlet of the boiler steam flow that maintains a specific steam temperature by using water injection through a control valve.

Superheated steam is steam that has been heated to a very high temperature and a majority of the moisture content has been removed.

BOILER TYPES

Boilers come in a variety of types. The most common types are fire tube, water tube, and process heat.

Fire Tube Boiler

Fire tube boilers are heat exchanger-type devices that pass hot combustion gases through the tubes to heat water on the shellside of the exchanger.

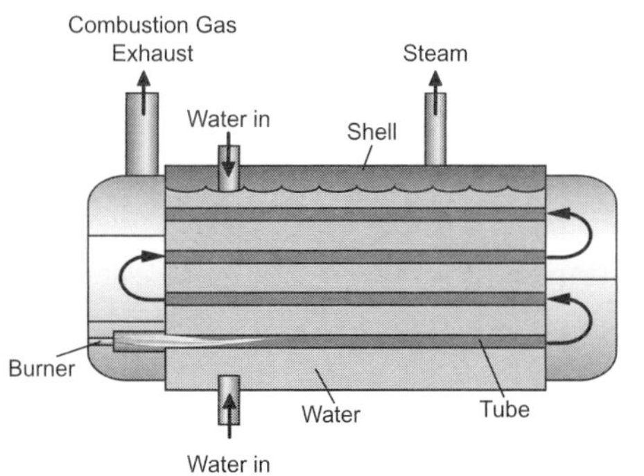

Figure 25-3: Fire tube boiler

Figure 25-3 shows an example of a fire tube boiler.

In this type of boiler, combustion gases are directed through the tubes, while water is directed through the shell. As the water begins to boil, steam is formed. This steam directed out of the boiler to other parts of the process, and makeup water is added to compensate for the fluid loss.

In this type of system, the water level within the shell must always be maintained so that the tubes are covered. Otherwise, the tubes could overheat and become damaged.

Water Tube Boiler

Water tube boilers are one of the most common types of boilers. In a water-tube boiler, water flows through tubes heated externally by combustion gases, and steam is collected above in a drum.

Figure 25-4 shows an example a water tube boiler and its components.

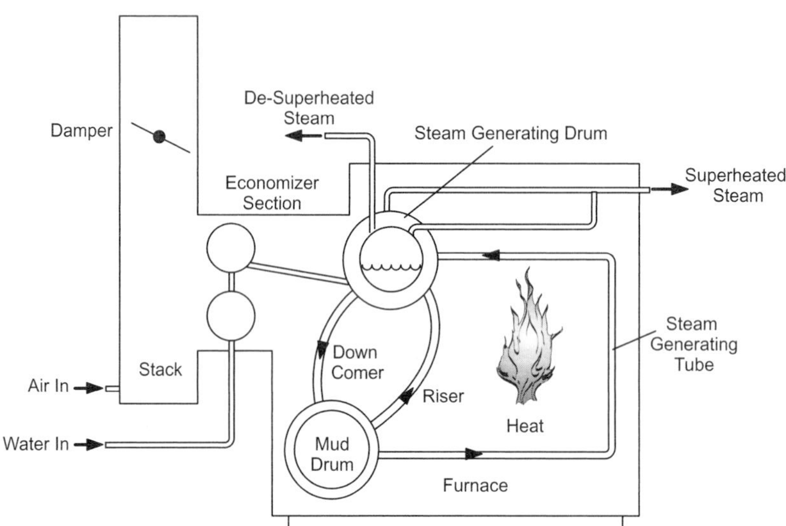

Figure 25-4: Water tube boiler

Water tube boilers have an upper and lower drum connected by tubes. The upper drum is called the steam drum, and the lower drum is called the mud drum. Proper water level in a water tube boiler is controlled in the steam drum. It is important for this water level to be maintained for safety and operating purposes. If the water level gets too low, the boiler could be damaged. If the water gets too high, the steam could become contaminated with the chemicals used to maintain the boiler. Some chemicals enter the boiler system through the boiler water feed pumps. Chemicals are used in boilers to prevent fouling, control conductivity, prevent corrosion, and more. Combustion occurs at the burners located in the firebox. This, in turn, heats the water in the tubes.

MONITORING AND MAINTENANCE ACTIVITIES

When monitoring and maintaining boilers technicians must always remember to look, listen and feel for the following:

Look	Listen	Feel
■ Check firebox for flame impingement on tubes ■ Check burner flame color ■ Check for wall hotspots (external and internal) ■ Check draft balance (pressures) ■ Check temperature gradient ■ Check firing efficiency (CO_2 and O_2 in the stack) ■ Check burner balance (fuel and air) ■ Check controlling instruments (water level, fuel flow, feed water flow, pressure, steam pressure and temperature)	■ For abnormal noise (fans, burners, water leaks, steam leaks, or external alarms)	■ Check for excessive vibration (fans and burners)

Failure to perform proper maintenance and monitoring could impact the process and result in equipment damage.

HAZARDS ASSOCIATED WITH IMPROPER OPERATION

There are many hazards associated with boilers. Table 25-1 lists some of these hazards and their impacts.

TABLE 25-1: Hazards associated with boiler operations

IMPROPER OPERATION	POSSIBLE IMPACTS			
	Individual	Equipment	Production	Environment
Opening header drains and vents	Burns and eye injuries			
Failing to purge firebox (startup)	Possible burns or injuries	Explosion in firebox, damage to boiler internals	Facility upset, lost steam production, down time for repairs	Exceeding EPA opacity limits
Poor control of excess air and draft control		Flame impingement	Boiler efficiency	Exceeding EPA opacity limits

TABLE 25-1:
Continued

IMPROPER OPERATION	POSSIBLE IMPACTS			
	Individual	Equipment	Production	Environment
Loss of boiler feed water		Tube rupture Loss of downstream equipment use	Lost production due to downtime for repairs	
Loss of fuel gas or oil		Loss of downstream equipment use	Lost steam production	

BOILER SYMBOLS

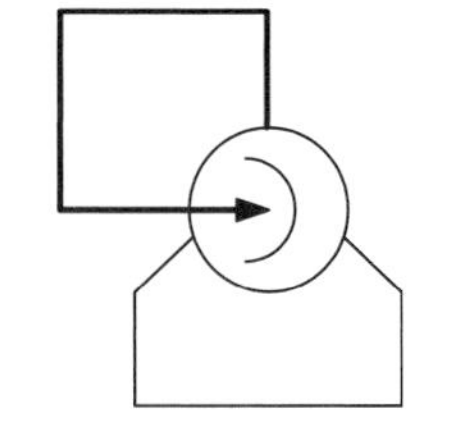
Figure 25-5: Boiler symbol

In order to accurately locate boilers on a piping and instrumentation diagram (P&ID) process technicians need to be familiar with the different types of boilers and their symbols. These symbols

Figure 25-5 shows examples of a boiler symbol. However, these symbols may vary from facility to facility.

SUMMARY

Boilers are vessels that are used to create steam. Steam has many applications in the process industries. It is used to heat and cool process fluids, fight fires, purge and drive equipment, and facilitate reactions.

Boilers use the principle of differential density for the purpose of water circulation. In order for boilers to work properly, they must have adequate amounts of heat and water flow. Factors that affect boiler operation include pressure, temperature, water level, air flow, and differences in water density.

Boilers contain many components. The main components include a firebox, stack, burners, a steam drum, a lower (mud) drum, an economizer, and a superheater. The three most common types of boilers are fire tube, water tube, and process heat.

Improperly operating a boiler can result lead to safety and process issues, injury to personnel and equipment damage.

Technicians should always monitor boilers for flame impingement, hotspots, pressures drops, excessive temperatures, abnormal noises, excessive vibrations, or other abnormal conditions, and conduct preventive maintenance as needed.

CHECKING YOUR KNOWLEDGE

1. Define the following key terms:
   a. Downcomer
   b. Riser
   c. Lower (mud) drum
   d. Economizer
   e. Superheater
   f. Steam drum

2. A fire tube boiler is called a fire tube boiler because:
   a. The tubes are painted to look like flames.
   b. The tubes are in direct contact with the burner.
   c. The tubes are surrounded by hot fluids.
   d. The heat source is located on the tube side.

3. In a water tube boiler, the upper (steam) drum is connected to the lower (mud) drum through tubes called _______________ and ________________.

4. List three things steam is used for in the process industries.

5. List at least three things a process technician should look, listen, and feel for during normal boiler monitoring and maintenance.

6. List at least three hazards associated with improper boiler operation, and explain the impacts these hazards might have on individuals, equipment, production, and the environment.

ACTIVITIES

1. Given a picture of a water tube boiler, identify the following components:
   a. Firebox
   b. Stack
   c. Damper
   d. Air register
   e. Burner (gas or oil)
   f. Downcomer(s)
   g. Riser(s)
   h. Fan
   i. Lower (mud) drum
   j. Economizer
   k. Superheater
   l. Steam drum

2. Explain the differences between fire tube, water tube, and process heat boilers.

3. Given a Piping and Instrumentation Diagram (P&ID), identify all of the boilers.

Chapter 26
Distillation

OBJECTIVES

Upon completion of this chapter you will be able to:

1. Describe the distillation process.
2. Identify the primary parts and support systems of a typical tray-type distillation column.
3. Describe the use of packing as it pertains to distillation.
4. Discuss the hazards associated with the improper operation of a distillation column.
5. Describe the monitoring and maintenance activities associated with distillation column operations.
6. Identify the symbols used with distillation columns and associate equipment.

KEY TERMS

- **Bottoms**—see Heavy Ends.
- **Condenser**—a heat exchanger that is used to condense vapor to a liquid.
- **Distillation**—the separation of the constituents of a liquid mixture by partial vaporization of the mixture and separate recovery of vapor and residue.
- **Downcomer**—a conduit or pipe which allows liquid to pass from one distillation tray to another.
- **Feed Tray**—the tray located immediately below the feed line in a distillation tower. This is where liquid is introduced to or enters the column.
- **Flash Zone**—a term used to refer to the section located between the rectifying and stripping sections in a distillation column.
- **Fraction**—a portion of distillate having a particular boiling range separated from other portions in the fractional distillation of petroleum products.
- **Heavy Ends (bottoms)**—materials in a distillation column which boil at the highest temperature; usually found at the bottom of a distillation column.
- **Light Ends**—materials in a distillation column which boil at the lowest temperature; usually found at the top of a distillation column.
- **Overhead**—the product or stream which comes off the top of a fractionation tower (also called distillate or overhead takeoff).
- **Packed Tower**—a tower that is filled with specialized packing material instead of trays.
- **Packing**—material used in a distillation column to effect separating.
- **Pre-heater**—a heat exchanger used to warm liquids before they enter a distillation tower.

- **Reboiler**—a tubular heat exchanger placed at the bottom of a distillation column or stripper that is used to supply the necessary column heat.
- **Rectifying Section**—that portion of a fractionating tower located between the feed tray and the top of the tower. The area in which the overhead product is purified.
- **Reflux**—the condensed portion of the lighter liquid that is pumped back into the distillation column to cool and condense the rising vapors.
- **Risers**—tubes in a column that allow rising vapors to move into upper trays.
- **Stripping Section**—the section of a distillation tower below the feed tray where the heavier components with higher boiling points are located.
- **Tray**—a part of a fractionation tower containing a liquid level. Vapors rise through the liquid to form a dynamic equilibrium mixture. The number of trays varies with the degree of fractionation required.
- **Weir**—a flat or notched dam or barrier that is used to maintain a given depth of liquid on the trays of a distillation column.

INTRODUCTION

Distillation is a commonly used technique in the process industries. Through distillation, process technicians use heat, pressure, and boiling point to separate liquid mixtures into their different components.

Distillation columns are the devices used to facilitate distillation. These columns come in a variety of shapes and sizes, and may contain trays or packing. They are often used in conjunction with pre-heaters, condensers, and reboilers.

PURPOSE OF DISTILLATION

Distillation is a method for separating a liquid mixture using the different boiling points of each component. During the distillation process, a liquid mixture is separated into two or more components through partial vaporization and the separate recovery of the vapors and residue (usually through the process of condensation).

The distillation process has a wide array of uses throughout the process industries. For example, it is used in water treatment (e.g., removing particulates from water), food and beverage manufacturing (e.g. creating alcoholic beverages), oil refining (e.g. creating gasoline and other products from crude oil), chemical manufacturing (e.g., purifying chemicals), paper and pulp manufacturing, and the manufacturing of pharmaceuticals.

HOW THE DISTILLATION PROCESS WORKS

During the distillation process, a liquid mixture is heated in a distillation column until it reaches its boiling point and the various components of the mixture begin to vaporize into fractions. A **fraction** is a portion of the distillate that has a particular boiling point that is different from the boiling point of the other fractions in the column.

> **DID YOU KNOW?**
>
> The distillation process has been in use for a very long time.
>
> COGNAC
>
> The earliest recorded large scale distillation of Cognac was in 1411.

The following is a generalized overview of the distillation process:

1. Feed comes into the process unit.
2. Feed is heated.
3. Heat vaporizes all volatile components.
4. The vapor and liquid is fed into distillation column.
5. Hot vapor passes up the column (rectifying section).
6. Cooler liquids pass down the column (stripping section).
7. Contact between the vapor and liquid occurs on each tray. The lower boiling compounds migrate to the vapor phase and the higher boiling compounds to the liquid phase.

It is important for process technicians to understand the concept of distillation and the chemical properties of the substances they are distilling. By understanding the distillation processes that are taking place, process technicians are better able to troubleshoot and work more safely and effectively to produce cleaner, purer products.

TRAY-TYPE DISTILLATION COLUMNS

Condenser

Pre-Heater

Reboile

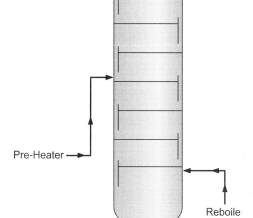

Figure 26-1: Distillation column with associated equipment

Distillation columns come in a variety of shapes and sizes. One type is the tray-type distillation column. The most common components of this type of column are:

- Tower
- Trays
- Condenser
- Reboiler

Figure 26-1 shows an example of a tray-type distillation column.

In this example there is a tower, a pre-heater, a condenser, and a reboiler. The tower is where the vaporization (fractionation) of the liquid occurs.

In some processes, if the heat of the fluid is not sufficient before the fluid enters the tower it must first go through a pre-heater. The **pre-heater** is a heat exchanger that warms the liquid before it enters the tower. This pre-heating process helps the tower maintain a more constant temperature, and reduces the amount of time required for the substance to reach its boiling point once it enters the tower.

As the fluids are circulating through the distillation column, a **reboiler** is used to apply additional heat to the base of the column. This additional heat helps the column maintain a proper temperature and causes the substances at the bottom to vaporize.

After the fluids have completed the vaporization process and moved to the top of the column they are routed to a condenser. A **condenser** is a heat exchanger that is used to condense vapors to a liquid. Some of these liquids are pumped back into the distillation column as reflux to cool, condense and improve separation of rising vapors. The remaining is liquid is transported to other parts of the process.

Tower Sections

Distillation towers have three sections: a flash zone, a rectifying (fractionating) section, and a stripping section.

Figure 26-2 shows an example of each of these sections.

Rectifying Section

Flash Zone

Stripping Section

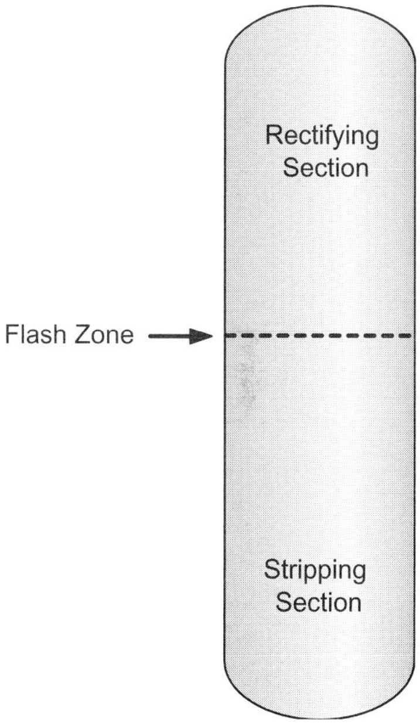

Figure 26-2: Rectifying and stripping sections in a distillation column

The **flash zone** is the entry point or feed tray. The portion of the tower below the feed point is the stripping section. The portion above the feed point is the rectifying section.

The **stripping section** is the section of a distillation tower where the bottoms products and heavy ends are located. **Bottoms** products are the materials collected at the bottom of the distillation column. These materials move down the distillation column as a liquid called heavy ends. **Heavy ends** are the substances that boil at the highest temperature.

The **rectifying (fractionating) section** is where the overhead product is purified. **Overhead** products are the materials collected at the top of the distillation column. They are also referred to as light ends. **Light ends** are the substances that boil at the lowest temperature.

Trays

Distillation trays are components within a tower that provide a contact point between the liquid and vapor. The liquid flows across the tray and down the tower. The vapor flows up thru holes in the tray. A liquid level is maintained on each tray (about 2 inches) using a weir. The liquid is routed from tray to tray using a downcomer.

There are various types of tray designs. The three most common are the sieve tray, valve tray, and bubble cap tray. A sieve tray contains small holes punched in the tray deck. A valve tray contains Valve lifters, similar to

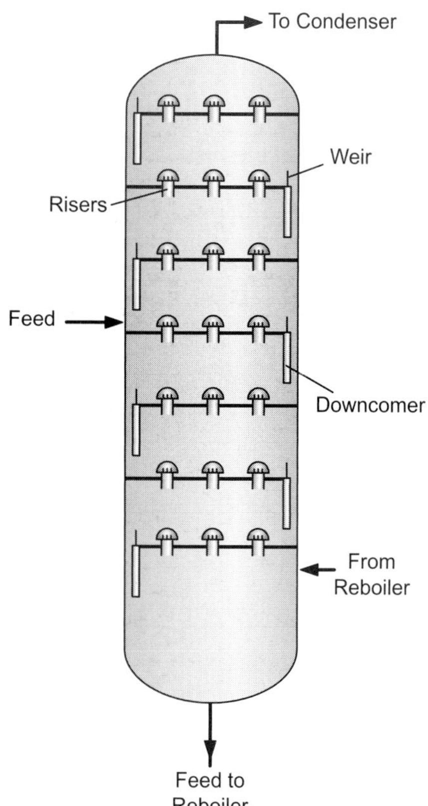

Figure 26-3: Distillation column with bubble cap trays

those in a car engine. The valve is lifted higher and higher as the vapor rate increases. A bubble cap tray contains risers with slotted caps on top of the risers.

Figure 26-3 shows an example of a bubble cap tray tower.

Tray towers are tend to be larger than packed towers (discussed below), and are used for larger scale operations, such as separating gasoline from crude oil, or refining and purifying a chemical product.

The **feed tray** is the tray located immediately below the feed line in a distillation tower. This is where liquid is introduced to or enters the column.

Vapors move up the column through openings in the tray such as bubblecaps, perforations or valves. Liquid flows down the column to lower trays through **downcomers**.

Weirs (flat or notched dams or barriers) are used to maintain a given depth of liquid on the trays.

PACKED TOWERS

Another type of tower used in the distillation process is a packed tower. **Packed towers** are distillation columns that are filled with specialized packing material instead of trays. In the process industries, packed columns are used less often than tray towers. They also tend to be smaller in size.

Packed towers use the same principles as a tray tower. However, packed towers use packing to separate liquids instead of trays. **Packing** is the material used to maximize the contact area for the liquid and gas and effect separation. Packing comes in many different shapes and sizes.

Figure 26-4 shows examples of some different types of packing.

Figure 26-4: Examples of structured packing

Packed towers consist of packing materials, packing supports, liquid and gas distributors, supports, and redistributors. The type of packing and the packing support depends on the material being processed in the tower. The packing support must be designed for free flow of liquid and gas.

The supports must be able to support the packing and the weight of the liquid flowing through the packing.

Packed towers can be used for:

- Absorption of gases in liquid
- Stripping
- Evaporation of liquids
- Condensation of liquids
- Drying operations

MONITORING AND MAINTENANCE ACTIVITIES

When monitoring and maintaining distillation columns, technicians must always remember to look, listen and feel for the following:

Look	Listen	Feel
■ Look for leaks anywhere in the system. ■ Monitor column pressure and temperature ■ Monitor reboiler shell pressure ■ Monitor levels of bottoms products and reflux in drums ■ Monitor condenser temperature drops ■ Check reboiler condensate trap for malfunctions ■ Conduct material balance check ■ Monitor tower/column differential pressure and temperature ■ Monitor feed, reflux and product distribution flows	■ Listen for any unusual noises.	■ Feel for temperature changes in equipment. ■ Feel for unusual vibration in equipment.

Failure to perform proper maintenance and monitoring could impact the process and result in equipment damage.

HAZARDS ASSOCIATED WITH IMPROPER OPERATION

When working with distillation columns, process technicians should always be aware of potential hazards such as overheating, leakage, and over pressurization:

IMPROPER OPERATION	POSSIBLE IMPACTS			
	Individual	Equipment	Production	Environment
Introduction of steam to reboiler (too fast or too much) during start-up	Exposure to chemicals Possible burn or injury	Damage to trays or packing due to vapor rush up the column from the sudden vaporization of chemicals Overpressure Damage to the tower	Extend startup time and increase startup cost Loss of production	Possible atmospheric release of product, and/or flaring Release of vapors to the environment
Tower not vented	Exposure to chemicals Possible burn or injury	Overpressure Damage to the tower	Extend startup time and increase startup cost Loss of production	Release of vapors to the environment
No cooling liquid on condenser	Exposure to chemicals Possible burn or injury	Overpressure Damage to the tower	Extend startup time and increase startup cost Loss of production	Release of vapors to the environment
Rapid changes in column control and set points	Employee discipline and possible retraining	Upset conditions	Product contamination Loss of production	Possible atmospheric release of product, and/or flaring
Operation of column bottoms level too high or too low		Tray damage (high level). Overheat tubes in reboiler (low level).	Increased cost of operation Off-spec product Loss of instrumentation	
Opening a bleed valve on a vacuum column		Upset conditions due to loss of vacuum. Air in column—possible flammable mixture in system	Loss of production Incorrect instrumentation readings	
Opening a bleed valve on a positive pressure column	Exposure to chemicals Possible burn or injury		Loss of production	Spill to the environment

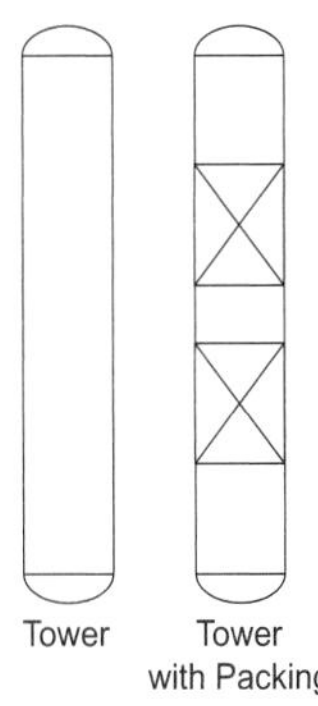

Figure 26-5: Distillation column symbols

DISTILLATION COLUMN SYMBOLS

In order to accurately locate distillation columns on a piping and instrumentation diagram (P&ID) process technicians need to be familiar with the different types of columns and their symbols.

Figure 26-5 shows examples of distillation column symbols. However, these symbols may differ from facility to facility.

SUMMARY

Distillation is a commonly used technique in the process industries. Through distillation, process technicians are able to use heat, pressure, and boiling point to separate volatile liquids into their different components or fractions (portions of the distillate that have a boiling point that is different from the boiling point of other fractions in the distillation column).

Distillation is used throughout the process industries, for processes like water treatment, food and beverage manufacturing, oil refining, paper and pulp manufacturing, and the manufacturing of pharmaceuticals.

Distillation columns are the devices used to facilitate distillation. These columns come in a variety of shapes and sizes and are often used in conjunction with pre-heaters, condensers, and reboilers. They may also use trays or packing to effect separation of the various fractions.

Distillation columns are divided into two sections: the stripping section (bottom) and the rectifying section (top). The stripping section contains the substances with the highest boiling point which are called heavy ends. The rectifying section contains the substances with the lowest boiling point, which are called overhead products or light ends.

When monitoring and maintaining distillation towers, process technicians must always remember to look, listen, and feel to ensure that there are no leaks, that the pressures and temperatures are within normal operating range, and that the equipment is not producing abnormal noises, or unusual vibrations.

There are many hazards associated with improper distillation column operation. These include overheating, leakage, and over pressurization.

CHECKING YOUR KNOWLEDGE

1. Define the following key terms:
   a. Bottoms
   b. Distillation
   c. Downcomer
   d. Fraction
   e. Heavy Ends
   f. Light Ends
   g. Overhead
   h. Packing
   i. Rectifying Section
   j. Riser
   k. Stripping Section
   l. Tray

2. Which of the following is *not* a characteristic of bottoms products:
   a. Are located in the stripping section of the distillation column
   b. Are referred to as heavy ends
   c. Have the lowest boiling point of all the fractions in the column

3. List at least three things a process technician should look, listen, and feel for during normal distillation column monitoring and maintenance.

4. List at least three hazards associated with improper distillation column operation, and explain the impacts these hazards might have on individuals, equipment, production, and the environment.

5. *(True or False)* Rasching rings and intalox saddles can be used to effect fluid separation in a packed tower.

ACTIVITIES

1. List and explain the steps in the distillation process.

2. Draw a simple diagram of a distillation column a pre-heater, a condenser, and a reboiler. Identify where each item is located in the process, and explain what each component is used for.

3. Given a picture of a tray-type column, identify the following components:
   a. Feed tray
   b. Riser
   c. Downcomer
   d. Weir

4. List five things packed towers can be used for.

5. Given a Piping and Instrumentation Diagram (P&ID), identify the distillation column, and any associated equipment.

Chapter 27

Process Utilities

OBJECTIVES

Upon completion of this chapter you will be able to:

1. Discuss the different types of process utilities and their applications:
   - Water
   - Steam
   - Air
   - Nitrogen
   - Gas
   - Electricity
2. Describe the different types of equipment associated with each of the utility systems found in the process industries:
   - Pumps
   - Tanks
   - Distribution

KEY TERMS

- **Boiler Feed Water**—water that is sent to a boiler in order to produce steam.
- **Condensate**—liquid resulting from cooled or condensed vapor. Frequently refers to condensed steam.
- **Cooling Water**—Water that is sent from a cooling tower to unit heat exchangers in order to cool process fluids, then returned to the cooling tower for cooling and reuse.
- **Fire Water**—pressurized water that is used to extinguish fires.
- **Fuel Gas**—gas used as fuel in boilers and other types of furnaces.
- **Instrument Air**—plant air that has been filtered and/or dried for use by instruments.
- **Nitrogen**—an inert (non-reactive) gas that is used to purge explosive gases and air from process systems and equipment.
- **Plant Air**—the supplied air system used for pneumatic tools, air hoists, and other air operated equipment. Also called, "service" or "utility air."
- **Water**—water that is safe to drink and can be used for cooking.
- **Process Water**—water that is used in any level of the manufacturing process; not suitable for consumption.
- **Service Water**—general-purpose water which may or may not have been treated.
- **Treated Water**—water that has been filtered, cleaned, and chemically treated.

- **Utility Air**—compressed air that is used to power equipment and tools.
- **Waste Water**—water that contains waste products in the form of dissolved or suspended solids.

INTRODUCTION

Process utilities are integral to the daily operations of a facility. These utilities provide essentials such as water, steam, compressed air, nitrogen and other gases, fuel gas or electricity to a process unit.

Because these utilities are common throughout the process industries and play a vital role in almost every process, process technicians must understand these utilities, their purpose, the equipment associated with each utility, potential hazards, and how to monitor and maintain them.

TYPES OF PROCESS UTILITIES

There are many different types of process utilities systems. The most common utilities include:

- Water
- Steam
- Compressed air
- Nitrogen
- Gas
- Electricity

The following sections explain the applications of each utility. Some process facilities group some of these utilities together for general use, as part of centrally located stations that provide access to steam, plant air, nitrogen and service water.

WATER

Water is a critical utility in the process industries, with many applications. For example, water can be used for:

- Heating and cooling (e.g. boilers, cooling towers)
- Safety (e.g. fire protection, showers/eye washes)
- Power generation (e.g., steam used to drive turbines, and moving water used in the hydroelectric process)
- Drinking and cooking
- Washing and cleaning (solvent)
- Irrigation
- Transportation (e.g. slurry)
- Propulsion (e.g. hydraulic jets)

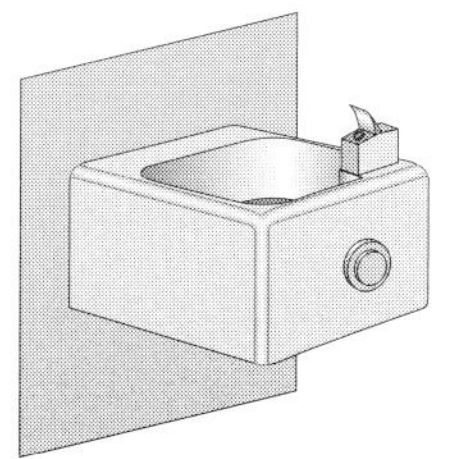

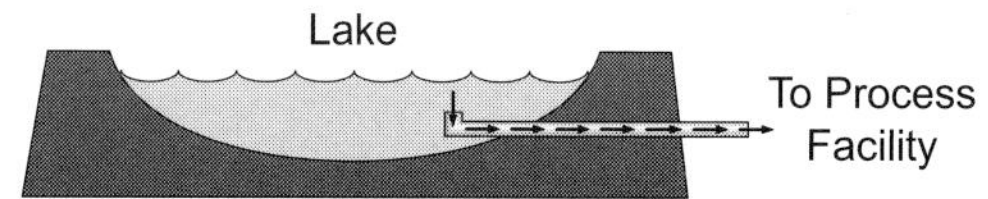

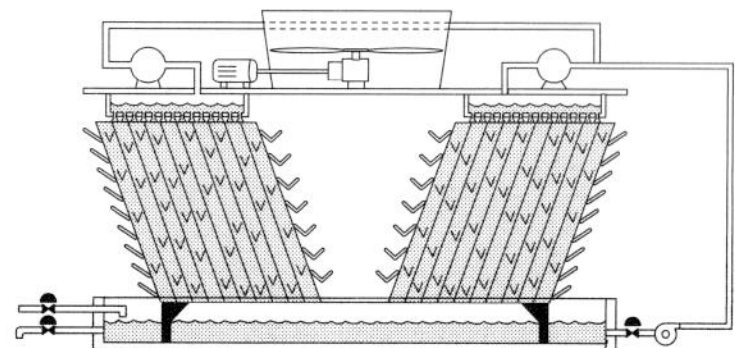

Figure 27-1 Uses of water in the process industries

The water used in a process facility comes from a source, such as a municipal supply or lake, and then is subjected to treatment, filtration, deaeration, pressurization or other changes as necessary. Different systems are used to distribute the water and make sure that it is in the proper form.

- **Service (raw) water**—general-purpose water which may or may not have been treated.
- **Fire water**—water that is used to extinguish fires.
- **Treated water**—water that has been filtered, cleaned, and chemically treated.
- **Potable**—water that is safe to drink and can be used for cooking.
- **Boiler feed water**—water that is sent to a boiler in order to produce steam.
- **Cooling water**—water that is sent to heat exchangers and cooling tower in order to cool process fluids.
- **Process water**—water that is used in any level of the manufacturing process.
- **Waste, sewer and storm water**—water that contains waste products in the form of dissolved or suspended solids.

Water quality is crucial for many uses in a facility. Water quality analysis and testing is key. Depending on how the water is used, different standards are needed. Following are some of the required water properties:

- Free of air or low in oxygen (to prevent corrosion).
- Free of dissolved substances such as chlorine or salts.
- Free of suspended solids.
- Free of hydrocarbons and minerals.
- pH controlled (pH level can vary based on the use of the water).
- Proper temperature and pressure range.

Service/Raw Water

Water comes into a process facility from a source (or sources) such as a municipal supply, lake or river, well(s) or the sea. This supply is called service or raw water, and is the primary source of water for multiple uses around a facility.

Service water is usually treated and filtered, either by a municipal source or the facility itself. Typically, service water is stored in a reservoir then pumped to other water utility systems throughout the facility. Water pressure is maintained at around 125-150 psig.

Fire Water

Fire water is used during safety-related incidents such as fires, spills and releases. The fire water system provides a high pressure emergency water supply (up to 300 psig). Foam can be added to the water, since water alone cannot extinguish a flammable liquid.

Fire water pumping systems can also use a combination of electric motors, natural gas engines, steam turbines, and diesel engines to run the pumps in order to have the ability to extinguish fires even during the loss of one of the major utilities. These pumps are designed to start automatically if water pressure drops below a pre-determined level (like 100 psig).

Fire water lines run underground typically, with drops to hydrants and hose stations, sprinklers and deluge systems, water wall curtains, sprinkler risers, and other fire fighting equipment. Some fire fighting systems automatically come on based on fire, smoke or gas detectors, while other systems can be manually tripped.

Treated Water

Water from the service system can be treated using various techniques, such as chemical treatment and/or filtration, to achieve a high quality of purity. This treated water, sometimes called filtered water, is used for purposes such as:

- Boiler feed water
- Process water
- Seal water for rotating equipment
- Cooling tower water

Potable Water

Potable water means water that is safe for drinking. If you see a sign labeling water as non-potable, you cannot drink it. Potable water is sanitary, so it can also come into contact with skin but not cause harm. It can be used for eye washes and safety showers (Personal Protective Equipment).

A facility can either purchase potable water (like from a municipal water source) or process it from other sources (e.g. a well) using a water treatment system.

The potable water system delivers water to drinking fountains, rest rooms, sinks, eye wash stations, emergency showers and other similar fixtures.

Boiler Feed Water

Boiler feed water is used as a feedstock for boilers and steam generators. Boiler feed water is treated and filtered to remove chemicals, solids, dissolved gases and impurities. The water quality must be consistent and fall within tight chemical parameters. Treatment prevents corrosion problems and increases the efficiency of operations and safety.

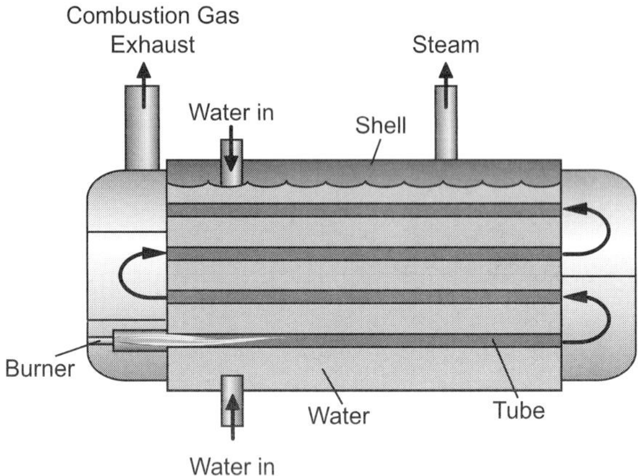

Figure 27-2: Boiler Feed Water

Water testing is essential to proper boiler performance, so frequent tests must be performed. Although testing varies depending on the facility process requirements, typical samples taken of boiler feed water check for:

- Conductivity
- Alkalinity
- Hardness
- pH
- Phosphates

Cooling Water

Cooling water is circulated through equipment components to help remove heat from different stages of a process. Cooling water systems consist of heat exchangers and cooling towers. Cooling water flows through heat exchangers to remove heat from the process. The heated water is then pumped to a cooling tower, which uses air to remove heat from the water. The water can then be re-circulated to the heat exchangers. See chapters 22 *Heat Exchangers* and 23 *Cooling Towers* for details.

Cooling water is monitored for tower basin level, tower discharge pressure, water pH and water impurities. Cooling water is filtered at the intake source to prevent fish and other wild life from entering the tower and, possibly, the unit equipment. Cooling water is treated with chemicals to prevent, or reduce, the buildup of algae, metal corrosion, and scale in the cooling tower and process unit equipment.

Process Water

Water can be used as part of the process. Process water is any water used to process materials into an end product. As mentioned previously, cooling water and boiler feed water play an integral part in processes. But water can also be used as feedstock. For example, water is integral to processes making:

- Cement
- Medicines
- Foods
- Pulp and paper
- Hydrocarbons

The ability of water to evaporate makes it versatile to many processes. Water is mixed with other feed stocks, then evaporated, forming an end product. Process water can also be used for cooling, being circulated through equipment like heat exchangers.

Waste Water

Since many processes rely on water, a lot of the water used becomes contaminated in some way. Wastewater is water that has been used by processes that affect the water quality, such as hydrocarbon cracking, distillation reforming, and alkalization.

Federal and state agencies regulate that wastewater must be treated before the water can be discharged back to its source. Water discharge permits must be obtained, to ensure that discharged water meets specific parameters such as temperature, bacteria, pH level and so on.

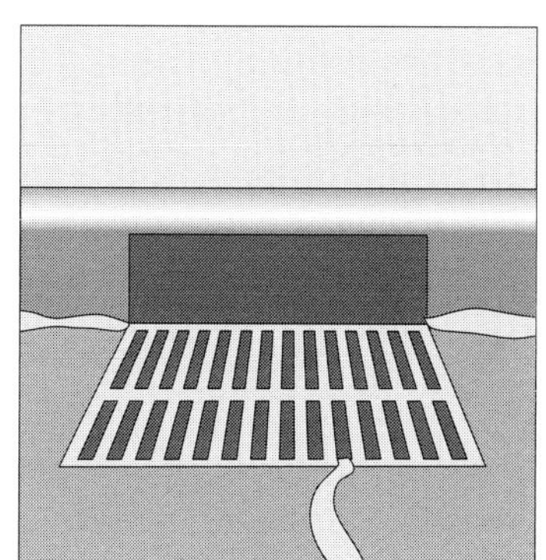

The Environmental Protection Agency (EPA), U.S. Fisheries and Wildlife and state water commissions set wastewater guidelines. Specific federal legislation addressing wastewater treatment includes the Clean Water Act and the Resource Conservation and Recovery Act (RCRA).

At the wastewater facility, the water is treated in various ways. For example, the water is filtered, pH is adjusted, suspended and dissolved solids are removed, the temperature is lowered, biological elements are adjusted, and clarification is performed

Wastewater treatment is usually handled in a closed system, isolated from the facility sewer systems.

Process Sewer

In a plant, industrial sewer systems collect water and other process fluids from drains around the facility. Once collected, this water is sent to the wastewater management facility.

Some process sewer water consists of oil and other hydrocarbons. Typically, these fluids are separated from the water, recovered, and reused if possible. Some plants direct their process water sewer into the wastewater treatment plant. Sanitary (septic) sewers can be handled through a municipal source or routed to the facility's wastewater management system.

Storm Sewer

Storm sewers are open-graded sewer systems that collect and distribute surface water run-off through a series of lift stations and pumping stations to large holding ponds until treatment can occur. Storm water must be analyzed and treated before returning to the water source.

Large volumes of storm water must be held and treated before being released into the navigable water system (e.g. lakes and rivers).

STEAM AND CONDENSATE SYSTEMS

Steam, or water in its vapor form, has many applications in the process industries. For example, steam can be used as a driver or mover for process equipment, such as turbines or compressors, or to provide heat for heat exchangers and reactors.

Steam is generated at a variety of temperatures and pressures, from low to super high, depending on the process requirements:

- Super high pressure: 1000-1600 psig
- High pressure: 400-800 psig
- Medium pressure: 180-200 psig
- Low pressure: 15-60 psig

As stated in the *Chemistry* chapter, pressure and temperature are related. So, high-pressure steam has a higher temperature than low-pressure steam.

Facilities generate and distribute steam using a system comprised of a boiler, steam header, piping and other equipment. Before water is sent to the boiler, water must be treated to remove impurities and prevent scaling and corrosion that can shorten the life of equipment. Steam condensate can also be collected and reused.

Condensate is a pure and valuable source of water in a process facility. Condensate is water that has condensed from steam equipment, unit processes, and steam distribution systems. It is collected, recovered, and reused since it requires minimal treatment.

COMPRESSED AIR

Compressed air can be used in a variety of ways at a process facility. For example, it can be used to power tools and equipment, operate control instrumentation, and process materials.

Compressed air is typically provided for the entire facility using compressors placed in a central location. Air produced in this central location is called plant air. The air is typically filtered and the water is removed. The air is then compressed to a pressure of around 125 psig. Plant air is used in one of two main ways: as utility air (for power tools and general use) or instrument air (for control instrumentation).

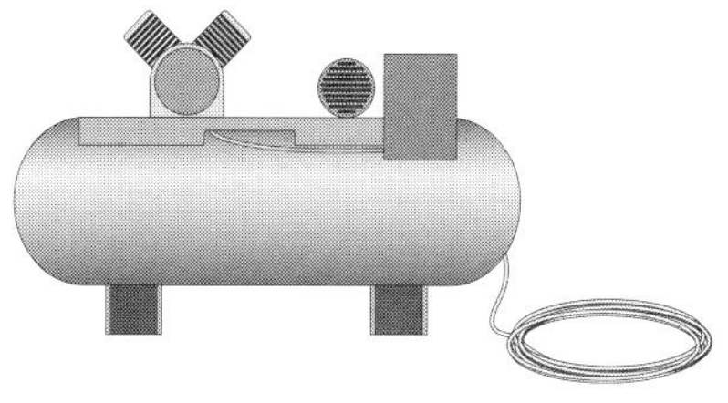

Often, the plant air compressors are configured so that loss of power does not affect the critical instrument air system.

Utility Air

Utility air is used to power equipment and tools (e.g. cyclone air movers and pneumatic power tools), dry equipment, and off-load delivery transports in certain situations (not recommended for unloading of liquid hydrocarbons, due to the hazards this creates). Utility air is not pressure controlled or dried the same as instrument air. It is also used to handle process needs and provide air to utility stations.

Instrument Air

Instrument air provides pneumatic power for process control instrumentation (e.g. valves). Instrument air is cleaned and dried repeatedly, depending on facility usage.

The pressure is maintained at all times to ensure that instruments are provided a continuous supply for operation. It can also be used as clean, dry, purge air. Some plants use a nitrogen system as a backup in the event the instrument air system fails.

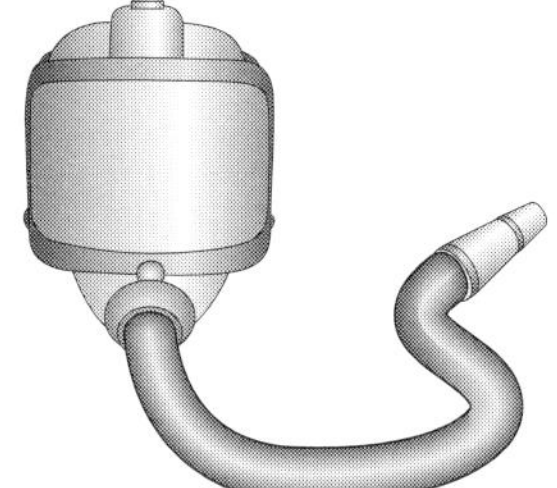

BREATHING AIR

A separate breathing air system provides an air supply to personnel working in oxygen-deficient or high particulate environments. Hoses from the breathing system connect to face masks and tanks that the employees wear. Purity and oxygen concentration (i.e., percentage of oxygen) are important to breathing air.

NITROGEN SYSTEM

Nitrogen has various uses in a process facility. It can be used to purge or remove explosive gases from process systems and equipment (e.g. furnaces, reactors), as a backup for instrument air (in emergencies), as an inert, blanket gas for storage tanks and lines (see the chapter on Tanks, Drums and Vessels), and for cooling and cryogenics.

A facility can generate nitrogen or purchase it from a vendor. Nitrogen can be delivered by high-pressure pipeline, truck or rail. Nitrogen is delivered at lower pressure to units for use at utility hose stations or in equipment. Nitrogen-rich environments can result in asphyxiation (suffocation).

FUEL GAS

Fuel gas is used to operate fired equipment (e.g. furnaces, boilers). This gas can also be used as a blanket gas or as a supply to the pilot light on a flare system. Some process facilities purchase natural gas from a supplier who delivers the gas by pipeline, truck or rail. Other facilities produce their own fuel gas (e.g., as byproducts of their processes) and supplement it with natural gas.

ELECTRICITY

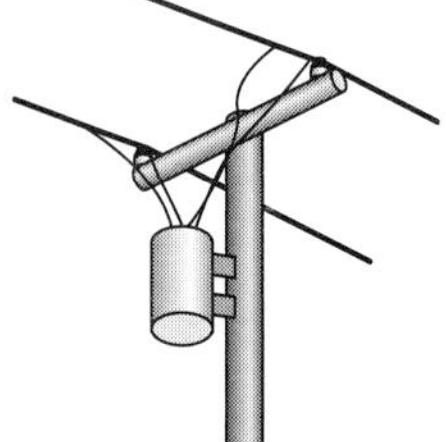

The electrical power and distribution system provides operating power to numerous systems and equipment throughout a facility. Electricity can be provided by a variety of sources including:

- Power generation company
- Steam (driving turbines)
- Co-generation (using waste heat from processes to create steam)
- Diesel gas generators
- Emergency generators
- Portable generators
- Battery systems

See the *Electricity and Motors* chapter for more information.

TYPES OF EQUIPMENT ASSOCIATED WITH UTILITY SYSTEMS

Following are some of the most common types of equipment associated with utility systems. Specific details about each system are beyond the scope of this textbook.

Water

- Tanks and drums
- Pumps
- Instrumentation and control (pressure, level, flow, temperature)
- Pipes and hoses
- Valves
- Filtration system
- Chemical addition & other treatment systems (chlorine, clarifiers, pH treatment, deaerators)
- Cooling Towers (cooling water systems)
- Heat Exchangers (cooling water systems)
- Boilers (boiler feed water)
- Fire fighting equipment and systems

Steam and Condensate

- Tanks
- Pumps
- Boilers
- Turbines (steam)
- Collection systems
- Pipes and hoses
- Valves
- Instrumentation

Compressed Air

- Compressors
- Tanks
- Dryers, moisture separators
- Filters
- Pipes
- Valves
- Instrumentation
- Breathing equipment (e.g. hoses, masks, suits)

Nitrogen

- Compressors
- Pipes
- Valves
- Instrumentation

Gas

- Compressors
- Pipes
- Valves
- Instrumentation

Electricity

- Turbines
- Boilers
- Generators
- Motors
- Motor control centers and circuit breakers
- Instrumentation

HAZARDS ASSOCIATED WITH IMPROPER PROCESS UTILITY OPERATIONS

There are a variety of hazards associated with the improper operations of process utility systems. This section describes general hazards. You can find specific hazards listed in the appropriate equipment chapters of this textbook.

Water

- Pressure loss can damage equipment and impact the process.
- High-pressure water can cause injuries or death, damage equipment and impact the process.
- Hot water can cause burns, injury or death.
- Low pressure in the fire water system can impact fire fighting activities.
- Connecting low pressure system to a high pressure system without a check valve can cause injuries or death, damage equipment and impact the process
- Scaling and corrosion can damage equipment and impact the process.
- Leaks can cause slips and injuries, damage equipment and impact the process.
- Collected water can result in drowning.
- Crossing any non-potable water system with the potable water system can cause illness or death.
- Chemicals used in water treatment can cause respiratory problems, illness or death.

Steam/Condensate

- Water hammer (slugs of condensate in steam lines) can damage equipment and impact the process.

- High pressure or high heat steam can cause injuries or death, damage equipment and impact the process.
- Loss of steam can damage equipment and impact the process.
- Inefficient fuel burn can impact the process and cause air pollution.
- Closed or stuck valves can result in a rupture or explosion, which can cause injuries or death, damage equipment, and impact the process.

Compressed Air

- Low pressure can damage equipment and impact the process.
- Moisture or condensate can affect control instrumentation, which can damage equipment and impact the process.
- Compressor failure can damage equipment and impact the process.
- Unfiltered air can damage equipment and impact the process.
- Ruptured or loose hoses under air pressure can result in hose whipping, which can cause injuries or death.
- Using utility air lines improperly (e.g. horseplay) can cause injuries or death.

Breathing Air

- Breathing air with less oxygen purity than necessary can result in injuries or death.
- Cracks and leaks to breathing air hoses and masks can result in injuries or death.

Nitrogen

- Nitrogen-rich environments can result in asphyxiation (suffocation).
- Coming into contact with cooled nitrogen can result in freeze-related burns.
- Cross connection of nitrogen line to another type of system line can result in explosions, which can cause injuries or death, damage equipment and impact the process.

Gas

- Cross connections with other systems can result in fires or explosions, which can cause injuries or death, damage equipment and impact the process.
- Gas leaks can cause injuries or death.

- Buildup of Hydrogen Sulfide (a gas known for its rotten egg odor) can occur, causing injuries or death, damaging equipment and impacting the process.
- Fire or explosion hazards can cause injuries or death, damage equipment and impact the process.

Electricity

- Loss of power can damage equipment and impact the process.
- Over or under voltage can damage equipment such as circuits or motors.
- Short circuits or arcing can cause injuries or death, damage equipment and impact the process.

MONITORING AND MAINTENANCE ACTIVITIES

Scheduled maintenance activities are an essential part of keeping process utility systems in good condition and preventing failure. The most common monitoring and maintenance activities are described in the specific chapters related to the equipment listed under the section ***Types of Equipment Associated with Utility Systems***. Any other activities are advanced and beyond the scope of this textbook.

SUMMARY

Process utilities are integral to the daily operations of a facility. These utilities provide essentials such as water, steam, compressed air, nitrogen and other gases, fuel gas or electricity to a process unit. The most common utilities include water, steam, compressed air, nitrogen, gas and electricity.

Water can be service or raw water, fire water, treated water, potable, boiler feed water, cooling water, process water and waste/sewer/storm water. Steam, or water in its vapor form, has many applications in the process industries. For example, steam can be used as a driver or mover for process equipment, such as turbines or compressors, or to provide heat for heat exchangers and reactors.

Compressed air can be used in a variety of ways at a process facility. For example, it can be used to power tools and equipment, operate control instrumentation, and process materials. Compressed air is typically provided for the entire facility using compressors placed in a central location. Air produced in this central location is called plant air. Plant air is used in one of two main ways: as utility air (for power tools and general use) or instrument air (for control instrumentation).

A separate breathing air system provides an air supply to personnel working in oxygen-deficient or high particulate environments. Hoses from the breathing system connect to face masks and tanks that the employees wear.

Nitrogen has various uses in a process facility. It can be used to purge or remove explosive gases from process systems and equipment (e.g. furnaces, reactors), as a backup for instrument air (in emergencies), as an inert, blanket gas for storage tanks and lines (see the chapter on Tanks, Drums and Vessels), and for cooling and cryogenics.

Fuel gas is used to operate fired equipment (e.g. furnaces, boilers). This gas can also be used as a blanket gas or as a supply to the pilot light on a flare system.

The electrical power and distribution system provides operating power to numerous systems and equipment throughout a facility.

There is a wide variety of equipment associated with utility systems. Process technicians should understand the hazards associated with utility systems and the associated equipment. Each type of equipment must be properly monitored and maintained.

CHECKING YOUR KNOWLEDGE

1. Define the following key terms:
   a. Condensate
   b. Cooling water
   c. Fire water
   d. Fuel gas
   e. Instrument air
   f. Potable water
   g. Process water
   h. Service water
   i. Treated water
   j. Utility air
   k. Waste water

2. Describe the term service water.

3. Which of the following statements best applies to potable water?
   a. Safe to drink
   b. Can be used for fire deluge systems
   c. Is used for cooling towers
   d. Is collected as condensate

4. Define the term process water.

5. ________ water is water used by a process that has become contaminated.

6. (*True of False*) The air pressure for utility air is regulated.

7. Which is more critical, utility air or instrument air, and why?

8. Name three uses for nitrogen in a process facility.

9. Name five types of equipment and components associated with water utilities.

10. If a person comes into contact with a nitrogen-rich environment and dies, what is the most likely cause of death?
    a. Chemical burns
    b. Asphyxiation
    c. Explosion
    d. Skin irritation

11. Which of the following is NOT a hazard of compressed air?
    a. Low pressure
    b. High pressure
    c. Freeze burns
    d. Moisture

12. Name three electrical hazards.

ACTIVITIES

1. Pick one of the process utility systems and write a one-page paper describing its uses, equipment, and hazards.

2. What is the purpose of plant air at a process facility? Describe it, along with utility air and instrument air, including how they are used and why they are important.

3. Select a utility system, and make a list of five hazards based on the equipment used with the system (you will need to refer to the specific equipment chapter for details).

Chapter 28

Process Auxiliaries

OBJECTIVES

Upon completion of this chapter you will be able to:

1. Describe the purpose or function of the different process auxiliary systems and their applications:
   - Flare and relief systems
   - Refrigeration systems
   - Lubrication systems
   - Hot oil systems
2. Discuss the equipment associated with flare and relief systems found in the process industries.
3. Discuss the components associated with refrigeration systems found in the process industries.
4. Discuss the components associated with lubrication systems found in the process industries.
5. Discuss the components associated with hot oil systems found in the process industries.

KEY TERMS

- **Burner**—a device used to introduce, distribute, mix, and burn a fuel.
- **Condenser**—a heat exchanger that is used to condense vapor to a liquid.
- **Evaporator**—a heat exchanger that uses a refrigerant to remove heat.
- **Expansion Valve**—a valve that reduces the pressure of a refrigerant, causing it to cool.
- **Flare**—a device designed to safely burn excess hydrocarbons. It is generally a pipe located some specified distance from a process unit.
- **Flare Header**—a pipe that connects several vents to the flare system.
- **Hot Oil System**—a system that heats heavy oils as a fuel source for furnaces, boilers, reboilers, and exchangers
- **Knock-Out Drum**—vessel located between the flare header and the flare. It is designed to separate liquid substances from vapors being sent to the flare for burning.
- **Lubricant**—any substance interposed between two surfaces in motion for the purpose of reducing the friction, heat, and/or wear between them.
- **Lubrication System**—a system that supplies oil or grease to equipment parts such as bearings, gearboxes, sleeve bearings, and seals.
- **Pilot**—an initiating device, such as a pilot burner on a furnace burner, or a pilot valve on a pressure safety valve.

- **Receiver**—a tank that stores refrigerant once it leaves the condenser.
- **Refrigerant**—a fluid with a low boiling point circulated throughout a refrigeration system.

INTRODUCTION

Process auxiliary systems, which are secondary support systems, play an important role in the daily operations of a facility. Process auxiliary systems include flares, refrigeration, lubrication and hot oil. Process technicians must understand these auxiliaries, their purpose, what equipment is associated with each auxiliary system, potential hazards, and how to monitor and maintain them.

TYPES OF PROCESS AUXILIARIES

The most common types of process auxiliary systems include:

- **Flare and relief systems**—prevent system overpressure and remove excess flammable substances from a process by burning it to the atmosphere.
- **Refrigeration systems**—cool process feed or products to a desired temperature.
- **Lubrication systems**—supply oil or grease to equipment parts such as bearings, gearboxes, and seals.
- **Hot oil systems**—heat heavy oils used as a heat source for furnaces, boilers, reboilers, and heat exchangers.

The following sections explain the use and equipment/components of each auxiliary system.

FLARE AND RELIEF SYSTEMS AND ASSOCIATED EQUIPMENT

Flare and relief systems prevent system overpressure and provide a safety relief path for a process. Note that to burn off flammable excess (or waste) substances, an incinerator is used instead of a flare system.

Flares are generally tall, vertical pipes (although they can also be horizontal) called "stacks" located a safe distance from a process unit. Flares burn flammable and toxic substances from a process, converting them into environmentally acceptable substances, such as carbon dioxide (CO_2) and water (H_2O).

Figure 28-1 shows an example of a flare.

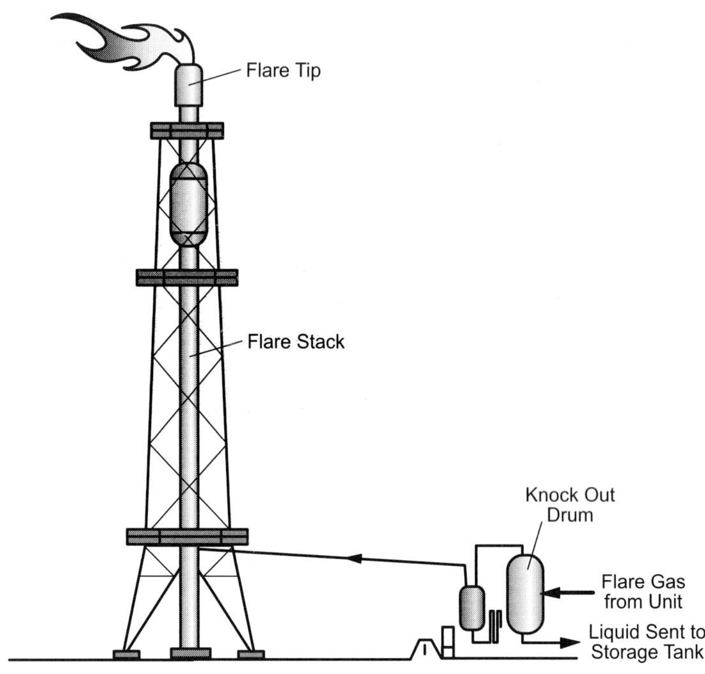

Figure 28-1: Example of a Flare with Knockout Drum

In a flare system, flammable substances are routed from a process unit to a **flare header**, a pipe system that connects process unit vents to the flare. A **knock-out drum** (a vessel located between the flare header and the flare) separates liquids from the vapors being sent to the flare stack. These liquids are then pumped out to a storage tank.

In vertical flares, vapors travel up the **flare stack** (the riser section or body of the flare that supports the steam ring and burner) and are ignited by a **pilot** burner (an ignition source, located at the flare tip, which is used for lighting the flare) as they flow out the flare tip.

Figure 28-2 shows an example of a flare tip with a pilot burner.

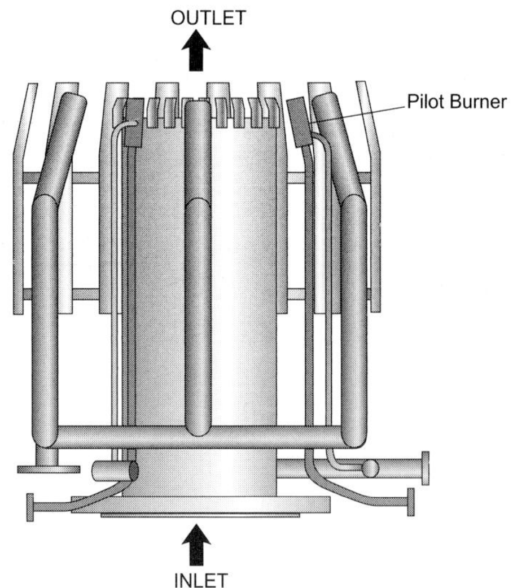

Figure 28-2: Flare Tip with Pilot Burner

As the substances are combusted, steam is distributed through a **steam ring** on the tip of the flare. The steam ring helps create a draft, reduce smoke, and protect the flare tip from the flame. Also, a continuous gas purge prevents the flame from flowing back down the flare.

Figure 28-3 shows an example of a flare with a steam ring.

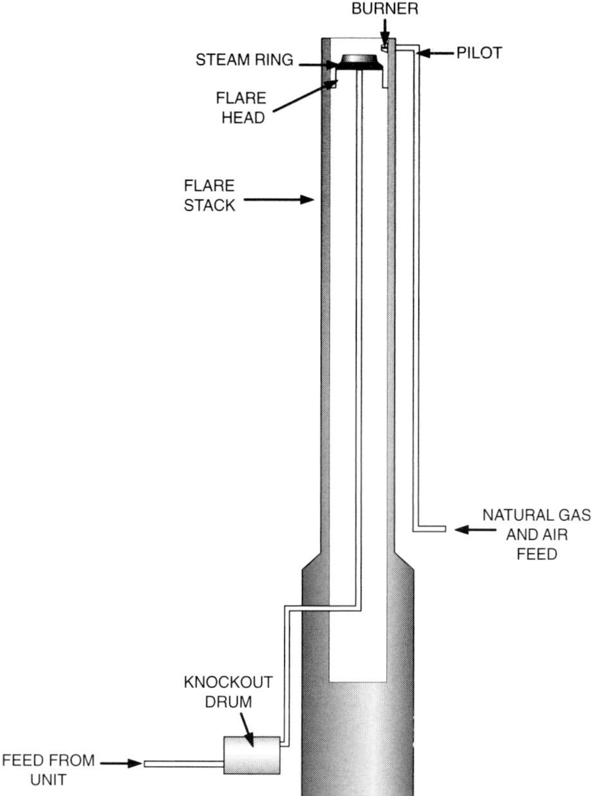

Figure 28-3: Flare with Steam Ring

In addition to vertical flares, there are also ground flares that do not have a vertical flare stack.

Flares can be designated as either hot service or cold service depending on the temperature of substance being flared. Hot service flares collect fluids above 32° F, while cold service flares collect fluids below that temperature. This is important because materials of construction are designed for specific temperature ranges.

Flare and relief systems are essential during facility startups, shutdowns, and emergency shutdowns. They are also essential in the event of process upsets.

Flare and relief systems are connected to various process unit safety relief valves. If a process upset occurs that results in a dangerous pressure buildup, safety relief valves open and vent to the flare and relief system.

Flare and relief systems are regulated by clean air and safety laws set forth by various government agencies. Because of this, flares must be located a safe distance from the process unit and the flame cannot endanger people, the surrounding area, or equipment. In addition, substances must be burned with a minimum amount of smoke. Flares can only burn during short emergency periods.

REFRIGERATION SYSTEMS AND ASSOCIATED COMPONENTS

DID YOU KNOW?

The first patent for a mechanical refrigeration system was issued in Great Britain in 1834 to the American inventor Jacob Perkins.

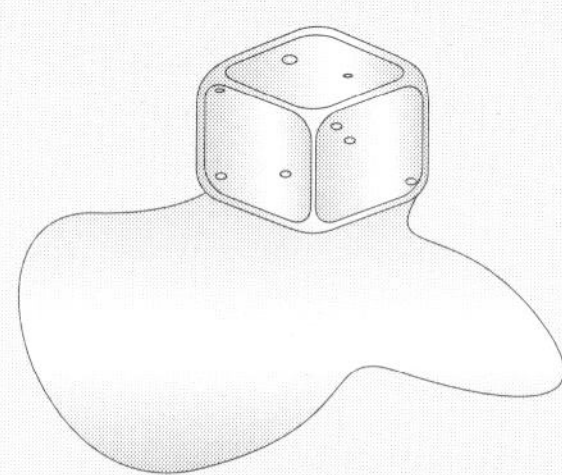

Mechanical refrigeration systems are based on the principle that absorption of heat by a fluid (refrigerant) as it changes from a liquid to a gas lowers the temperature of the objects around it.

Cooling is a major element in many processes. Some processes require feed and/or product temperatures to be lower than what can be reached using cooling methods such as air or cooling water. In these instances, refrigeration systems are used to cool feed or products. The two most common types of refrigeration 7systems are mechanical and absorption.

Mechanical systems use a **refrigerant** (a fluid that boils at low temperatures) for cooling. In this type of system, refrigerant is cycled through a closed loop system that contains a low pressure cold side (evaporator) and a high pressure hot side (condenser). A compressor and expansion valve maintain the system pressure. Car and home air conditioners are examples of a mechanical refrigeration system.

Absorption systems use water and a heat absorbent material (typically lithium bromide, LiBr) that is also cycled through a closed loop system. The loop still has a low pressure side (evaporator) and a high pressure side (condenser), but adds a low pressure absorber and high pressure generator with a heater to complete the cycle. This type of system is used less often than mechanical systems.

Components of a Mechanical Refrigeration System

Mechanical refrigeration systems typically consist of a closed loop process that includes components for compression, condensation, and evaporation. The components associated with mechanical refrigeration systems include:

- **Refrigerant**—a fluid with a low boiling point circulated throughout the refrigeration system.
- **Evaporator**—a heat exchanger that uses the refrigerant to remove heat; in the process, the refrigerant evaporates.
- **Compressor**—a device used to increase the pressure of a refrigerant in the gas phase.
- **Condenser**—a device used to cool the gas phase refrigerant from the compressor and causes it to condense back to a liquid phase.
- **Receiver**—a tank that stores the liquid phase refrigerant once it leaves the condenser.
- **Expansion valve**—a valve that reduces the pressure of the liquid refrigerant, causing it to cool.

LUBRICATION SYSTEMS AND ASSOCIATED COMPONENTS

Lubricants are substances used between two surfaces to make them slide more smoothly against each other and reduce friction. Oil and grease are common lubricants. Lubrication systems supply oil or grease to equipment parts such as bearings, gearboxes, sleeve bearings, and seals.

Some process units have a centrally-located lubrication system that atomizes (reduces to a fine spray) lubricants as they are distributed to equipment parts, thereby reducing heat. The components associated with a typical lubrication systems include:

- **Lube oil reservoir tank**—a vessel that contains lubrication for a system.
- **Cooler**—a device that removes heat from lubricant circulating through a system.
- **Filter**—a device that removes contaminants from lubricant circulating through a system.
- **Pump**—a device used to move lubricant through a system, to process equipment, and back again.

Other ways of supplying lubricants to equipment are:

- **Grease guns**—hand-held devices used to apply grease to equipment components (e.g., bearings).
- **Glass bottle oiler cup**—a cup that connects an oil reservoir to a part so it remains lubricated.
- **Automatic grease lubricator**—(also called a grease cup) a device which slowly forces grease from a reservoir into a part so it remains lubricated.

DID YOU KNOW?

According to the American Petroleum Institute (API), motor oil doesn't wear out, it just gets dirty.

Used motor oil can be cleaned and reprocessed for use in furnaces for heat, or in power plants to generate electricity. It can also be recycled into lubricating oils that meet the same specifications as virgin motor oil.

HOT OIL SYSTEMS AND ASSOCIATED COMPONENTS

Hot oil systems are used to heat heavy oils as a heat source for furnaces, boilers, reboilers, and exchangers. The components associated with hot oil systems are:

- **Storage tank**—a vessel that contains the oil.
- **Oil circulation and heating system**—a component that distributes oil from the tank to the heaters then on to the process equipment and back again.
- **Oil pump**—equipment that moves the oil through the circulation and heating system.
- **Oil burner**—a device that atomizes oil and air for ignition in process equipment.

HAZARDS ASSOCIATED WITH IMPROPER OPERATION

Process technicians must understand the hazards associated with the improper operation of process auxiliaries:

IMPROPER OPERATION	POSSIBLE IMPACTS			
	Individual	Equipment	Production	Environment
Flare pilot or burner not lit	Possible exposure to toxic fumes	Fire or explosion	Process upset	Potential environmental release
Flare knockout drum allowed to fill with liquid and spill out of the flare stack	Possible exposure to toxic fumes and/or liquids Potential for burns and injury	Back pressure on the flare can create dangerous system pressure	Process upset	Potential environmental release Spill, fire
Failure to line up cooling water to condenser in refrigeration system		Equipment damage	Process upset	Potential environmental release
Loss of lubrication on equipment		Bearing failure	Process upset	Potential environmental release
Hot oil pump drains and vents left open	Potential for burns and injury	Equipment failure	Process upset	Potential environmental release
Hot oil header drains left open	Potential for burns and injury	Equipment failure	Process upset	Potential environmental release

MONITORING AND MAINTENANCE ACTIVITIES

When monitoring and maintaining process auxiliary systems, technicians must always remember to perform the following tasks:

Flares	Refrigeration	Lubrication	Hot Oil
■ Maintain level in the knockout drum. ■ Check knockout drum pump operation. ■ Check pilot gas regulator for proper flow. ■ Check that the pilot is lit.	■ Check the cooling water in the condenser. ■ Check compressor operation. ■ Check the expansion valve. ■ Check the system temperature profile.	■ Maintain proper oil levels in reservoirs. ■ Check bearing lubrication. ■ Check grease fittings. ■ Check inlet and outlet temperature for excess heat. ■ Check lubrication flow to equipment. ■ Check filter differential pressure. ■ Check pump for proper operation.	■ Check pump for proper operation. ■ Check tank temperature. ■ Check oil circulating temperature. ■ Check oil tank level. ■ Check circulating valves for proper alignment.

Failure to perform proper maintenance and monitoring could impact the process and result in equipment damage.

SUMMARY

Process auxiliary systems include flares/relief systems, refrigeration, lubrication and hot oil.

Flare and relief systems provide two main functions at a facility. These systems prevent over pressurization and provide a safe relief path for process materials. These systems also burn off flammable, excess (or waste) substances. Equipment associated with flare systems includes: flare header, knock-out drum, pilot, steam ring, and flare tip.

Cooling is a major element in many processes. Refrigeration systems are used to cool feed or products to required temperatures during the process or for storage. Mechanical refrigeration systems use a refrigerant (a fluid that boils at low temperatures) for cooling. In this type of system, refrigerant is cycled through a closed loop system that contains a low pressure cold side (evaporator) and a high pressure hot side (condenser). A compressor and expansion valve maintain the system pressure.

Lubricants allow two surfaces to slide more smoothly against each other, thereby reducing friction. Common types of lubricants are oil and grease. Lubrication systems supply oil or grease to equipment parts such as bearings, gearboxes, sleeve bearings, and seals. The components associated with lubrication systems are the lube reservoir tank, cooler, filter and pump.

Hot oil systems are used to heat heavy oils as a fuel source for furnaces, boilers, reboilers, and exchangers. The components associated with hot oil systems are the storage tank, oil circulation and heating system, oil pump, and oil burner.

Technicians must always remember to perform monitoring and maintaining tasks on process auxiliary systems. Technicians must also understand the hazards associated with the improper operation of process auxiliaries.

CHECKING YOUR KNOWLEDGE

1. Define the following key terms:
   a. Flare
   b. Refrigerant
   c. Condenser
   d. Compressor
   e. Evaporator
   f. Expansion Valve

2. Which of the following flare system components is a vessel designed to separate liquid hydrocarbons from vapors being sent to the flare for burning?

   a. Flare Head

   b. Knockout Drum

   c. Flare Stack

   d. Steam Ring

3. The __________ is the ignition source for lighting the flare, located at the flare tip.

4. Which of the following does an expansion valve do in a refrigeration system?

   a. Acts a heat exchanger that uses the refrigerant to remove heat

   b. Cools the gas phase refrigerant from the compressor and causes it to condense back to a liquid phase.

   c. Stores the liquid phase refrigerant once it leaves the condenser.

   d. Reduces the pressure of the liquid refrigerant, causing it to cool.

5. Which of the following refrigeration system components is to a tank that stores liquid refrigerant after it leaves the condenser?

   a. Evaporator

   b. Receiver

   c. Filter

   d. Compressor

6. *(True or False)* Lubrication systems are used to heat heavy oil for a fuel source

   a. True

   b. False

7. Which component of a lubrication system removes contaminants?

   a. Filter

   b. Lubricator

   c. Scrubber

   d. Cooler

8. Name three types of equipment that use a hot oil system.

9. Which of the following is NOT part of a hot oil system?
   a. Storage tank
   b. Oil pump
   c. Oil burner
   d. Refrigerant

10. *(True or False)* It is not necessary to check tank temperature during monitoring and maintenance activities for a hot oil system.
    a. True
    b. False

ACTIVITIES

1. Write 5-6 paragraphs on TWO of the following process auxiliary systems. Describe each system, how it is used, and why it is important to the process industries:
   - Flare system
   - Refrigeration system
   - Lubrication system
   - Hot oil system

2. Match the diagram labels with the components listed below.

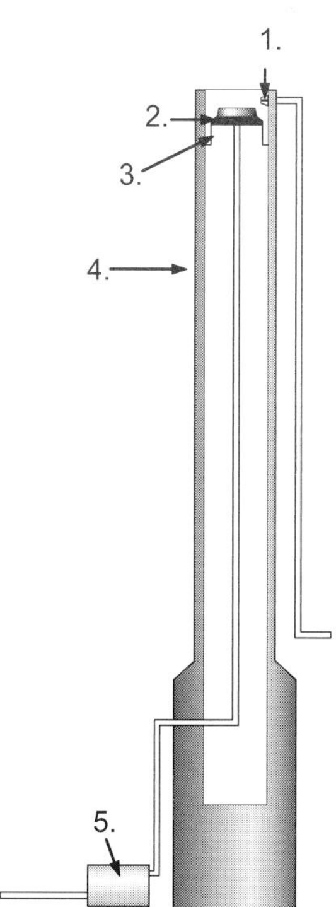

   a. ___ Flare Tip

   b. ___ Knock-out drum or sump system

   c. ___ Pilot

   d. ___ Steam Ring

   e. ___ Flare Stack

3. List the monitoring and maintenance activities for flare systems, refrigeration systems, lubrication systems, and hot oil systems.

Chapter 29
Instrumentation

OBJECTIVES

Upon completion of this chapter you will be able to:

1. Describe the purpose or function of process control instrumentation in the process industries.
2. Identify the key variables which are controlled by process control instrumentation:
   - Temperature
   - Pressure
   - Level
   - Flow
   - Analytical
3. Identify typical process control instruments, their applications and functions:
   - Transmitter
   - Indicator
   - Recorder
   - Controller
   - Thermocouples
   - Sensor
   - Transducer
   - Alarms
   - Control valve
   - Actuators
   - Positioner
4. Describe process control instrumentation signals.
5. Define a generic control loop and provide an example.
6. Describe distributive control systems and how they are applied in the process industries.
7. Describe the monitoring and maintenance activities associated with process control instrumentation.
8. Discuss the hazards associated with process control instruments.
9. Identify symbols used to represent process control instruments.

KEY TERMS

- **Actuator**—an electrical, pneumatic, or hydraulic device that moves a control device, usually a valve.
- **Alarm**—a device, such as a horn, flashing light, whistle or bell, which provides visual and/or audio cues that alert technicians of conditions outside the normal operating range.
- **Analytical**—a measurement of the chemical or physical properties of a substance.
- **Control Valve**—a valve normally equipped with an actuator to control valve stem movement; these valves are used as the final control element for controlling flow, level, temperature or pressure in a process.
- **Controller**—An instrument device which controls a process variable at a specific condition.

- **Distributed Control System (DCS)**—a computer-based system that is used to monitor and control a process.
- **Electronic Instruments**—use electricity as the power source.
- **Flow Rate**—a quantity of fluid that moves past a specific point within a given amount of time; usually expressed in volume units per unit of time, such as Gallons Per Minute (GPM) or Cubic Feet per minute (CFM).
- **Hydraulic Instruments**—use liquid (hydraulic fluid) as the power source.
- **Indicator**—shows the current condition of a process variable, usually through a visual representation such as needle on a scale in a gauge or a digital readout.
- **Instrumentation**—devices used to measure and control or indicate flow, temperature, level pressure, and analytical data.
- **Level**—a position of height or depth along a vertical axis. In the process industries, level usually means the height of the surface of a material compared to a zero reference point.
- **Local**—instrumentation considered to be in close proximity to the process, generally found outside the control room.
- **Pneumatic instruments**—use air pressure or a gas as the power source.
- **Positioner**—a device that ensures a valve is positioned properly in reference to the signal received from the controller.
- **Pressure**—the force exerted on a surface divided by its area.
- **Process Variable**—a varying operational condition that is associated or goes with a chemical processing operation such as temperature, pressure, flow rate, level, and composition.
- **Recorder**—a mechanical recording instrument used to measure variables such as pressure, speed, flow, temperature, level and electrical units; also provides a history of the process over a given period of time.
- **Remote**—instrumentation that is located away from the process (e.g., in a control room).
- **Sensor**—a device, in direct contact with the process variable to be measured, that senses or measures, then transmits a signal to another instrument.
- **Temperature**—The degree of hotness or coldness measured on a definite scale.
- **Thermocouple**—a sensing device consisting of two dissimilar metals joined at a junction that when exposed to heat, the metal junction generates an electrical signal proportional to the change in temperature.
- **Transducer**—convert one instrumentation signal type to another type (e.g., electronic to pneumatic).
- **Transmitter**—a device that receives its signal from the sensing element and "transmits" its signal to the controller or indicator.

INTRODUCTION

This chapter introduces the process technician to instrumentation that he or she will use to monitor and control processes in a facility. Instrumentation is a vital part of the process industries, allowing companies to produce higher quality products, more efficiently, in a safer environment. Without instrumentation, the process industries would need numerous workers to monitor and control even the simplest process.

Automation plays an ever-increasing role in modern facilities, changing the technology behind instrumentation rapidly. Processes that were once manually monitored and controlled are now done so automatically, allowing a process technician to oversee increasingly complex processes from a single source.

THE PURPOSE OF PROCESS CONTROL INSTRUMENTATION

The process industries use instrumentation to monitor and control conditions in processes, such as temperature and pressure. These conditions are called Process Variables (PVs).

Instrumentation helps process technicians measure, compare, compute and correct process variables. Instrumentation can be considered "the eyes and ears" of a process, constantly reporting the status of variables. It also plays a role in the decision-making involved with keeping a process operating efficiently and effectively, often times directly controlling it.

Instrumentation is critical to processes, because with even a simple process there can be multiple process variables to monitor and control. In a complex process, there can be dozens of process variables.

KEY PROCESS VARIABLES CONTROLLED BY INSTRUMENTATION

The five most common (and thus key) process variables that instrumentation can control are pressure, temperature, level, flow, and analytical. However, speed, vibration and a range of other variables can be monitored and controlled, depending on the process requirements.

Pressure

Pressure is the force exerted on a surface divided by its area. A common unit of pressure measurement is pounds per square inch (PSI). A common term associated with pressure is differential. Differential is the difference between measurements taken from two separate points, (the difference between two related pressures). Differential pressure devices are commonly used in the process industries. Two ways to express differential is the Greek letter delta (Δ) with a P (ΔP) following it (or the letter combination D/P).

Pressure-related instruments include Bourdon tubes, pressure gauges, manometers, and D/P (Differential Pressure) cells. Figures 29-1 through Figure 29-3 show examples of some of these devices.

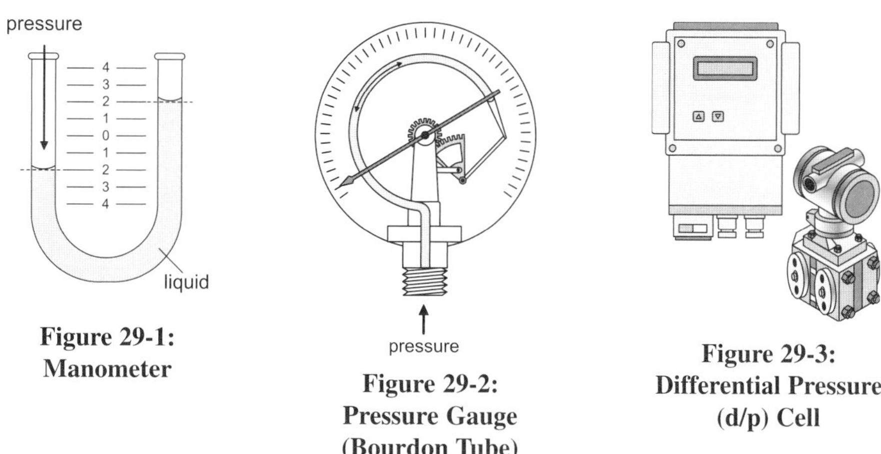

Figure 29-1: Manometer

Figure 29-2: Pressure Gauge (Bourdon Tube)

Figure 29-3: Differential Pressure (d/p) Cell

Temperature

Temperature is a specific degree of heat or cold as indicated on a reference scale. The two most common temperature measurements scales are Fahrenheit (°F) and Celsius (°C). A common term associated with pressure is differential. Differential is the difference between measurements taken from two separate points, (the difference between two related temperatures). Differential temperature devices are commonly used in the process industries. Two ways to express differential is the Greek letter delta (Δ) with a T (ΔT) following it (or the letter combination D/T).

Temperature-related instruments include thermocouples, Resistance Temperature Devices (RTDs), thermometers and filled systems. Figure 29-4 and Figure 29-5 show examples of some of these devices.

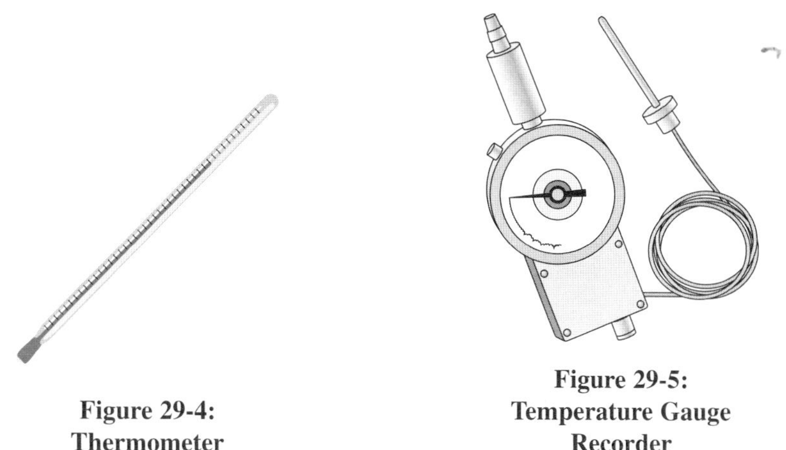

Figure 29-4: Thermometer

Figure 29-5: Temperature Gauge Recorder

Level

Level is a position of height or depth along a vertical axis. In the process industries, level usually means the height of the surface of a material compared to a zero reference point. Level-related instruments include sight glasses, float and tape devices, and displacers (buoyancy devices). Figures 29-6 and Figure 29-7 show examples of some of these devices.

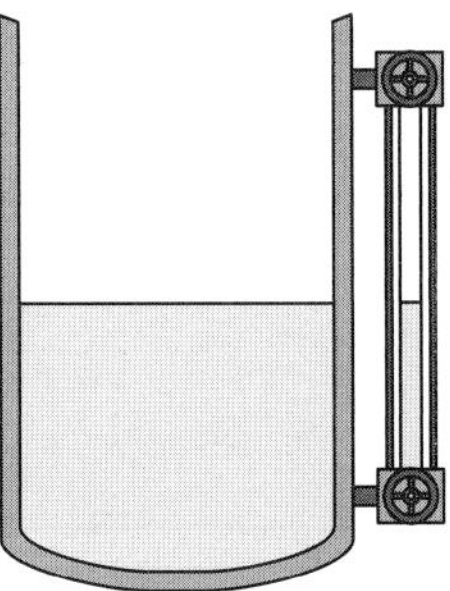

Figure 29-6: Tubular Type Sight Glass

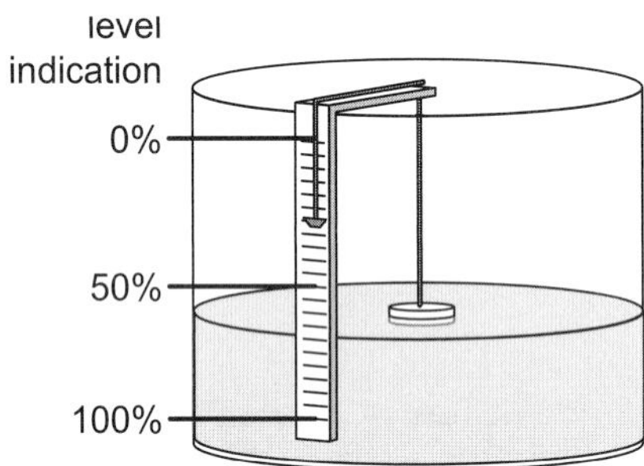

Figure 29-7: Tape Gauge

Flow

Flow (also flow rate) is a quantity of fluid that moves past a specific point within a given amount of time; usually expressed in volume units per unit of time, such as Gallons Per Minute (GPM) or Cubic Feet per minute (CFM). Flow-related instruments include orifice plates, Venturi tubes that use a D/P (differential pressure) cell, vortex, magnetic flow meter, and coriolis. Orifice plates and Venturi tubes do not actually measure anything, they create a condition that a sensor can measure. Figure 29-8 and Figure 29-9 show examples of some of these devices.

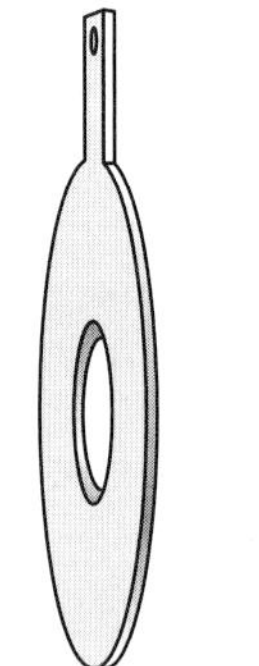

Figure 29-8: Orifice Plate

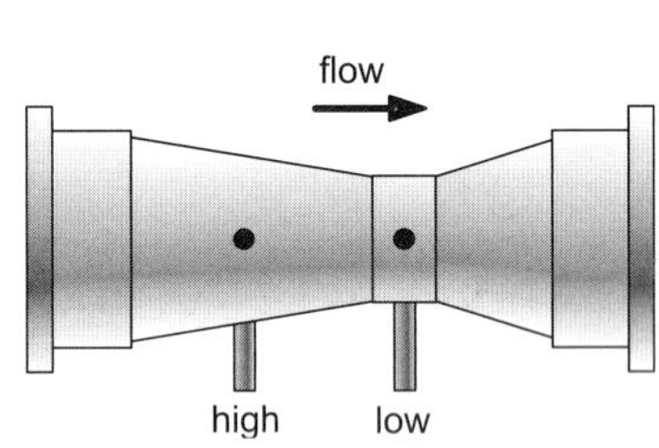

Figure 29-9: Venturi Tube

Analytical

Analytical is a measurement of chemical or physical properties of a substance. Some analytical-related instruments include pH meters, chromatographs and viscosity meters. Figures 29-10 and 29-11 show some examples of these devices.

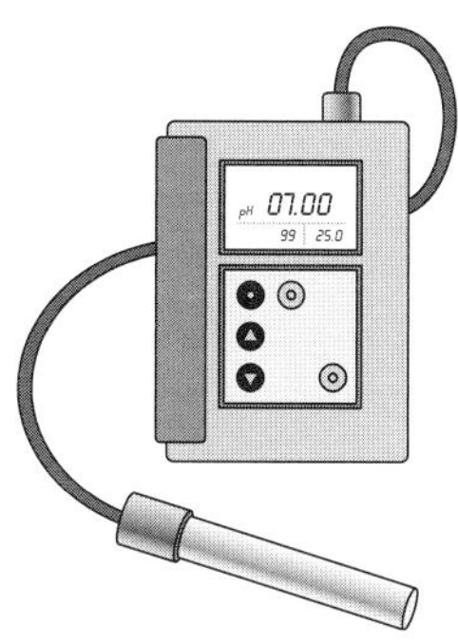

Figure 29-10: pH Meter

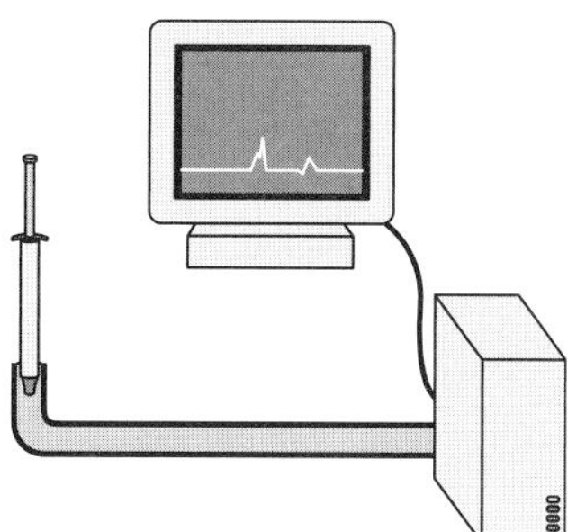

Figure 29-11: Gas Chromatograph

TYPICAL PROCESS CONTROL INSTRUMENTS AND THEIR APPLICATIONS

Process Control Instrumentation can be categorized in three different ways:

- Location
- Function
- Power source

An instrument can fall into more than one of these categories. For example, a pneumatic (power source) pressure transmitter (function) can be local (location) to the process.

The type of signal produced can also be used to categorize instrumentation. Refer to the Process Control Instrumentation Signals section for details.

Location

Location is where the instrument is located with respect to the actual process it is measuring and/or controlling. An instrument can be categorized as local or remote. **Local** (also referred to as field) is considered to be in close proximity to the process and generally located outside the control room, while **remote** is usually in a control room.

Function

An instrument can function as a sensor, transmitter, indicator, recorder, or controller.

Sensors are devices that sense and measure a variable. A **transmitter** is a device that receives its signal from the sensing element and "transmits" its signal to the controller or indicator. Sensors and transmitters work together to measure and transmit data. Some devices combine sensors and transmitters. For example, a sensor/transmitter can measure pressure and then transmit a signal representing that measurement.

Indicators show the current condition of a process variable, usually through a visual representation such as needle on a scale in a gauge or a digital readout. Sometimes, transmitters, recorders and controllers incorporate an indicator.

A **Recorder** is a device that registers a process variable by recording it in some form (e.g. stored as data in a microprocessor or written to a log). Recorders can show historical data which is helpful for troubleshooting or optimizing a process.

Controllers are devices which maintain a process variable at a specific value. Controllers take input from a transmitter, compare that input to the setpoint, and send output signals to a final control element such as a valve or damper.

Power source

A power source is what provides the instrumentation with the power to work. Power sources can be pneumatic, hydraulic, electronic, or a combination of these sources.

Pneumatic uses air pressure or a gas such as nitrogen as the power source. This type of power source is low cost

Hydraulic uses liquid as the power source. This source is used when heavy equipment, such as large valves, must be controlled.

Electronic uses electricity as the power source (grouped as analog, digital, or a hybrid of the two). Analog uses a continuously variable electrical quantity to measure and generate an output signal. Digital uses microprocessors (mini computers) to measure and generate an output signal; digital is used more frequently due to the increased use of automation in the process industries. Signals can also use a electromagnetic source.

Figure 29-12 and Figure 29-13 show examples of analog and digital instruments.

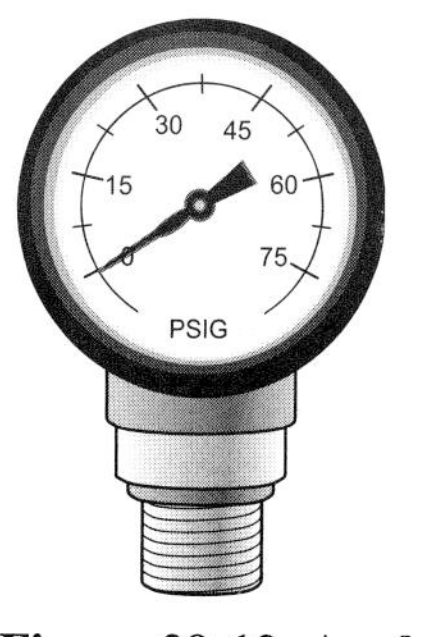

Figure 29-12: Analog Instrument

Figure 29-13: Digital Instrument

OTHER COMPONENTS

Other typical components in a process control instrumentation system include thermocouples, sensors, transducers, alarms, control valves, actuators, valve positioners, and signal control elements.

Thermocouples are devices used to measure the temperature of a process.

Sensors are devices that provide a reading from a process (e.g., such as a strain gauge for weight or fire eye for flame).

Transducers convert one instrumentation signal type to another type (e.g., electronic to pneumatic).

Alarms are devices that provide a visual or audio alert to technicians of a condition that needs attention, such as a tank reaching capacity. An alarm can be a computer display, horn, flashing light, whistle or bell or other similar device.

A **control valve** is a final control element that physically controls flow in a process. **Actuators** are electronic or pneumatic device that move control valves.

A **valve positioner** is a device that assures a valve is positioned correctly based on the signal it received.

PROCESS CONTROL INSTRUMENTATION SIGNALS

Instrumentation can also be categorized by the type of signal it produces. The two main types of electronic signals are analog and digital. An **analog** signal is a continuously variable representation of a process variable. A **digital** signal uses the language of computers, called binary (on/off switches that represent data) to represent continuous values or distinct (discrete) states.

DID YOU KNOW?

Process technicians should always ask permission before using a camera with a flash around fire detection equipment.

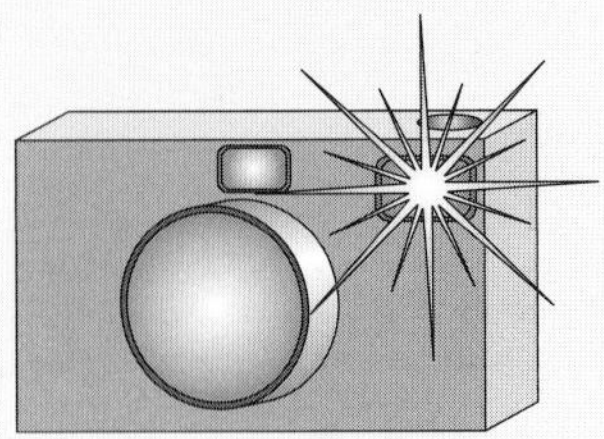

Camera flashes can interfere with fire detection sensors, potentially causing the fire suppression system to be activated.

Although digital signals are used more frequently, analog signals continue to be used in process industries for instrumentation applications. Some digital-based instruments (with a microprocessor inside) can generate either a digital or analog signal.

Fiber optics, which uses light frequency signals in either analog or digital mode, is gaining in industry usage.

Pneumatic signals are common in process facilities, also. These signals are generated using air or a compressed gas (like nitrogen).

Some plants may even have mechanical methods of transmitting signals, such as levers, cables and pulleys. Hydraulic systems can also be used to transmit signals.

Devices in an instrumentation system, called **transducers**, can convert one type of signal to another.

INSTRUMENTATION AND CONTROL LOOPS

A **control loop** is a collection of instruments that work together in a system to monitor and control a process. At a fundamental level, a control loop consists of a process variable, a device to measure it, and a device to control it.

The following describes how a simple control loop works:

1. The process variable is measured and that measurement is converted to a proportional pneumatic or electronic signal that represents the process.
2. The controller has a setpoint, or optimum value for that variable, that is then compared to the process requirement (note: the difference between the setpoint and the current process variable value is called the error).
3. The controller produces an output signal that is sent to the final control element (e.g. a valve or speed governor) to eliminate the error. Most control valves are actuated pneumatically, even if the loop signal is electronic. If the signal is electronic, a transducer converts the signal to pneumatic.
4. As the error is corrected, the process variable returns to match the setpoint.

To illustrate a control loop, think of a toilet. The variable is the water level in the tank. A float (measurement device) senses the height of the water (the process variable), until it reaches a pre-determined point (a setpoint). Once the float reaches that point, the valve (control element) shuts off water flow into the tank.

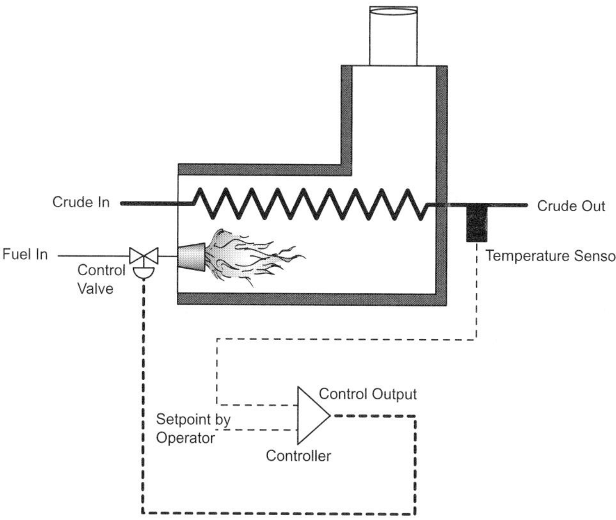

Figure 29-14: Simple Temperature Control Loop Example

Figure 29-14 shows another example of a simple control loop.

In this example, there is a heater that heats crude oil to 400∞ F. A sensor in the heater measures how hot the crude oil is and sends the temperature to a controller. The controller compares the temperature to the setpoint value. If the temperature is below the setpoint, the controller opens the fuel valve a little to add more fuel. If the temperature is above the setpoint, the controller closes the fuel valve a little to reduce the fuel. The valve will rarely be fully opened or fully closed.

A control loop can either be two position control (also called discrete, e.g. using on/off, open/shut, start/stop) or continuous (modulated). If the loop is on/off, the control element can only be on or off (such as a heating system in a home). If the loop is continuous, control is variable. The control element can be placed at any position between on and off (like a valve).

DISTRIBUTED CONTROL SYSTEMS AND THEIR APPLICATION

A **Distributed Control System (DCS)** is a computer-based system that is used to monitor and control a process. Because they are computer-based, DCS systems can be connected to other DCSs and related systems, providing flexibility, accuracy, precision and reliability.

The modern DCS can process and store data quickly and efficiently, sharing the workload among systems and offering redundancy (backup or fail-safe systems).

Although DCS terminology varies based on the company using it and the manufacturer that provided it, some common components to all systems are:

- **Input/output (I/O) devices**—field instrumentation
- **Network**—a data highway that connects the systems
- **Workstations**—computers that allow users to monitor and control processes
- **Central processor**—a computer used to configure the DCS
- **Accessory devices**—devices such as printers, data loggers/recorders and storage devices

HAZARDS ASSOCIATED WITH IMPROPER OPERATION

Process technicians must understand the hazards associated with the improper operation of instrumentation:

IMPROPER OPERATION	POSSIBLE IMPACTS			
	Individual	Equipment	Production	Environment
Opening bypass valve around control valve			Loss of process control	Potential release of a hazardous substance
Adjusting leaking control valve packing	Possible burns from product	The valve may not open, close, or operate properly; Over tightening may cause the valve to hang or stick	Could result in unit shutdown and loss of production Loss of process control	Potential release of a hazardous substance
Attempting to manually open a motor controlled valve	Hand and/or back injuries	Damage to valve and motor	Loss of process control	Potential release of a hazardous substance
Bypassing alarms	Injury due to malfunctioning equipment	Possible shutdown of process unit	Loss of production	Potential release of a hazardous substance
Failure to respond to an alarm	Injury due to malfunctioning equipment	Possible shutdown of process unit	Loss of production	Potential release of a hazardous substance
Turning off air supplies to control devices	Disciplinary action	Equipment shutdown	Could result in unit shutdown and loss of production Loss of process control	Potential release of a hazardous substance
Opening housings on electronic transmitters	Possible shock hazard	Equipment damage (e.g. blown fuses)	Possible ignition source for an explosion	

MONITORING AND MAINTENANCE ACTIVITIES

When monitoring and maintaining instrumentation, technicians must always remember to look, listen, and feel for the following:

Look	Listen	Feel
■ Monitor process variables. ■ Check for proper control status. ■ Check setpoint readings. ■ Watch for alarms. ■ Control process variables as required. ■ View historical data to analyze trends. ■ Check instrumentation calibration. ■ Check control valve packing. ■ Inspect instrumentation for broken parts. ■ Check for broken gauges. ■ Look at wiring and connections for potential problems. ■ Test alarms. ■ Inspect auxiliary equipment.	■ Check for noisy valve seats and disks. ■ Check for control line leaks (pneumatic or hydraulic). ■ Test alarms.	■ Check analog gauges to make sure they are not stuck. ■ Feel for valve and instrumentation vibration.

Failure to perform proper maintenance and monitoring could impact the process and result in instrumentation and equipment damage.

SYMBOLS FOR PROCESS CONTROL INSTRUMENTS

To accurately locate instrumentation on a piping and instrumentation diagram (P&ID) process technicians must be familiar with the symbols that represent different types of instrumentation.

Line Symbols	Line Type
	Pneumatic Signal
	Electrical Signal
	Electromagnetic or Sonic Signal
	Capillary Tubing
	Hydraulic Signal
	Software Link
	Connection to Process; Secondary Line; Utility Line
	Process Line
	Mechanical Link

Figure 29-15: Standard Line Symbols

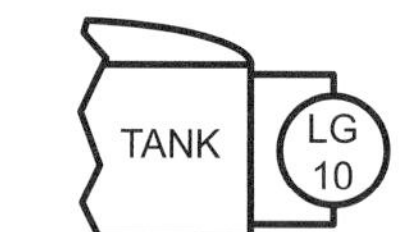

Figure 29-16: Balloon Symbol Example

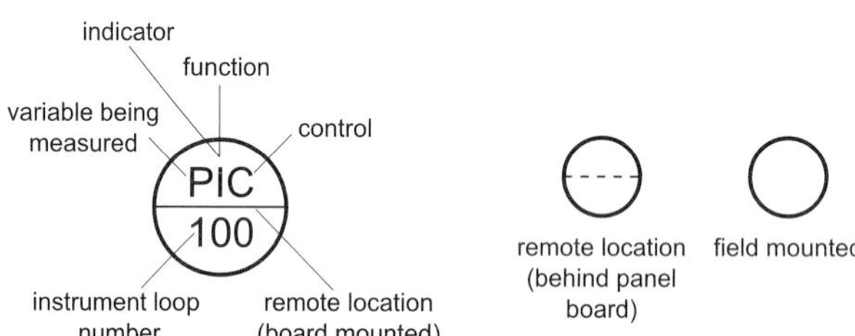

Figure 29-17: Instrument Symbol Interpretation Key

Figure 29-15 through Figure 29-17 show examples of instrumentation symbols:

SUMMARY

Instrumentation is a vital part of the process industries, allowing companies to produce higher quality products, more efficiently, in a safer environment. The process industries use instrumentation to monitor and control conditions in processes, such as temperature and pressure. These conditions are called Process Variables (PVs).

The five most common types of process variables are: pressure, temperature, level, flow, and analytical.

Pressure is the force exerted on a surface divided by its area. A common unit of pressure measurement is PSI, or pounds per square inch.

Temperature is a specific degree of heat or cold as indicated on a reference scale. The two most common temperature measurements scales are Fahrenheit (°F) and Celsius (°C).

Level is a position of height or depth along a vertical axis. In the process industries, level usually means the height of the surface of a material compared to a zero reference point.

Flow (also flow rate) a quantity of fluid that moves past a specific point within a given amount of time; usually expressed in volume units per unit of time, such as Gallons Per Minute (GPM) or Cubic Feet per minute (CFM).

Analytical is a measurement of the chemical or physical properties of a substance. Some analytical-related instruments include pH meters, chromatographs and viscosity meters.

Differential is the difference between measurements taken from two separate points. Differential pressure and differential temperature devices are commonly used in the process industries.

Process Control Instrumentation can be categorized three different ways: by location, function, or power source. An instrument can be categorized as local (field) or remote. An instrument can function as a sensor/transmitter, indicator, recorder, or controller. A power source can be pneumatic, hydraulic, or electronic.

Other typical components in an instrumentation system are thermocouples, sensors, transducers, alarms, control valves, actuators, valve positioners, and signal control elements.

Process control signals can be pneumatic, electronic (analog or digital), hydraulic, or electromagnetic.

A control loop is a collection of connected instruments that work together in a system to monitor and control a process.

A Distributed Control System (or DCS) is a computer-based system that is used to monitor and control a process.

When monitoring and maintaining instrumentation, technicians must always remember to look, listen, and feel for a variety of conditions. Process technicians must understand the hazards associated with the improper operation of instrumentation.

To accurately locate instrumentation on a piping and instrumentation diagram (P&ID) process technicians must be familiar with the symbols that represent different types of instrumentation.

CHECKING YOUR KNOWLEDGE

1. Define the following key terms:
   a. Transmitter
   b. Indicator
   c. Recorder
   d. Controller
   e. Thermocouples
   f. Sensor
   g. Transducer
   h. Alarms
   i. Control valve
   j. Actuators
   k. Positioner

2. What do the initials PV stand for?
   a. Pneumatic Value
   b. Process Variation
   c. Pressure Value
   d. Process Variable

3. Which of the following is NOT a key process variable?
   a. Pressure
   b. Temperature
   c. Flow
   d. Surface Area

4. ___________ is a process variable that measures the chemical or physical properties of a substance.

5. Differential is:
   a. The difference between measurements taken from two separate points.
   b. The similarity between measurements taken from two separate points.
   c. The difference between measurements taken from the same point at different times.
   d. The similarity between measurements taken from the same point at different times.

6. What are three ways that an instrument can be categorized?

7. An instrument is considered______ if it is in close proximity to the process, but considered ______ if it is located away from the process, like in a control room.

8. Which of the following power sources for an instrument uses air pressure, vacuum, or a gas such as nitrogen?
   a. Hydraulic
   b. Electronic
   c. Digital
   d. Pneumatic

9. *True or False:* A transmitter is a device that shows the current condition of a process variable, usually through a visual representation.

10. A __________ __________ is a collection of instruments that work together in a system to monitor and control a process.

11. What do the initials DCS stand for?
    a. Dynamic Collaboration System
    b. Dual Control Sensors
    c. Distributed Control System
    d. Disturbed Control Sensor

12. Which of the following is NOT a potential hazard if someone attempts to manually open a motor controlled valve?
    a. Damage to valve and motor
    b. An alarm will sound
    c. Loss of process control
    d. Potential release.

13. What do the following symbols represent?

A. —//———//—

B. – – – – – – –

C. —∿———∿—

ACTIVITIES

1. List the five most common process control variables, describe each one, and list any related instrumentation.

2. Write a paragraph that describes the functions of each of the following: sensor/transmitter, indicator, recorder, and controller.

3. Describe how a simple control loop works.

Glossary

Absolute Pressure (psia)—gauge pressure plus atmospheric pressure; pressure referenced to a total vacuum (zero PSIA). ***(Chapter 10)***

Absorb—the process of drawing inward. ***(Chapter 7)***

Acid—substances with a pH less than 7.0 that release hydrogen (H+) ions when mixed with water. ***(Chapter 11)***

Actuate—to put into action. ***(Chapter 20)***

Actuator—an electrical, pneumatic, or hydraulic device that moves a control device, usually a valve. ***(Chapter 29)***

Administrative Controls—the implementation programs (e.g., policies and procedures) and activities to address a hazard. ***(Chapter 12)***

Adsorb—the process of sticking together. ***(Chapter 7)***

Aftercoolers—a heat exchanger located on the discharge side of a compressor with the function of removing excess heat from the system created during compression. ***(Chapter 22)***

Air Pollution—the contamination of the atmosphere, especially by industrial waste gases, fuel exhausts, smoke or particulate matter (finely divided solids). ***(Chapter 12)***

Air Register—air intake devices, located at the burner, which are used to adjust the primary and secondary airflow to the burner. Air registers are the main source of air to the furnace. ***(Chapters 24, 25)***

Alarm—a device, such as a horn, flashing light, whistle or bell, which provides visual and/or audio cues that alert technicians of conditions outside the normal operating range. ***(Chapter 29)***

Alchemy—a medieval practice that combined occult mysticism and chemistry. ***(Chapter 3)***

Alkaline—a term used to refer to a base (e.g., substances with pH greater than 7). ***(Chapter 11)***

Alloy—a material composed of two or more metals that are mixed together when molten to form a solution, not a chemical compound (e.g., bronze is an alloy of copper and tin). ***(Chapter 16)***

Alternating Current (AC)—an electric current that reverses direction periodically. This is the primary type of electrical current used in the process industries. ***(Chapters 5, 21)***

Ammeter—device used to measure current. ***(Chapter 21)***

Amperes (Amps)—a unit of measure of the electrical current flow in a wire; similar to "gallons of water" flow in a pipe. ***(Chapter 21)***

Analytical—a measurement of the chemical or physical properties of a substance. ***(Chapter 29)***

Antibiotics—substances derived from mold or bacteria that inhibit the growth of other microorganisms (e.g., bacteria or fungi). ***(Chapter 8)***

API Gravity—the American Petroleum Institute (API) standard used to measure the density of hydrocarbons. ***(Chapter 10)***

Apothecary—a person who studies the art and science of preparing medicines; in modern times we call these individuals pharmacists. ***(Chapter 8)***

Articulated Drain—hinged drains, attached to the roof of a floating roof tank that raise and lower as the roof and the fluid levels raise and lower. ***(Chapter 17)***

Assignable Variation—states that when a product's variation goes beyond the limits of a natural variation, it is the result of a cause that can be identified. ***(Chapter 13)***

Atmospheric Pressure—the pressure at the surface of the earth (14.7 PSIA at sea level). ***(Chapter 10)***

Atmospheric Tank—enclosed vessels in which atmospheric pressure is maintained; usually cylindrical in shape and equipped with either a fixed or floating roof. ***(Chapter 17)***

Atom—the smallest particle of an element that still retains the properties and characteristics of that element. ***(Chapter 5, 11)***

Atomic Number—the number of protons found in the nucleus of an atom. ***(Chapter 11)***

Atomic Weight—the sum of protons and neutrons in the nucleus of an atom. ***(Chapter 11)***

Attitude—a state of mind or feeling with regard to some issue or event. ***(Chapter 12)***

Attributes—also called discrete data, or data that can be counted and plotted as distinct or unconnected events (such as percentage of late shipments or number of mistakes made during a process). ***(Chapter 13)***

Axial Compressor—a dynamic-type compressor, which uses a series of blades with a set of stator blades between each rotating, wheel. In this type of compressor, the gas flow is axial, or straight through, parallel to the compressor shaft. ***(Chapter 19)***

Axial Pump—a dynamic pump that uses a propeller or row of blades to propel liquids axially along the shaft. ***(Chapter 18)***

Back Flush—to wash by reversing the normal flow. ***(Chapter 22)***

Baffle—1) a metal plate, placed inside tanks or vessel, which is used to alter the flow of chemicals or facilitate mixing. ***(Chapter 17*** 2) partitions located inside a shell and tube heat exchanger that increase turbulent flow and reduce hot spots. ***(Chapter 22)***

Balanced Draft Furnace—a furnace that uses two fans to facilitate airflow, one inducing flow out of the firebox (induced draft) and one providing positive pressure to the burners (forced draft). ***(Chapter 24)***

Ball Valve—a flow control element shaped like a hollowed out ball used to start and stop flow; a ball valve only requires a quarter turn to get from fully open to fully closed. ***(Chapter 16)***

Base—a substance with a pH greater than 7.0 that releases hydroxyl (OH-) anions when dissolved in water. ***(Chapter 11)***

Basin—a compartment, located at the base of a cooling tower, which is used to store water until it is pumped back into the process. ***(Chapter 23)***

Baume Gravity—the industrial manufacturing measurement standard used to measure the gravity of non-hydrocarbon materials. ***(Chapter 10)***

Behavior—an observable action or reaction of a person under certain circumstances. ***(Chapter 12)***

Bernoulli's Law—a physics principle that states as the speed of a fluid increases, the pressure inside the fluid, or exerted by it, decreases. ***(Chapter 10)***

Bin/Hopper—vessels, which typically hold dry solids. ***(Chapter 17)***

Biological Hazard—any hazard that comes from a living, or once living, organism such as viruses, mosquitoes, or snakes, which can cause a health problem. ***(Chapter 12)***

Biologicals—products (e.g., vaccines) derived from living organisms that detect, stimulate or enhance immunity to infection. ***(Chapter 8)***

Blanketing—the process of putting nitrogen into the vapor space above the liquid in a tank to prevent air leakage into the tank (often referred to as a "nitrogen blanket"). ***(Chapter 17)***

Block Flow Diagram (BFD)—a very simple drawing that shows a general overview of a process, indicating the parts of a process and their relationships. ***(Chapter 15)***

Blowdown—removing water from a cooling tower basin or boiler drum in order to reduce the level of impurity concentration and control the chemical ratio. ***(Chapters 23, 25)***

Blower—a limited discharge compressor (usually below 100 PSI) that is used to move (airvey) powders or pellets from one point to another. ***(Chapter 19)***

Boiler—a closed vessel in which water is boiled and converted into steam under controlled conditions. ***(Chapter 25)***

Boiler Feed Water—water that is sent to a boiler in order to produce steam. ***(Chapter 27)***

Boiling Point—the temperature at which liquid physically changes to a gas at a given pressure. ***(Chapters 10, 11)***

Boot—a section at the bottom of a process drum where water is collected and drained to waste. This space is the lowest portion of the drum. ***(Chapter 16)***

Bottoms—see Heavy Ends. ***(Chapter 26)***

Boyle's Law—a physics principle that states, at a constant temperature, as the pressure of a gas increases, the volume of the gas decreases. ***(Chapter 10)***

Braze—to solder together using a hard solder with a high melting point. ***(Chapter 16)***

British Thermal Unit (BTU)—the amount of heat energy required to raise the temperature of one pound of water one degree Fahrenheit. ***(Chapter 10)***

Buoyancy—the principle that a solid object will float if its density is less than the fluid in which it is suspended; the upward force exerted by the fluid on the submerged or floating solid is equal to the weight of the fluid displaced by the solid object. ***(Chapter 10)***

Burner—a device used to introduce, distribute, and mix air, fuel and flame in the firebox. ***(Chapters 24, 25, 28)***

Butt Weld—a type of weld used to connect two pipes of the same diameter that are butted against each other. ***(Chapter 16)***

Butterfly Valve—a flow regulating device that uses a disc-shaped flow control element to increase or decrease flow; requires a quarter-turn to go from fully open to fully closed. ***(Chapter 16)***

Calorie—the amount of heat energy required to raise the temperature of one gram of water by one degree Celsius. ***(Chapter 10)***

Carbon Seal—a sealing system that utilizes carbon rings surrounded by springs. As the shaft heats, these carbon seals help prevent the escape of steam from the turbine casing. ***(Chapter 20)***

Casing—the housing around the internal components of a turbine. The component of a steam turbine, which holds all moving parts, including the rotor, bearings, seals, and is connected to the driven end equipment. ***(Chapter 20)***

Catalyst—a substance used to change the rate of a chemical reaction without being consumed into the reaction. ***(Chapter 11, 17)***

Catalytic Cracking—the process of adding heat and a catalyst to facilitate a chemical reaction. ***(Chapter 11)***

Caustic—a term used to describe a substance capable of destroying or eating away human tissues or other materials by chemical action; also a process industry term used to refer to a strong base. ***(Chapter 11)***

Cavitation—a condition inside a pump wherein the liquid being pumped partly vaporizes due to factors such as temperature and pressure drop; occurs when the pressure on the eye of a pump impeller falls below the boiling pressure of the liquid being pumped; can be identified by noisy operation and erratic discharge pressure; can cause excessive wear on the impeller and case; often remedied by increasing the suction pressure on the pump, usually by raising the level of liquid in the suction line. ***(Chapter 18)***

Cellulose—the principal component of the cell walls in plants. ***(Chapter 6)***

Centrifugal Compressor—a dynamic type compressor using a series of impellers in which the gas flows from the inlet located near the shaft to the outer tip of the impeller blade. Flow is then routed from the outer edge of one stage back to the inlet port of the next stage. ***(Chapter 19)***

Centrifugal Pump—A pump that imparts velocity to liquid by centrifugal force and then converts some of the velocity to pressure. ***(Chapter 18)***

Chain Reaction—a series of reactions in which each reaction is initiated by the energy produced in the preceding reaction.)

Charles' Law—a physics principle that states, at constant pressure, as the temperature of a gas increases, the volume of the gas also increases. ***(Chapter 10)***

Check Valve—a type of valve that only allows flow in one direction and is used to prevent reversal of flow in a pipe. ***(Chapter 16)***

Chemical—a substance with a distinct composition that is used in, or produced by, a chemical process. ***(Chapter 3)***

Chemical Change—a reaction in which the properties of a substance do change and a new substance is produced. ***(Chapter 11)***

Chemical Formula—a short-hand, symbolic expression that represents the elements in a substance and the number of atoms present in each molecule (e.g., water, H_20, is two hydrogen atoms and one oxygen atom bonded together). ***(Chapter 11)***

Chemical Hazard—any hazard that comes from a solid, liquid or gas element, compound, or mixture that could cause health problems or pollution. ***(Chapter 12)***

Chemical Property—a property of elements or compounds that is associated with a chemical reaction. ***(Chapter 11)***

Chemical Reaction—a chemical change or rearrangement of chemical bonds to form a new product. ***(Chapter 11)***

Chemical Symbol—one or two letter abbreviations for elements on the periodic table. ***(Chapter 11)***

Chemical Treatment—chemicals added to cooling water that are used to control algae, sludge, and fouling of exchangers and cooling equipment. ***(Chapter 23)***

Chemistry—the science that describes matter, its chemical and physical properties, the chemical and physical changes it undergoes, and the energy changes that accompany those processes. ***(Chapters 3, 11)***

Chiller—a device used to cool a fluid to a temperature below ambient temperatures; chillers generally use a refrigerant as a coolant. ***(Chapter 22)***

Circuit—a system of one or many electrical components connected together to accomplish a specified purpose. ***(Chapter 21)***

Coal—an organic, energy-producing mineral, consisting primarily of hydrogen and carbon. ***(Chapter 5)***

Cogeneration Station—A utility plant that produces both electricity and steam (for heating). ***(Chapter 5)***

Commodity Chemicals—basic chemicals that are typically produced in large quantities, and in large facilities. Most of these chemicals are inexpensive and are used as intermediates. ***(Chapter 3)***

Compound—A pure and homogeneous substance that contains atoms of different elements in definite proportions, and that usually has properties unlike those of its constituent elements. ***(Chapter 11)***

Compounding—mixing two or more substances or ingredients to achieve a desired physical form. ***(Chapter 8)***

Compression Ratio—The ratio of discharge pressure (psia) to inlet pressure (psia). ***(Chapter 19)***

Compressor—mechanical device used to compress gases and vapors for use in a process system that requires a higher pressure. ***(Chapter 19)***

Condensate—liquid resulting from cooled or condensed vapor. Frequently refers to condensed steam. ***(Chapter 27)***

Condenser—a heat exchanger that is used to condense vapor to a liquid.

Conduction—the transfer of heat through matter via vibrational motion. ***(Chapters 10, 22)***

Conductor—a substance or body that allows a current of electricity to pass continuously along it. ***(Chapter 21)***

Containment Wall—a wall used to protect the environment and people against tank failures, fires, runoff and spills. ***(Chapter 17)***

Continuous Reaction—a reaction in which raw materials (reactants) are continuously being fed into the reactor and products are continuously being formed and removed. ***(Chapter 17)***

Control Valve—a valve normally equipped with an actuator to control valve stem movement; these valves are used as the final control element for controlling flow, level, temperature or pressure in a process. ***(Chapters 16, 29)***

Controller—an instrument device, which controls a process variable at a specific condition.

Convection—the transfer of heat through the circulation or movement of a liquid or gas. ***(Chapters 10, 22)***

Convection Section—the upper portion of a furnace where heat transfer is primarily through convection. *(Chapter 24)*

Convection Tubes—furnace tubes, located above the shock bank, that receive heat through convection. *(Chapter 24)*

Cooler—a heat exchanger that uses a cooling medium to lower temperature of a process material. *(Chapter 23)*

Cooling Tower—a structure designed to lower the temperature of a water stream by evaporating part of the stream (latent heat of evaporation); these towers are usually made of plastic and wood and are designed to promote maximum contact of the water with the air. *(Chapter 23)*

Cooling Water—Water that is sent from a cooling tower to unit heat exchangers in order to cool process fluids, then returned to the cooling tower for cooling and reuse. *(Chapter 27)*

Criticism—a serious examination and judgment of something; criticism can be positive (constructive) or negative (destructive). *(Chapter 14)*

Cyber Security—security measures intended to protect information and information technology from unauthorized access or use. *(Chapter 12)*

Cylinder—vessels that can hold extremely volatile or high pressure materials. *(Chapter 17)*

Dalton's Law—a physics principle that states the total pressure of a mixture of gases is equal to the sum of the individual partial pressures. *(Chapter 10)*

Damper—a movable plate or adjustable louvers used to regulate the flow of air or draft in a furnace. *(Chapters 24, 25)*

Density—the ratio of an object's mass to its volume. *(Chapter 10)*

Deposit—a natural accumulation of ore. *(Chapter 4)*

DESCC Conflict Resolution Model—a model for resolving conflict, comprised of the following steps: Describe, Express, Specify, Contract and Consequences. *(Chapter 14)*

Diaphragm Valve—a flow regulating device that use a chemical resistant, rubber-type diaphragm to control flow instead of a typical flow control element. *(Chapter 16)*

Dike—a wall (earthen, shell or concrete) built around a piece of equipment to contain any liquids should the equipment rupture or leak. *(Chapter 17)*

Direct Current (DC)—an electrical current that always travels in the same direction. *(Chapter 5, 21)*

Discharge—normally refers to the outlet side of a pump, compressor, fan or jet. *(Chapter 19)*

Disinfecting—destroying disease-causing organisms (e.g., through washing, irradiation, or ultraviolet exposure). *(Chapter 8)*

Disinfection—the process of killing pathogenic organisms. *(Chapter 7)*

Dissolved Solids—solids, which are held in suspension indefinitely. *(Chapter 7)*

Distillation—the separation of the constituents of a liquid mixture by partial vaporization of the mixture and separate recovery of vapor and residue. *(Chapters 22, 26)*

Distributed Control System (DCS)—a computer-based system that is used to monitor and control a process. *(Chapter 29)*

Diversity—the presence of a wide range of variation in qualities or attributes; in the workplace, it can also refer to anti-prejudice training. *(Chapter 14)*

DOT—U.S. Department of Transportation; a U.S. government agency with a mission of developing and coordinating policies to provide efficient and economical national transportation system, taking into account need, the environment and national defense. *(Chapter 12)*

Downcomer—a conduit or pipe, which allows liquid to pass from one distillation tray to another. *(Chapters 25, 26)*

Draft Gauges—gauges, calibrated in "inches of water," that measure the firebox pressure, airflow, and differential pressure between the outside of the furnace and the flue gas inside. *(Chapter 24)*

Drift Eliminators—devices that prevent water from being blown out of the cooling tower; the main purpose of a drift eliminator is to minimize water loss. *(Chapter 23)*

Drugs—substances used as medicines or narcotics. *(Chapter 8)*

Drum—specialized types of storage tank. *(Chapter 17)*

Dynamic Compressor—a compressor that uses centrifugal or rotational force to move gases (as opposed to positive displacement compressors which use a piston to compress the gas). Dynamic compressors are classified as either centrifugal or axial. *(Chapter 19)*

Dynamic Pump—a non-positive displacement pump, classified as either centrifugal or axial, that converts centrifugal force to dynamic or flowing pressure to move liquids. *(Chapter 18)*

Dynamics—describes interpersonal relationships; how workers get along with each other and function together.)

Economizer—the section of a boiler used to preheat feed water before it enters the main boiler system. *(Chapter 25)*

Elasticity—an object's tendency to return to its original shape after it has been stretched or compressed. *(Chapter 10)*

Electrical Diagram—a drawing that show electrical components and their relationships. *(Chapter 15)*

Electricity—a flow of electrons from one point to another along a pathway, called a conductor. *(Chapters 5, 21)*

Electromotive Force—the force that causes the movement of electrons through an electrical circuit. *(Chapter 21)*

Electronic Instruments—use electricity as the power source. *(Chapter 29)*

Electrons—negatively charged particles that orbit the nucleus of an atom. *(Chapter 11)*

Elements—substances composed of like atoms that cannot be broken down further without changing its properties. *(Chapter 11)*

Elevation Diagram—a drawing that represents the relationship of equipment to ground level and other structures. *(Chapter 15)*

Endothermic—a chemical reaction that requires the addition or absorption of energy. *(Chapter 11)*

Engineering Controls—controls that use technological and engineering improvements to isolate, diminish, or remove a hazard from the workplace. *(Chapter 12)*

EPA—Environmental Protection Agency; a Federal agency charged with authority to make and enforce the national environmental policy. *(Chapter 12)*

Equilibrium—a point in a chemical reaction in which the rate of the products forming from reactants is equal to the rate of reactants forming from the products. *(Chapter 11)*

Equipment Location Diagram—a drawing that shows the relationship of units and equipment to a facility's boundaries. *(Chapter 15)*

Ergonomic Hazard—hazards that can create physical and psychological stresses because of forceful or repetitive work, improper work techniques, or poorly designed tools and workspaces. *(Chapter 12)*

Ethnocentrism—belief in the superiority of one's own ethnic group; belief that others should believe and interpret things exactly the way you do. *(Chapter 14)*

Evaporation—an endothermic process in which a liquid is changed into a gas. *(Chapter 23)*

Evaporator—a heat exchanger that uses a refrigerant to remove heat. *(Chapter 28)*

Exchanger Head (also called the channel head)—a device at the end of a heat exchanger that directs the flow of the fluids into and out of the tubes. *(Chapter 22)*

Exothermic—a chemical reaction that releases energy. *(Chapter 11)*

Expansion Valve—a valve that reduces the pressure of a refrigerant, causing it to cool. *(Chapter 28)*

Exploration—the process of locating oil and gas reservoirs by conducting surveys and studies, and drilling wells. *(Chapter 2)*

Facility—also called a plant. Something that is built or installed to serve a specific purpose. *(Chapter 1)*

Fan—a device used to force or draw air through a furnace or cooling tower. *(Chapter 23, 25)*

Feed Tray—the tray located immediately below the feed line in a distillation tower. This is where liquid is introduced to or enters the column. *(Chapter 26)*

Feedback—evaluative or corrective information provided to the originating source about a task or a process. ***(Chapter 14)***

Fiber—a long, thin substance resembling a thread. ***(Chapter 6)***

Fill—the material that breaks water into smaller droplets as it falls inside the cooling tower. ***(Chapter 23)***

Filtration—the process of removing particles from water by passing it through porous media. ***(Chapter 7)***

Fire Triangle/Tire Tetrahedron—the three elements (fuel, oxygen and heat) required for a fire to start and sustain itself; a fire tetrahedron adds a fourth element: a chemical chain reaction. ***(Chapter 12)***

Fire Water—pressurized water that is used to extinguish fires. ***(Chapter 27)***

Firebox—the portion of a boiler/furnace where burners are located and radiant heat transfer occurs. ***(Chapters 24, 25)***

Firewall—earthen banks or concrete walls built around oil storage tanks to contain the oil in case of a spill or rupture. Also called bund. ***(Chapter 17)***

Fission—the process of splitting the nucleus, the positively charged central part of an atom, that results in the release of large amounts of energy. ***(Chapter 5)***

Fitting—a piping system component used to connect two or more pieces of pipe together. ***(Chapter 16)***

Fixed Bed Reactor—a reactor in which the catalyst bed is stationary as the reactants are passed over it; in this type of reactor, the catalyst occupies a fixed position and is not designed to leave the reactor with the process. ***(Chapter 17)***

Fixed Blades—blades inside a steam turbine that remain stationary when steam is applied. ***(Chapter 20)***

Flame Impingement—a condition in which the flames from a burner touch tubes in a furnace. ***(Chapter 24)***

Flange—a type of pipe connection that is bolted together. ***(Chapter 16)***

Flare—a device designed to safely burn excess hydrocarbons. It is generally a pipe located some specified distance from a process unit.)

Flare Header—a pipe that connects several vents to the flare system. ***(Chapter 28)***

Flash Zone—a term used to refer to the section located between the rectifying and stripping sections in a distillation column. ***(Chapter 26)***

Floating Roof—a type of roof (steel or plastic), used on storage tanks, which floats upon the surface of the stored liquid and is used to decrease the vapor space and reduce the potential for evaporation. ***(Chapter 17)***

Flow—the movement of fluids. ***(Chapter 10)***

Flow Rate—a quantity of fluid that moves past a specific point within a given amount of time; usually expressed in volume units per unit of time, such as Gallons Per Minute (GPM) or Cubic Feet per minute (CFM). ***(Chapter 29)***

Fluid—substances, usually liquids or vapors, that can be made to flow. ***(Chapter 10)***

Fluidized Bed Reactor—a reactor in which finely divided solids are suspended by an upward flow of gas. ***(Chapter 17)***

Foam (Chamber—a reservoir and piping that contain chemical foam used to extinguish fires within a tank. ***(Chapter 17)***

Forced Draft Furnace—a furnace that use fans or blowers to force air into the air registers. ***(Chapter 24)***

Forced Draft Tower—cooling towers that have fans or blowers at the bottom of the tower that force air through the equipment. ***(Chapter 23)***

Fourdriniers—a papermaking machine, developed by Henry and Sealy Fourdrinier, which produces a continuous web of paper. ***(Chapter 6)***

Fraction—a portion of distillate having a particular boiling range separated from other portions in the fractional distillation of petroleum products. ***(Chapter 26)***

Friction—the resistance encountered when one material slides against another. ***(Chapter 10)***

Fuel—any material that burns; can be a solid, liquid or gas. ***(Chapter 12)***
Fuel Gas—gas used as fuel in boilers and other types of furnaces. ***(Chapter 27)***
Fuel Gas Valve—a valve that controls fuel gas flow and pressure to burners. ***(Chapter 24)***
Furnace—an apparatus in which heat is liberated and transferred directly or indirectly to a fluid mass for the purpose of increasing the temperature of a process fluid. ***(Chapter 24)***
Furnace Purge—a method of removing combustibles from a furnace firebox in preparation for lighting burners before startup. ***(Chapter 24)***
Gas (vapor)—substance with a definite mass but no definite shape, whose molecules move freely in any direction and completely fill any container it occupies, and which can be compressed to fit into a smaller container. ***(Chapter 10)***
Gas Turbine—a device that consists of an air compressor, combustion chamber, and turbine. Hot gases produced in the combustion chamber are directed towards the turbine blades causing the rotor to move. The rotation of the connecting shaft can be used to operate other equipment. ***(Chapter 20)***
Gasket—a flexible material placed between flanges to seal against leaks. ***(Chapter 16)***
Gate Valve—a positive shutoff valve utilizing a gate or guillotine which when moved between two seats causes tight shutoff. ***(Chapter 16)***
Gauge Hatch—an opening on the roof of a tank that is used to check tank levels and obtain samples of the product or chemical. ***(Chapter 17)***
Gauge Pressure (psig)—pressure measured with respect to the Earth's surface at sea level (zero PSIG). ***(Chapter 10)***
General Gas Law (Combined Gas Law)—relationship between pressure, volume and temperature in a closed container; pressure and temperature must be in absolute scale ($P_1V_1/T_1 = P_2V_2/T_2$) ***(Chapter 10)***
Generator—a device that converts mechanical energy into electrical energy. ***(Chapter 5, 21)***
Geology—the study of the earth and its history as recorded in rocks.
Geothermal—a power generation source that uses steam produced by the earth to generate electricity. ***(Chapter 5)***
Globe Valve—a type of valve that uses a plug and seat to regulate the flow of fluid through the valve body, which is shaped like a sphere or globe. ***(Chapter 16)***
Governor—a device used to control the speed of a piece of equipment such as a turbine. ***(Chapter 20)***
Ground Fault Circuit Interrupter (GFCI)—a safety device that detects the flow of current to ground and opens the circuit to interrupt the flow. ***(Chapter 21)***
Grounding—the process of using the earth as a return conductor in a circuit. ***(Chapter 21)***
Hazardous Agent —the substance, method, or action by which damage or destruction can happen to personnel, equipment, or the environment. ***(Chapter 12)***
HDPE—High Density Polyethylene; a plastic material used to create water pipes and drains. ***(Chapter 16)***
Heat—1) the transfer of energy from one object to another as a result of a temperature difference between the two objects. ***(Chapter 10)*** 2) added energy that causes an increase in the temperature of a material (sensible heat) or a phase change (latent heat); the energy required by the fuel to generate enough vapors for the fuel to ignite. ***(Chapter 12)***
Heat Exchanger—a device used to exchange heat from one substance to another. ***(Chapter 22)***
Heat Tracing—a coil of heated wire or tubing that is adhered to or wrapped around a pipe in order to increase the temperature of the process fluid, reduce fluid viscosity, and facilitate flow. ***(Chapter 10)***
Heavy Ends (bottoms)—materials in a distillation column, which boil at the highest temperature; usually found at the bottom of a distillation column. ***(Chapter 26)***
Heterogeneous—matter with properties that are not the same throughout. ***(Chapter 11)***
Homogeneous—matter that is evenly distributed or consisting of similar parts or elements. ***(Chapter 11)***

Hot Oil System—a system that heats heavy oils as a fuel source for furnaces, boilers, reboilers, and exchangers. ***(Chapter 28)***

Hot Spot—a furnace tube or area within a furnace that gets too hot. ***(Chapter 24)***

Hunting—a term used to describe the condition when a turbine's speed fluctuates while the governor/controller is searching for the correct operating speed. ***(Chapter 20)***

Hydraulic Instruments—use liquid (hydraulic fluid) as the power source. ***(Chapter 29)***

Hydraulic Turbine—a turbine that is moved, operated, or effected by liquid. ***(Chapter 20)***

Hydrocarbon—organic compounds that contain only carbon and hydrogen that are most often found occurring in petroleum, natural gas and coal. ***(Chapters 2, 10, 11)***

Hydroelectric—a power generation source that uses flowing water to generate electricity. ***(Chapter 5)***

Hydrometer—an instrument designed to measure the specific gravity of a liquid. ***(Chapter 10)***

Ideal Gas Law—the mathematical expression PV = nRT which expresses the simplest relationship between temperature, pressure, volume and moles of a gas. ***(Chapter 10)***

Indicator—shows the current condition of a process variable, usually through a visual representation such as needle on a scale in a gauge or a digital readout. ***(Chapter 29)***

Induced Draft Furnace—a furnace that uses fans, located in the stack, to induce airflow from the firebox. ***(Chapter 24)***

Induced Draft Tower—cooling towers that have fans at the top of the tower that pull air through the tower. ***(Chapter 23)***

Inhibitors—substances, which slow or stop a chemical reaction. ***(Chapter 17)***

Inlet—the point where something enters. ***(Chapter 20)***

Inorganic Chemistry—the study of substances that do not contain carbon. ***(Chapter 11)***

Insoluble—describes a substance that does not dissolve in a solvent. ***(Chapter 11)***

Instrument Air—plant air that has been filtered and/or dried for use by instruments. ***(Chapter 27)***

Instrumentation—devices used to measure and control or indicate flow, temperature, level pressure, and analytical data. ***(Chapter 29)***

Insulator—a device made from a material that will not conduct electricity; the device is normally used to give mechanical support to electrical wire or electronic components. ***(Chapter 21)***

Interchanger (also called a cross exchanger)—one of the process-to-process heat exchangers. ***(Chapter 22)***

Intermediates—substances that are not made to be used directly, but are used to produce other useful compounds. ***(Chapter 3)***

Ions—charged particles. ***(Chapter 11)***

ISO—taken from the Greek word isos, which means equal, ISO is the International Organization for Standardization, which consists of a network of national standards institutes from over 140 countries. ***(Chapter 13)***

ISO 9000—an international standard that provides a framework for quality management by addressing the processes of producing and delivering products and services. ***(Chapter 12)***

ISO 14000—an international standard that addresses how to incorporate environmental aspects into operations and product standards. ***(Chapter 12)***

Isometric—a drawing that shows objects as they would be seen by the viewer (like a 3-D drawing, the object has depth). ***(Chapter 15)***

Kinetic Energy—energy associated with mass in motion. ***(Chapter 20)***

Knock-Out Drum—vessel located between the flare header and the flare. It is designed to separate liquid substances from vapors being sent to the flare for burning. ***(Chapter 28)***

Laminar Flow—a condition in which fluid flow is smooth and unbroken; viewed as a series of laminations or thin cylinders of fluid slipping past one another inside a tube. ***(Chapter 22)***

Latent Heat—heat that does not result in a temperature change but causes a phase change. ***(Chapter 10)***

Latent Heat of Condensation—the amount of heat energy given off when a vapor is converted to a liquid without a change in temperature. ***(Chapter 10)***

Latent Heat of Fusion—the amount of heat energy required to change a solid to a liquid without a change in temperature. ***(Chapter 10)***

Latent Heat of Vaporization—the amount of heat energy required to change a liquid to a vapor without a change in temperature. ***(Chapter 10)***

Level—a position of height or depth along a vertical axis. In the process industries, level usually means the height of the surface of a material compared to a zero reference point. ***(Chapter 29)***

Light Ends—materials in a distillation column, which boil at the lowest temperature; usually found at the top of a distillation column. ***(Chapter 26)***

Liquid—substances with a definite volume, but no fixed shape, that demonstrate a readiness to flow with little or no tendency to disperse, and are limited in the amount in which they can be compressed. ***(Chapter 10)***

Local—instrumentation considered to be in close proximity to the process, generally found outside the control room. ***(Chapter 29)***

Loop Diagram—a drawing that shows all components and connections between instrumentation and a control room. ***(Chapter 15)***

Lubricant—any substance interposed between two surfaces in motion for the purpose of reducing the friction, heat, and/or wear between them. ***(Chapter 28)***

Lubrication—a friction-reducing film placed between moving surfaces in order to reduce drag and wear. ***(Chapter 10)***

Lubrication System—a system that supplies oil or grease to equipment parts such as bearings, gearboxes, sleeve bearings, and seals. ***(Chapter 28)***

Makeup Water—water that is used to replace the water lost during blowdown and evaporation. ***(Chapter 23)***

Manufacturing—making a product from raw materials by hand or with machinery (e.g., cooking, decorating, grinding, milling, and mixing). ***(Chapter 8)***

Manway—an opening in a vessel that permits entry for inspection and repair. ***(Chapter 17)***

Mass—the amount of matter in a body or object measured by its resistance to a change in motion. ***(Chapter 10)***

Material Safety Data Sheet (MSDS)—a document that provides key safety, health and environmental information about a material. ***(Chapter 12)***

Matter—anything that takes up space and has mass. ***(Chapter 10)***

Mechanical Energy—the energy of motion that is used to perform work. ***(Chapter 20)***

Metal—chemical elements that have luster (ability to reflect light) and can conduct heat and electricity (e.g., copper, bauxite, iron, lead, gold, silver, zinc, nickel, and uranium). ***(Chapter 4)***

Mill—a facility where a raw substance is processed and refined to another form. ***(Chapter 6)***

Mine—a pit or excavation from which minerals are extracted. ***(Chapter 4)***

Minerals—naturally occurring, inorganic substances, which have a definite chemical composition and a characteristic crystalline structure. ***(Chapter 4)***

Mining—the extraction of valuable minerals or other geological materials from the earth. ***(Chapter 4)***

Mist eliminator—a device in a tank, composed of mesh, vanes or fibers that collect droplets of mist from gas. ***(Chapter 17)***

Mixer—a device used to mix chemicals or other substances. ***(Chapter 17)***

Mixture—occurs when two substances are mixed together but do not react chemically. ***(Chapter 11)***

Molecular Weight—a unit of measure for a substance equal to the sum of the atomic weights of the elements that are present in a substance. ***(Chapter 11)***

Molecule—two or more atoms held together by chemical bonds. ***(Chapter 11)***

Motor—a mechanical driver with rotational output; usually electrically operated. ***(Chapter 21)***

Mud drum—1) the lower drum of a boiler that is used as a junction area for boiler tubes. 2) a low place in a boiler where heavy particles in the water will settle out and can be blown down. ***(Chapter 25)***

Natural Draft Furnace—a furnace that has no mechanical draft or fans. Instead, the heat in the furnace causes draft. ***(Chapter 24)***

Natural Draft Tower—cooling towers that use temperature differences inside and outside the stack to facilitate air movement. ***(Chapter 23)***

Neutron—a neutrally charged particle found in the nucleus of an atom. ***(Chapter 11)***

Nitrogen—an inert (non-reactive) gas that is used to purge explosive gases and air from process systems and equipment. ***(Chapter 27)***

Non-metal—substances that conduct heat and electricity poorly, are brittle, waxy or gaseous, and cannot be hammered into sheets or drawn into wire (e.g., gems and precious stones, coal, gravel, sand, lime, stone, soda ash, phosphate rock, and clay). ***(Chapter 4)***

Nozzle—a small spout or extension on a hose or pipe that directs the flow of steam. ***(Chapter 20)***

NRC—Nuclear Regulatory Commission; a U.S. government agency that protects public health and safety through regulation of nuclear power and the civilian use of nuclear materials. ***(Chapter 12)***

Nuclear—a power generation source that uses the heat from splitting atoms to generate electricity. ***(Chapter 5)***

Ohm—a measurement of resistance in electrical circuits. ***(Chapter 21)***

Ore—a metal bearing mineral that is valuable enough to warrant mining (e.g., iron or gold). ***(Chapter 4)***

Organic Chemistry—the study of carbon-containing compounds. ***(Chapter 11)***

OSHA—Occupational Safety and Health Administration (OSHA); a U.S. government agency created to establish and enforce workplace safety and health standards, conduct workplace inspections and propose penalties for noncompliance, and investigate serious workplace incidents. ***(Chapter 12)***

Outlet—the point where something exits. ***(Chapter 20)***

Overhead—the product or stream, which comes off the top of a fractionation tower (also called distillate or overhead takeoff). ***(Chapter 26)***

Packed Tower—a tower that is filled with specialized packing material instead of trays. ***(Chapter 26)***

Packing—material used in a distillation column to effect separating. ***(Chapter 26)***

Pareto Principle—a quality principle, also called the 80-20 rule that states 80 percent of problems come from 20 percent of the causes. ***(Chapter 13)***

Pathogen—a disease causing microorganism. ***(Chapters 7, 8)***

Periodic Table—a chart of all known elements listed in order of increasing atomic number and grouped by similar characteristics. ***(Chapter 11)***

Personal Protective Equipment (PPE)—specialized gear that provides a barrier between hazards and the body and its extremities. ***(Chapter 12)***

Petrochemical—refers to chemicals derived from fossil fuels or petroleum products. ***(Chapter 3)***

Petroleum—a substance found in the earth, such as oil or gas, composed of chemical compounds consisting primarily of hydrogen and carbon. ***(Chapter 2)***

pH—a measure of the amount of hydrogen ions in a solution that can react and indicates whether a substance is an acid or a base. ***(Chapter 11)***

Pharmaceuticals—are man-made and natural drugs used to treat diseases, disorders, and illnesses. ***(Chapter 8)***

Phase Change—when a substance changes from one physical state to another, such as when ice melts to form water. ***(Chapter 10)***

Physical Change—an event in which the physical properties of a substance (e.g., how it looks, smells or feels) may change, the change may be reversible, and a new substance is not produced. ***(Chapter 11)***

Physical Hazard—any hazards that comes from environmental factors such as excessive levels of noise, temperature, pressure, vibration, radiation, electricity or mechanical hazards (note: this is not the OSHA definition of physical hazard). ***(Chapter 12)***

Physical Property—the properties of an element or compound that is observable and does not pertain to a chemical reaction. ***(Chapter 11)***

Physical Security—security measures intended to protect specific assets such as pipelines, control centers, tank farms, and other vital areas. ***(Chapter 12)***

Pilot—an igniting device used to light the primary burners in a furnace. ***(Chapters 24, 25, 28)***

Piping & Instrument Diagram (P&ID)—also called a Process & Instrument Drawing. A drawing that shows the equipment, piping and instrumentation of a process in the facility, along with more complex details than a Process Flow Diagram (see next definition). ***(Chapter 15)***

Plant Air—the supplied air system used for pneumatic tools, air hoists, and other air operated equipment. Also called, "service" or "utility air." ***(Chapter 27)***

Plasma—a gas that contains positive and negative ions. ***(Chapter 10)***

Plug Valve—a type of valve that uses a flow control element shaped like a hollowed out plug to start or stop flow; requires a quarter-turn to go from fully open to fully closed. ***(Chapter 16)***

Pneumatic Instruments—use air pressure or a gas as the power source. ***(Chapter 29)***

Positioner—a device that ensures a valve is positioned properly in reference to the signal received from the controller. ***(Chapter 29)***

Positive Displacement Compressor—compressors that use screws, sliding vanes, lobes, gears or diaphragms to deliver a set volume of gas; utilizes either reciprocating or rotary motion to trap a specific amount of gas and reduce its volume, thereby increasing the pressure at the discharge. ***(Chapter 19)***

Positive Displacement Pump—a pump that moves a constant amount of liquid through a system at a given pump speed. ***(Chapter 18)***

Potable Water—water that is safe to drink and can be used for cooking. ***(Chapter 27)***

PPM—Predictive/Preventive Maintenance, a program to identify potential issues with equipment and use preventive maintenance before the equipment fails. ***(Chapter 13)***

Pre-heater—a heat exchanger used to warm liquids before they enter a distillation tower. ***(Chapter 22, 26)***

Prejudice—attitude towards a group or its individual member based on stereotyped beliefs. ***(Chapter 14)***

Preserving—prepare for long-term storage (e.g., canning, drying, freezing, salting). ***(Chapter 8)***

Pressure—the force exerted on a surface divided by its area. ***(Chapter 10, 29)***

Pressurized Tank—enclosed vessels in which a pressure greater than atmospheric is maintained. ***(Chapter 17)***

Priming—filling the liquid end of a pump with liquid to remove vapors present and eliminate the tendency to become vapor bound or lose suction. ***(Chapter 18)***

Process—1) a system of people, methods, equipment, and structures that create products from other materials. ***(Chapter 1)*** 2) a method for doing something, which generally involves tasks, steps or operations, which are ordered and/or interdependent. ***(Chapter 14)***

Process Flow Diagram (PFD)—a basic drawing that shows the primary flow of product through a process, using equipment, piping and flow direction arrows. ***(Chapter 15)***

Process Industries—a broad term for industries that convert raw materials, using a series of actions or operations, into products for consumers. ***(Chapter 1)***

Process Technician—a worker in a process facility that monitors and controls mechanical, physical and/or chemical changes, throughout many processes, to produce either a final product or intermediate product, made from raw materials. ***(Chapter 1)***

Process Technology—a controlled and monitored series of operations, steps, or tasks that converts raw materials into a product. ***(Chapter 1)***

Process Variable—a varying operational condition that is associated or goes with a chemical processing operation such as temperature, pressure, flow rate, level, and composition. ***(Chapter 29)***

Process Water—water that is used in any level of the manufacturing process; not suitable for consumption. ***(Chapter 27)***

Product—also called output. The desired end components from a particular process. ***(Chapter 1)***

Production—1) output, such as material made in a plant, oil from a well, or chemicals from a processing plant. ***(Chapter 1)*** 2) the process of bringing oil and gas to the surface then preparing it for transport. ***(Chapter 2)***

Products—the substances that are produced during a chemical reaction. ***(Chapter 11)***

Proton—a positively charged particle found in the nucleus of an atom. ***(Chapter 11)***

Pulp—a cellulose fiber material, created by mechanical and/or chemical means from various materials (e.g. wood, cotton, recycled paper), from which paper and paperboard products are manufactured. ***(Chapter 6)***

PVC—Polyvinyl Chloride; a plastic type material that can be used to create cold water pipes and drains and other low-pressure applications. ***(Chapter 16)***

Quarry—an open excavation from which stones are extracted. ***(Chapter 4)***

Radiant Section—the lower portion of a furnace (firebox) where heat transfer is primarily through radiation. ***(Chapter 24)***

Radiant Tubes—tubes located in the firebox that receive heat primarily through radiant heat transfer. ***(Chapters 24, 25)***

Radiation—the transfer of heat energy through electromagnetic waves. ***(Chapters 10, 22)***

Raw Materials—also called feedstock or input. The material sent to a processing unit to be converted into a different material or materials. ***(Chapter 1)***

Reactants—the starting substances in a chemical reaction. ***(Chapter 11)***

Reaction Furnace—a reactor which combines a firebox with tubing to provide heat for a reaction that occurs inside the tubes. ***(Chapter 17)***

Reactor—a vessel in which chemical reactions are initiated and sustained. ***(Chapter 17)***

Reboiler—a tubular heat exchanger placed at the bottom of a distillation column or stripper to supply the necessary column heat. ***(Chapters 22, 26)***

Receiver—a tank that stores refrigerant once it leaves the condenser. ***(Chapter 28)***

Reciprocating Compressor—a type of positive displacement compressor that consists of a cylinder that contains a piston that travels back and forth (reciprocates) in a cylinder containing suction valves and discharge valves. ***(Chapter 19)***

Reciprocating Pump—a positive displacement pump that use the inward stroke of a piston or diaphragm to draw liquid into a chamber (intake) and then positively displaces (discharges) the liquid using an outward stroke. ***(Chapter 18)***

Recorder—a mechanical recording instrument used to measure variables such as pressure, speed, flow, temperature, level and electrical units; also provides a history of the process over a given period of time. ***(Chapter 29)***

Rectifying Section—that portion of a fractionating tower located between the feed tray and the top of the tower. The area in which the overhead product is purified. ***(Chapter 26)***

Refining—the process of purifying a crude substance into other products, such as petroleum being separated into gasoline, kerosene, gas oil. ***(Chapter 2)***

Reflux—the condensed portion of the lighter liquid that is pumped back into the distillation column to cool and condense the rising vapors. ***(Chapter 26)***

Refractory—a form of insulation used inside high temperature boilers, incinerators, heaters, reactors, and furnaces. ***(Chapter 24)***

Refrigerant—a fluid with a low boiling point circulated throughout a refrigeration system. ***(Chapter 28)***

Relief Valve—a safety device designed to open if the pressure of a liquid in a closed vessel exceeds a preset level. *(Chapter 16)*

Remote—instrumentation that is located away from the process (e.g., in a control room). *(Chapter 29)*

Riser—tubes that allow liquids or vapors to move upward in a vessel or a distillation column. *(Chapters 25, 26)*

Rotary Compressor—a type of positive displacement compressor that uses a rotating motion to move the gas. There are three basic compressor designs: screw, sliding vane, and lobe. *(Chapter 19)*

Rotary Pump—a positive displacement pump that moves liquids by rotating a screw or a set of lobes, gears or vanes. *(Chapter 18)*

Rotor—the rotating member or a motor or turbine. *(Chapters 20, 21)*

Safety Valve—a safety device designed to open if the pressure of a gas in a closed vessel exceeds a preset level. *(Chapter 16)*

Schematic—a drawing that shows the direction of current flow in a circuit, typically beginning at the power source. *(Chapter 15)*

Screwed (Threaded) Pipe—piping that is connected using male and female threads. *(Chapter 16)*

Sensible Heat—heat transfer that results in a temperature change. *(Chapter 10)*

Sensor—a device, in direct contact with the process variable to be measured, that senses or measures, then transmits a signal to another instrument. *(Chapter 29)*

Service Water—general-purpose water which may or may not have been treated. *(Chapter 27)*

Set Point—the point or place where the control index of a controller is set. *(Chapter 20)*

Settleable Solids—solids in wastewater that can be removed by slowing the flow in a large basin or tank. *(Chapter 7)*

Shaft—a metal rod, attached to an impeller, which is suspended by bearings. *(Chapter 20)*

Shell—the outer housing of a heat exchanger that covers the tube bundle. *(Chapter 22)*

Shell Inlets and Outlets—the openings that allow process fluids to flow into and out of the shell side of a shell and tube heat exchanger. *(Chapter 22)*

Shock Bank—tubes located directly above the firebox in a furnace that receive both radiant and convective heat. *(Chapter 24)*

Six Sigma—a relatively new TQM approach, Six Sigma is considered advanced quality management using a data driven approach and methodology to eliminate defects. *(Chapter 13)*

Socket Weld—a type of weld used to connect pipes and fittings when one pipe is small enough to fit snugly inside the other. *(Chapter 16)*

Soil Pollution—the accidental or intentional discharge of any harmful substance into the soil. *(Chapter 12)*

Solar—a power generation source that uses the power of the sun to heat water and generate electricity. *(Chapter 5)*

Solder—a metallic compound that is melted and applied in order to join and seal the joints and fittings together in tubing systems. *(Chapter 16)*

Solids—substances, with a definite volume and a fixed shape, that are neither liquid, nor gas, and that maintain their shape independent of the shape of the container. *(Chapter 10)*

Soluble—describes a substance that will dissolve in a solvent. *(Chapter 11)*

Solute—the substance being dissolved in a solvent. *(Chapter 11)*

Solution—a homogeneous mixture of two or more substances. *(Chapter 11)*

Solvent—the substance present in a solution in the largest amount. *(Chapter 11)*

Spacer rods—the rods that space the tubes in a tube bundle apart so they do not touch one another. *(Chapter 22)*

SPC—Statistical Process Control uses mathematical laws dealing with probability. Companies utilize SPC to gather data (numbers) and study the characteristics of processes, then use the data to make the processes behave the way they should. *(Chapter 13)*

Specialty Chemicals—chemicals that are produced in smaller quantities, are more expensive, and are used less frequently than commodity chemicals. *(Chapter 3)*

Specific Gravity—the ratio of the density of a liquid or solid to the density of pure water, or the density of a gas to the density of air at standard temperature and pressure (STP). *(Chapter 10)*

Specific Heat—the amount of heat required to raise the temperature of one gram of a substance one degree Celsius, or one pound of a substance one degree Fahrenheit. *(Chapter 10)*

Spherical Tank—a type of pressurized storage tank that is used to store volatile or highly pressurized material; also referred to as "round" tanks. *(Chapter 17)*

Stack—a cylindrical outlet, located on the top of a furnace, which remove flue gas from the furnace. *(Chapters 24, 25)*

Steam Drum—the top drum of a boiler where all of the generated steam gathers before entering the separating equipment. *(Chapter 25)*

Steam Turbine—a turbine that is driven by the pressure of steam discharged at high velocity against the turbine vanes. *(Chapter 20)*

Stereotyping—beliefs about individuals or groups based on opinions, habits of thinking, or rumors, which are then generalized to every member of a group. *(Chapter 14)*

Stirred Tank Reactor—a reactor that contains a mixer or agitator mounted to the tank. *(Chapter 17)*

Stripping Section—the section of a distillation tower below the feed tray where the heavier components with higher boiling points are located. *(Chapter 26)*

Suction—normally refers to the inlet side of a pump, compressor, fan or jet. *(Chapter 19)*

Suction Screens—usually a cone-shaped or flat metal strainer used to remove debris. *(Chapter 23)*

Sump—a pit or tank which receives and temporarily stores drainage at a low point. *(Chapter 17)*

Superheater—tubes, located toward the boiler outlet, which increase (superheat) the temperature of the steam flow. *(Chapter 25)*

Suspended Solids—solids that can not be removed by slowing the flow. *(Chapter 7)*

Symbol—figures used to designate types of equipment. *(Chapter 15)*

Synergy—the total effect of a whole is greater than the sum of its individual parts. *(Chapter 14)*

Synthetic—a substance resulting from combining components, instead of being naturally produced. *(Chapter 3)*

Tank—a large container or vessel for holding liquids and/or gases. *(Chapter 17)*

Task—a set of actions that accomplish a job. *(Chapter 14)*

Team—a small group of people, with complementary skills, committed to a common set of goals and tasks. *(Chapter 14)*

Temperature—the degree of hotness or coldness that can be measured by a thermometer and a definite scale. *(Chapters 10, 29)*

Thermocouple—a sensing device consisting of two dissimilar metals joined at a junction that when exposed to heat, the metal junction generates an electrical signal proportional to the change in temperature. *(Chapter 29)*

Throttle Valve—a valve that can be opened or closed to quickly (tripping) or slowly (throttling) to control flow. This valve works in conjunction with the governor to control the speed of the turbine. *(Chapter 20)*

Throttling—partially opening or closing a valve in order to restrict or regulate flow. *(Chapter 16)*

TPM—Total Productive Maintenance is an equipment maintenance program that emphasizes a company-wide effort to involve all levels of staff in various aspects of equipment maintenance. *(Chapter 13)*

TQM—Total Quality Management is a collection of philosophies, concepts, methods and tools used to manage quality; TQM consists of four parts: Customer Focus, Continuous Improvement, Manage by Data and Facts, and Employee Empowerment. *(Chapter 13)*

Transducer—convert one instrumentation signal type to another type (e.g., electronic to pneumatic). *(Chapter 29)*

Transmitter—a device that receives its signal from the sensing element and "transmits" its signal to the controller or indicator. *(Chapter 29)*

Transportation—the oil and gas industry segment responsible for moving petroleum from wells to processing facilities and finished products to consumers. Transportation methods include pipelines, watercraft, railways and trucks. *(Chapter 2)*

Tray—a part of a fractionation tower containing a liquid level. Vapors rise through the liquid to form a dynamic equilibrium mixture. The number of trays varies with the degree of fractionation required. *(Chapter 26)*

Treated Water—water that has been filtered, cleaned, and chemically treated. *(Chapter 27)*

Trip Valve—a safety valve that can be opened or closed quickly. *(Chapter 20)*

Tube Bundle—a group of fixed or parallel tubes, such as is used in a heat exchanger; the tube bundle includes the tube sheets with the tubes, the baffles and the spacer rods. *(Chapter 22)*

Tube Inlets and Outlets—the openings that allow process fluids to flow into and out of the tube bundle in a shell and tube heat exchanger. *(Chapter 22)*

Tube Sheet—a flat plate to which the tubes in a heat exchanger are fixed. *(Chapter 22)*

Tubing—small copper or stainless steel pipe used extensively in instrument work. Plastic tubing is also used. A seamless type of steel pipe is referred to as tubing. *(Chapter 16)*

Tubular Reactor—a tubular heat exchanger used to contain a reaction. *(Chapter 17)*

Turbidity—cloudiness caused by particles suspended in water or some other liquid. *(Chapter 7)*

Turbine—a machine for producing power. Activated by the expansion of a fluid (e.g., steam, gas, air) on a series of curved vanes on an impeller attached to a central shaft, which is used to create mechanical energy. *(Chapters 5, 20)*

Turbulent Flow—a condition in which the fluid flow pattern is disturbed so there is considerable mixing. *(Chapter 22)*

Unit—an integrated group of process equipment used to produce a specific product or products. All equipment contained in a department.)

Utilities Flow Diagram (UFD)—a drawing that shows the piping and instrumentation for the utilities in a process. *(Chapter 15)*

Utility Air—compressed air that is used to power equipment and tools. *(Chapter 27)*

Vacuum Pressure (psiv)—any pressure below atmospheric pressure. *(Chapter 10)*

Valve—a piping system component used to control the flow of fluids through a pipe. *(Chapter 16)*

Valve Seat—the internal component of a valve against which the sealing elements presses to stop flow. *(Chapter 16)*

Vane Separator—a device, composed of metal vanes, used to separate liquids from gases or solids from liquids. *(Chapter 17)*

Vapor Pressure—a measure of a substance's volatility and its tendency to form a vapor. *(Chapter 10)*

Vapor Recovery System—the process of capturing and recovering vapors. Vapors are captured by methods such as chilling or scrubbing. They are then purified and the vapors or products are either sent back to the process, sent to storage, or recovered. *(Chapter 17)*

Variables—also called continuous data, or data that can be measured and plotted on a constant scale (such as flow through a pipeline or liquid in a tank). *(Chapter 13)*

Velocity—the distance traveled over time or change in position over time. *(Chapter 10)*

Vessel—a container in which materials are processed, treated, or stored. *(Chapter 17)*

Viscosity—the measure of a fluid's resistance to flow. *(Chapter 10)*

Voluntary Protection Program (VPP)—an OSHA program designed to recognize and promote effective safety and health management. *(Chapter 12)*

Volute—a spiral casing for a centrifugal pump, designed so that speed will be converted to pressure without shock. ***(Chapter 18)***

Vortex—the cone formed by a swirling liquid or gas. ***(Chapter 17)***

Vortex breaker—a metal plate, or similar device, placed inside a cylindrical or cone-shaped vessel, that prevents a vortex from being created as liquid is drawn out of the tank. ***(Chapter 17)***

Waste Water—water that contains waste products in the form of dissolved or suspended solids. ***(Chapter 27)***

Water Distribution Header—a device that evenly distributes the water on top of the cooling tower through the use of distribution valves, and allows the water to evenly flow into the tower fill. ***(Chapter 23)***

Water Pollution—the introduction, into a body of water or the water table, any EPA listed potential pollutant that affects the chemical, physical or biological integrity of that water. ***(Chapter 12)***

Weight—a measure of the force of gravity on an object. ***(Chapter 10)***

Weir—a flat or notched dam or barrier to liquid flow that is normally used for either the measurement of fluid flows or to maintain a given depth of fluid as on a tray of a distillation column. ***(Chapters 17, 26)***

Wind Turbine—a device that converts wind energy into mechanical energy. ***(Chapter 20)***

Wiring Diagram—a drawing that shows electrical components in their relative position in the circuit and all connections in between. ***(Chapter 15)***

Work Group—a group of people organized by a logical grouping within a company, with a designated leader that handles routine tasks. ***(Chapter 14)***

Zero Defects—a quality practice with the objective of reducing defects, which can increase profits. ***(Chapter 13)***

INDEX

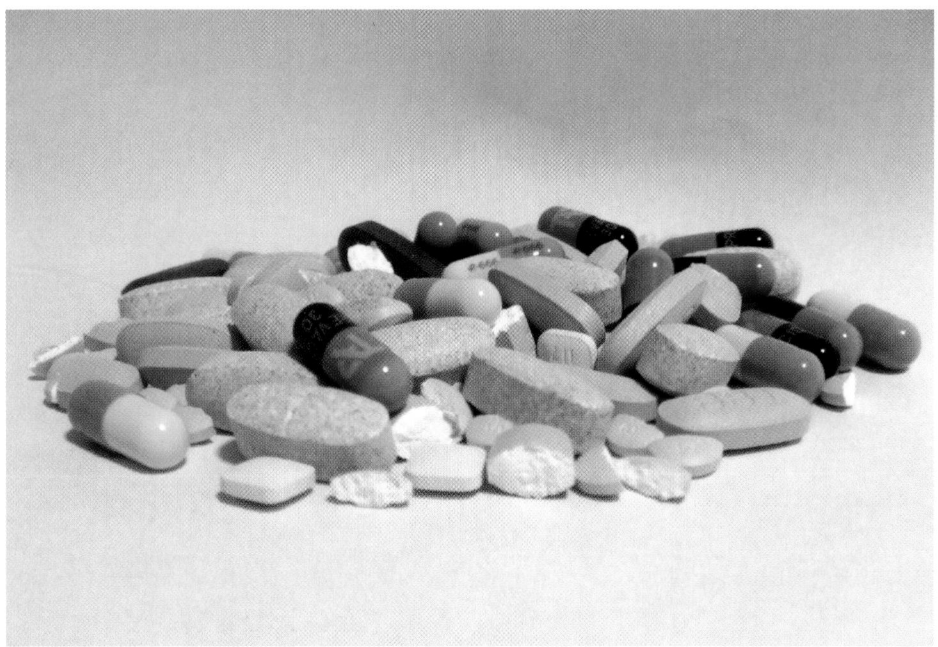

Pharmaceutical Industry

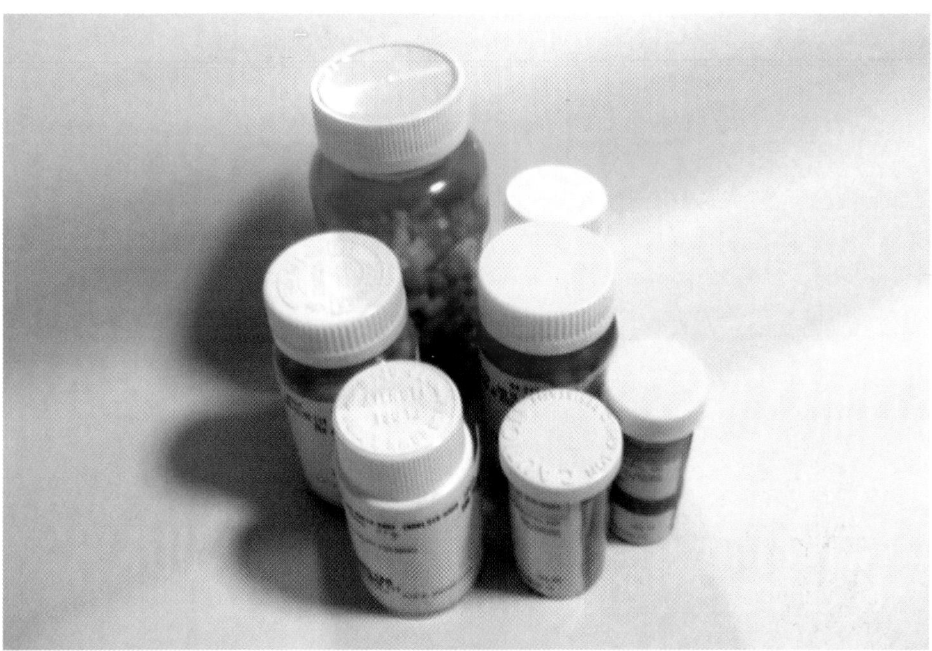

Pharmaceutical Industry

INDUSTRY OVERVIEW

Pharmaceutical Industry

Mining Industry

Courtesy of EyeWire/Getty Images, Inc.

INDUSTRY OVERVIEW

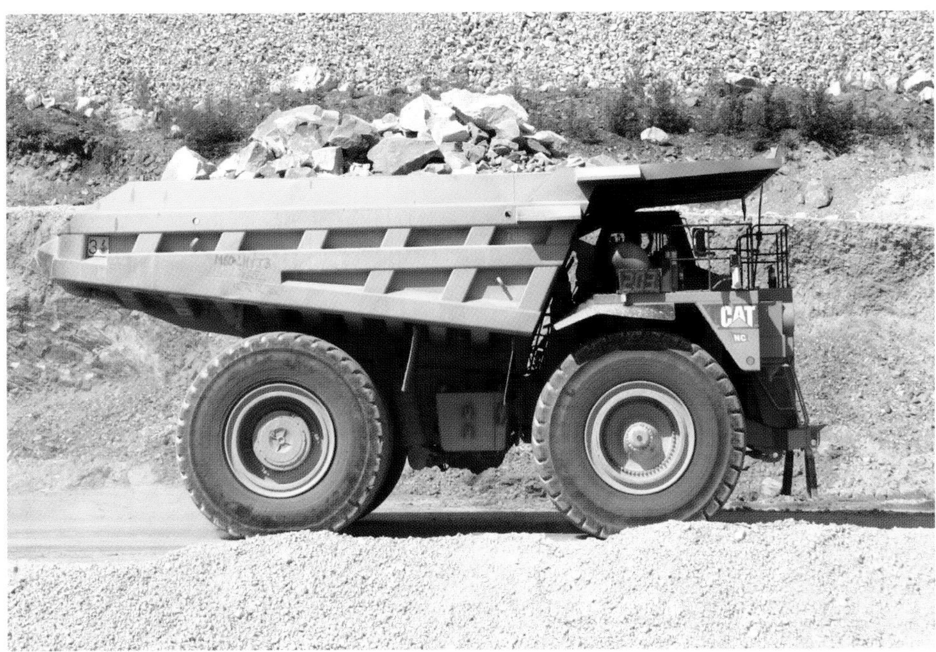

Mining Industry

Mining Industry

INDUSTRY OVERVIEW

Power Generation
Courtesy of Corbis Images.

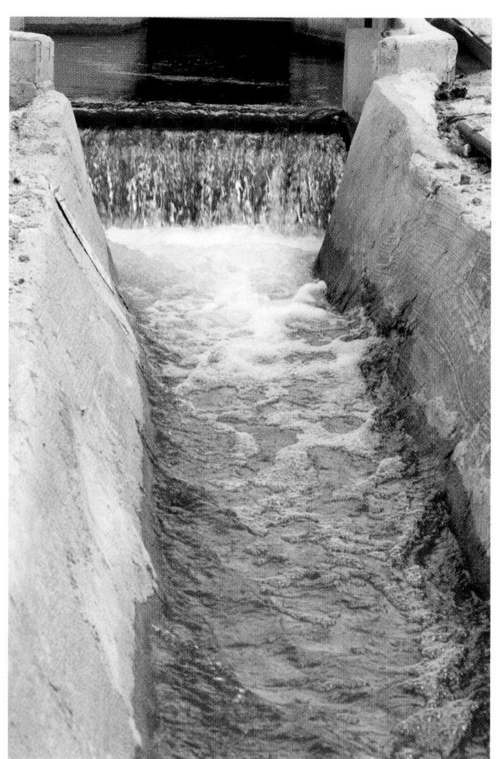

Water and Waste Water
Courtesy of EyeWire/Getty Images, Inc.

Power Generation
Courtesy of Corbis Images.

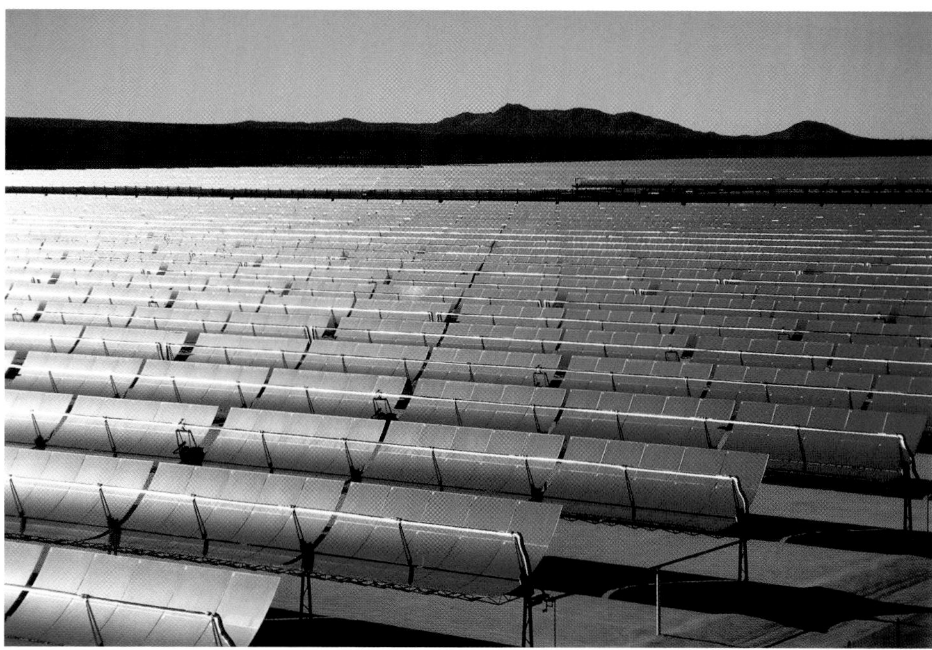

Power Generation
Courtesy of Corbis Images.

Power Generation
Courtesy of Corbis Images.

INDUSTRY OVERVIEW

Power Generation
Courtesy of Corbis Images.

Oil and Gas Production
(© 2005 BP Photo Resources.)

Oil and Gas Production

INDUSTRY OVERVIEW

Oil and Gas Production
(© 2005 BP Photo Resources.)

Oil and Gas Production
Courtesy of Corbis Images.

Oil and Gas Production
Courtesy of EyeWire/Getty Images, Inc.

Oil and Gas Production

Oil and Gas Production

Courtesy of EyeWire/Getty Images, Inc.

Pulp and Paper
(Photo Courtesy of Pacific Northwest National Laboratory)

Pulp and Paper
(Photo Courtesy of TISEC)

Ball Check Valve
(Photo Courtesy of Bayport Training & Technical Center)

Ball Valve
(Photo Courtesy of Bayport Training & Technical Center)

EQUIPMENT OVERVIEW

Gate Valve
(Photo Courtesy of Bayport Training & Technical Center)

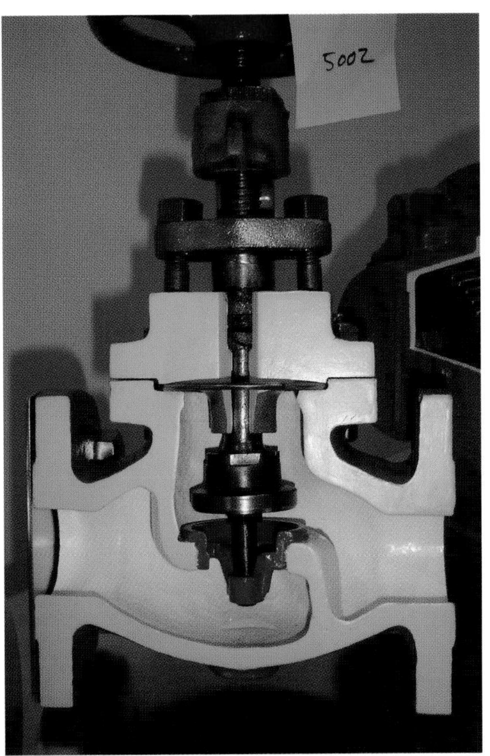

Globe Valve
(Photo Courtesy of Bayport Training & Technical Center)

Plug Valve
(Photo Courtesy of Bayport Training & Technical Center)\

EQUIPMENT OVERVIEW

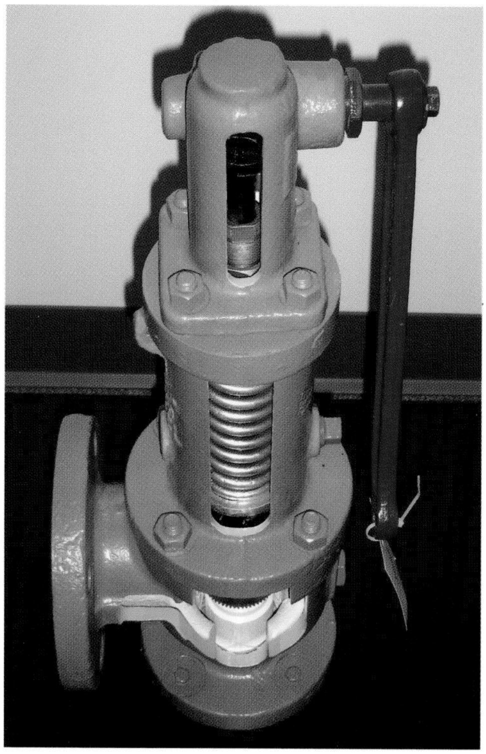

Relief Valve
(Photo Courtesy of Bayport Training & Technical Center)

Safety Relief Valve
(Photo Courtesy of Bayport Training & Technical Center)

Spring Check Valve
(Photo Courtesy of Bayport Training & Technical Center)

EQUIPMENT OVERVIEW

Swing Check Valve
(Photo Courtesy of Bayport Training & Technical Center)

Motor
(Photo Courtesy of Bayport Training & Technical Center)

EQUIPMENT OVERVIEW

Gear Pump
(Photo Courtesy of Bayport Training & Technical Center)

Rotary Vane Pump
(Photo Courtesy of Bayport Training & Technical Center)

EQUIPMENT OVERVIEW

Screw Pump
(Photo Courtesy of Bayport Training & Technical Center)

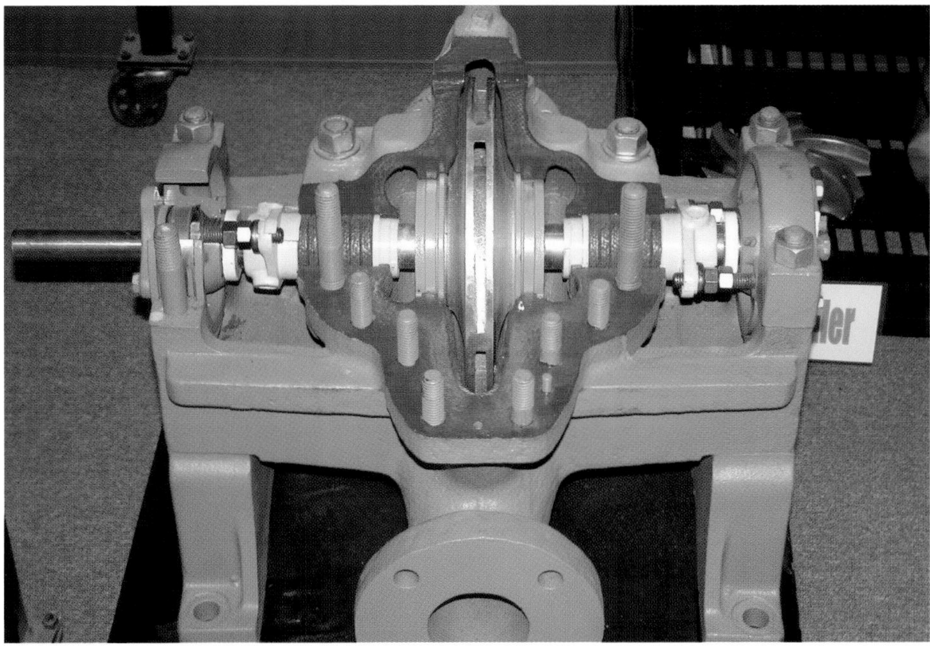

Centrifugal Compressor
(Photo Courtesy of Bayport Training & Technical Center)

EQUIPMENT OVERVIEW

Reciprocating Compressor
(Photo Courtesy of Bayport Training & Technical Center)

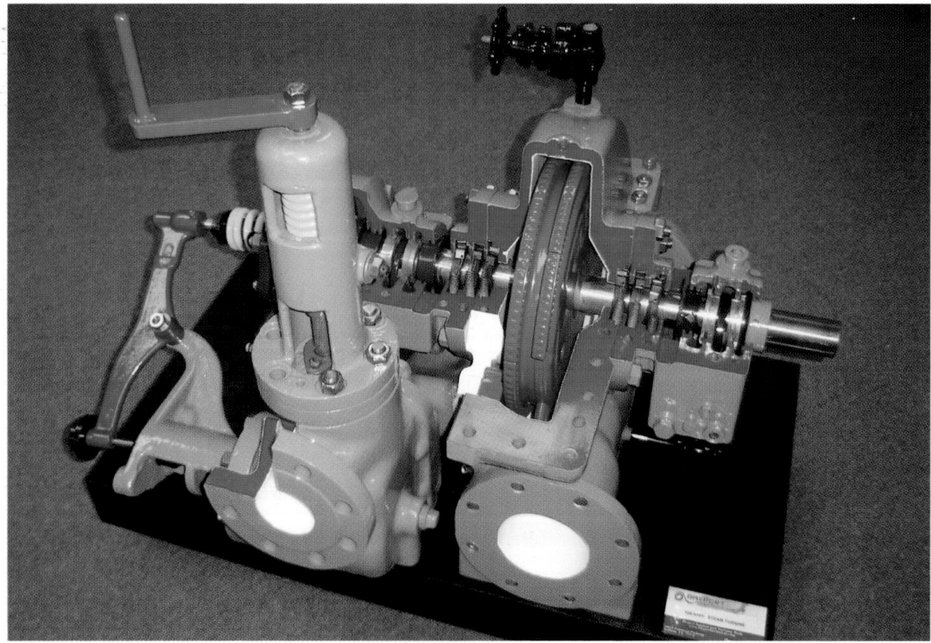

Steam Turbine
(Photo Courtesy of Bayport Training & Technical Center)

EQUIPMENT OVERVIEW

Heat Exchanger
(Photo Courtesy of Bayport Training & Technical Center)

Cooling Tower
(Photo Courtesy of Bayport Training & Technical Center)

EQUIPMENT OVERVIEW

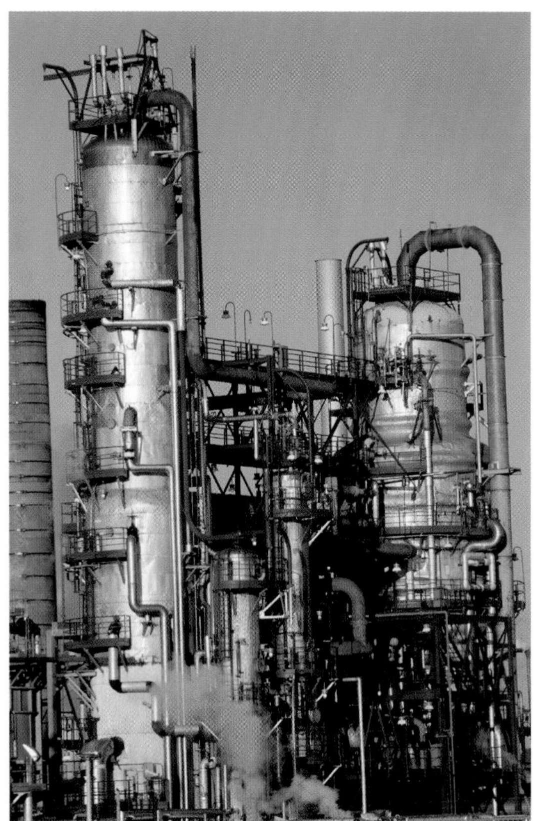

Distillation Column
Courtesy of EyeWire/Getty Images, Inc.

Distillation Column
Courtesy of Corbis Images.

EQUIPMENT OVERVIEW

Sphere Tanks
Courtesy of Corbis Images.

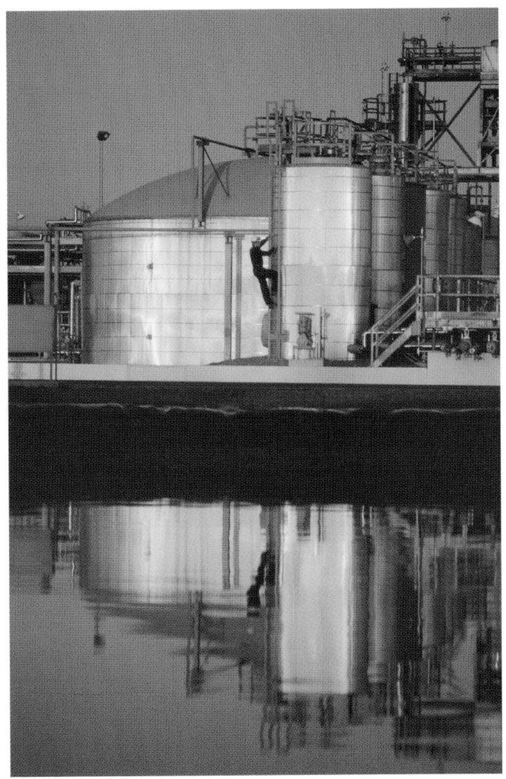

Storage Tanks
Courtesy of Corbis Images.

Tank Farm
Courtesy of EyeWire/Getty Images, Inc.

EQUIPMENT OVERVIEW

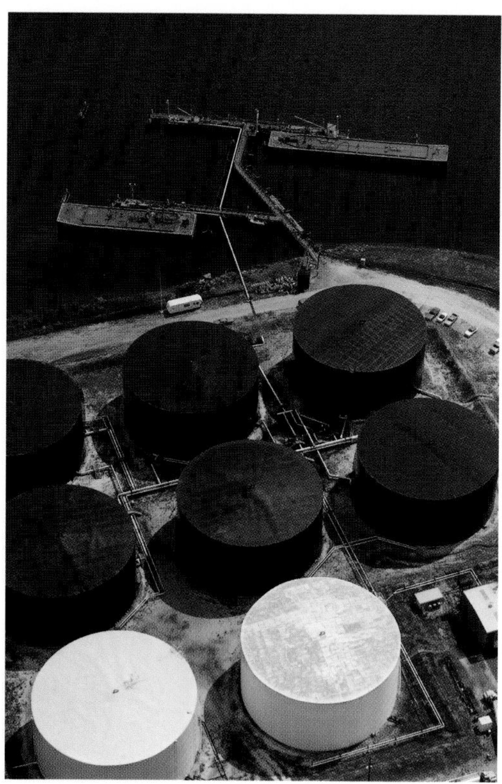

Tank Farm
Courtesy of Corbis Images.

Smoke Stacks
Courtesy of Corbis Images.

Flare
Courtesy of Corbis Images.

EQUIPMENT OVERVIEW

Furnaces
Courtesy of Corbis Images.

Monitoring Instrumentation
Courtesy of Corbis Images.

Drilling Rig
Courtesy of Corbis Images.

EQUIPMENT OVERVIEW

Drilling Pipe
Courtesy of Corbis Images.

Gauges
Courtesy of EyeWire/Getty Images, Inc.

ON THE JOB

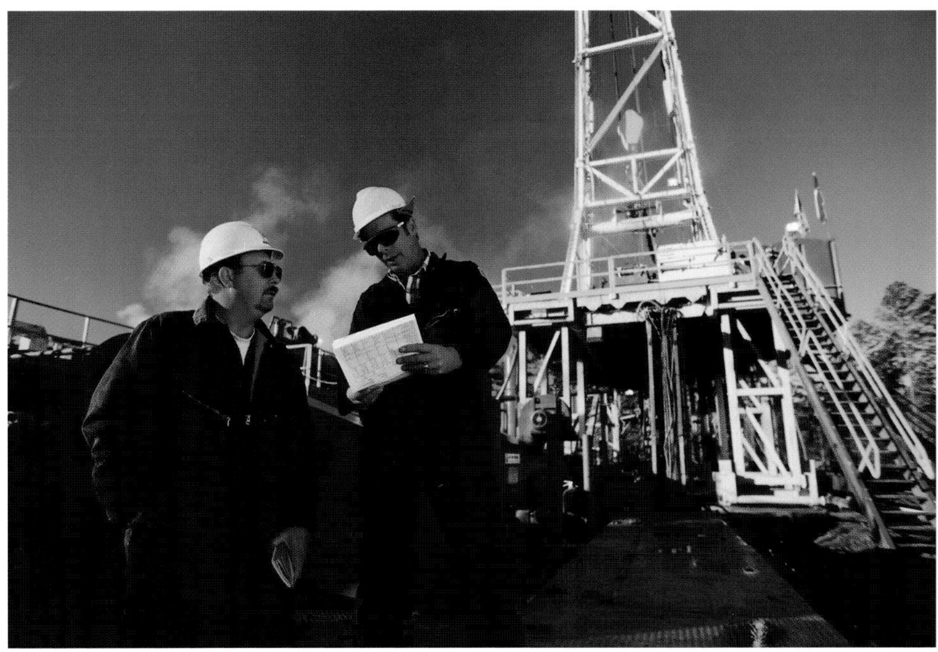

Print Reading

Communication

Courtesy of Corbis Images.

ON THE JOB

Well Head ("Christmas Tree")
Courtesy of Corbis Images.

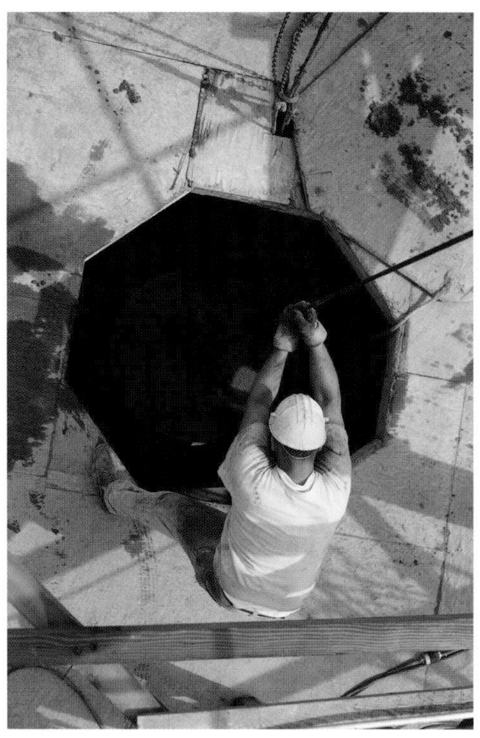

Drilling Well
Courtesy of Corbis Images.

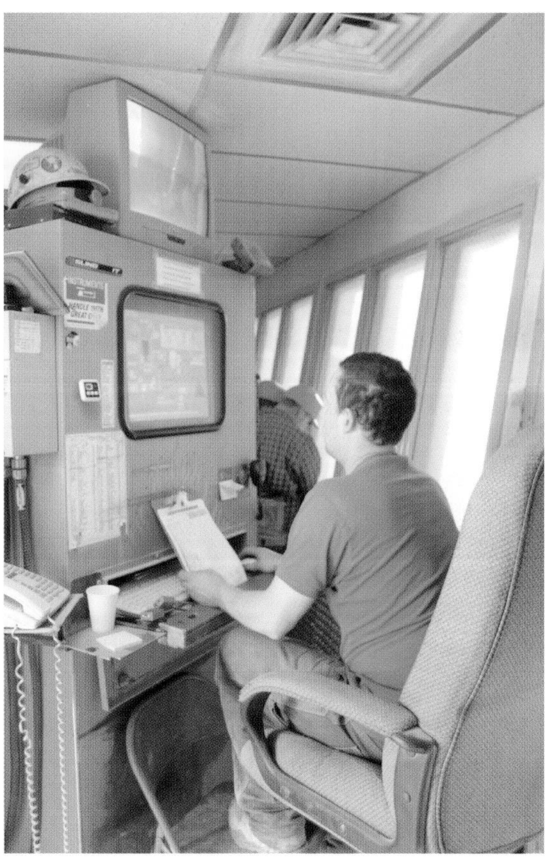

Process Technician at Mining Facility
Courtesy of Corbis Images.

ON THE JOB

Problem Solving
Courtesy of Corbis Images.

Inspection
Courtesy of Corbis Images.

Drilling Well
Courtesy of Corbis Images.

ON THE JOB

Shift Work
Courtesy of Corbis Images.

Monitoring
Courtesy of Corbis Images.